Britain Can Make It

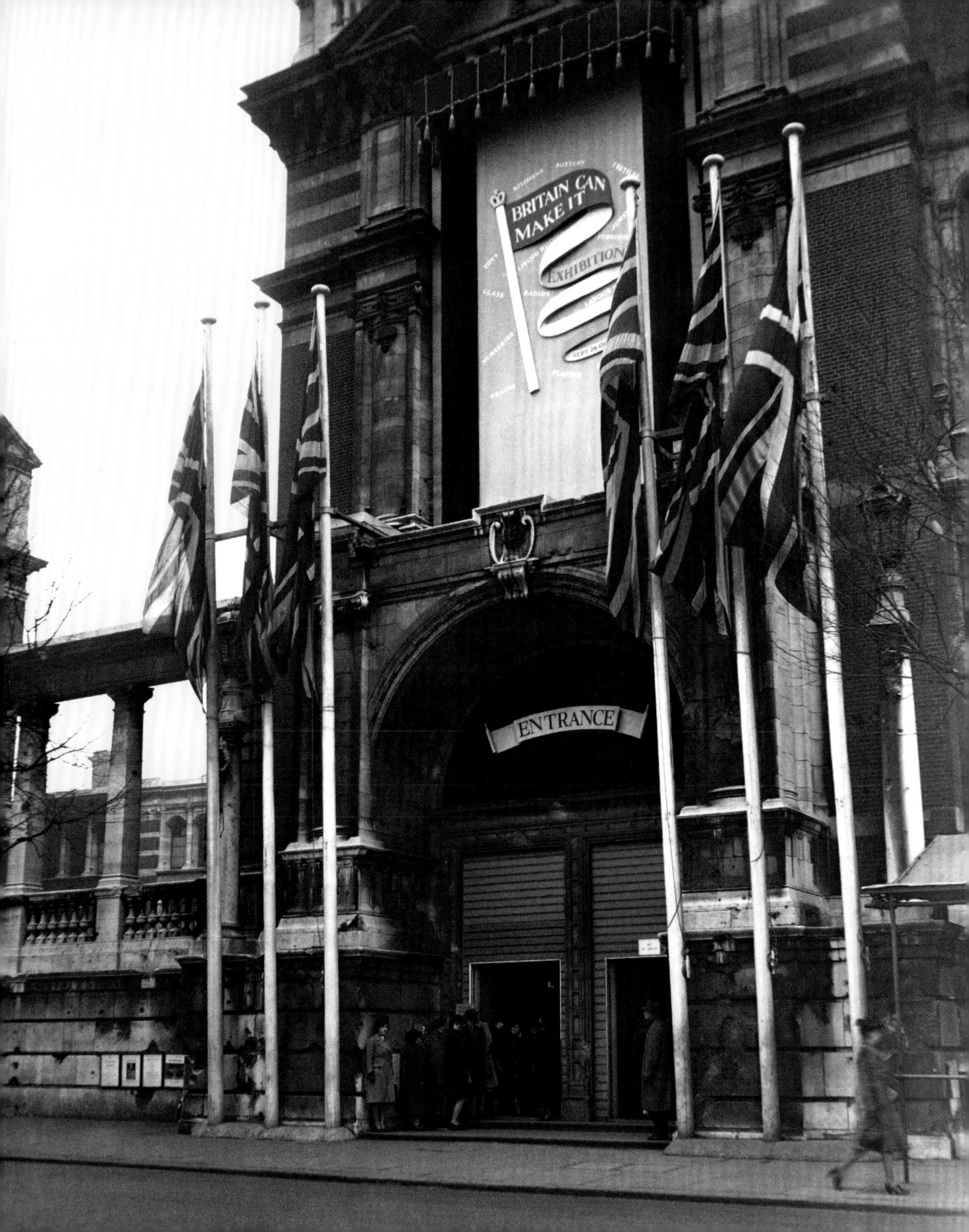
BRITAIN CAN MAKE IT
EXHIBITION
POTTERY
TOYS
GLASS
RADIOS
TRAINS
PLASTICS
ENTRANCE

Britain Can Make It

The 1946 Exhibition of Modern Design

CONTRIBUTING EDITOR
Diane Bilbey

PAUL HOLBERTON PUBLISHING
IN ASSOCIATION WITH V&A PUBLISHING
AND THE UNIVERSITY OF BRIGHTON DESIGN ARCHIVES

Supported by a Special Award from
the Royal Commission for the Exhibition of 1851

ISBN 978-1-911300-54-0

British Library Catalogue in Publishing Data

A CIP record of this publication is available from the British Library

Produced by Paul Holberton Publishing
89 Borough High St, London SE1 1NL
WWW.PAULHOLBERTON.COM

In association with

V&A Publishing
Victoria and Albert Museum
South Kensington, London SW7 2RL
WWW.VAM.AC.UK/PUBLISHING

Designed by Laura Parker
Printed by Gomer Press, Llandysul

Front cover Timber display from the 'What the Goods are made of' section
Back cover 'Mr Designer', a motif throughout the exhibition
Frontispiece The entrance to *Britain Can Make It* in Exhibition Road at 8am
Facing The CoID's 'Let's Look at Design' wall sheet

V&A Publishing

Supporting the world's leading museum of art and design, the Victoria and Albert Museum, London

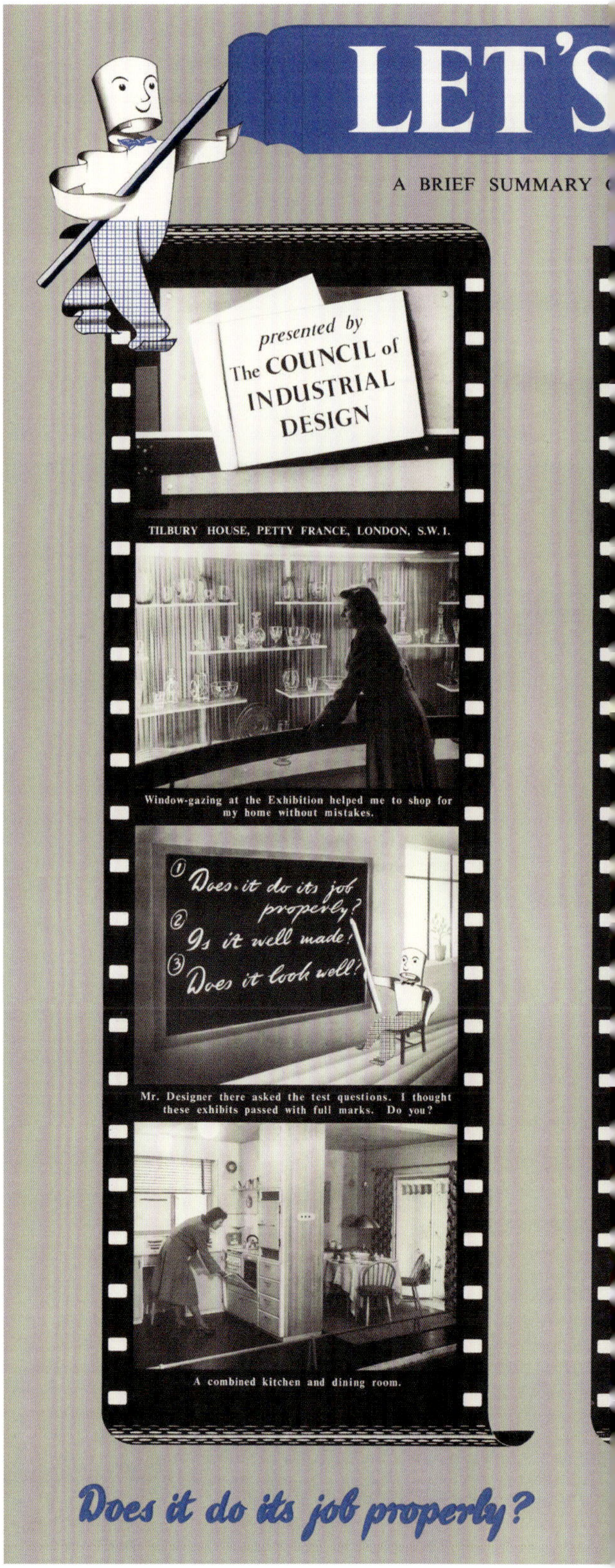

LOOK AT DESIGN

FILM STRIP RECORDING A HOUSEWIFE'S VISIT TO THE "BRITAIN CAN MAKE IT" EXHIBITION

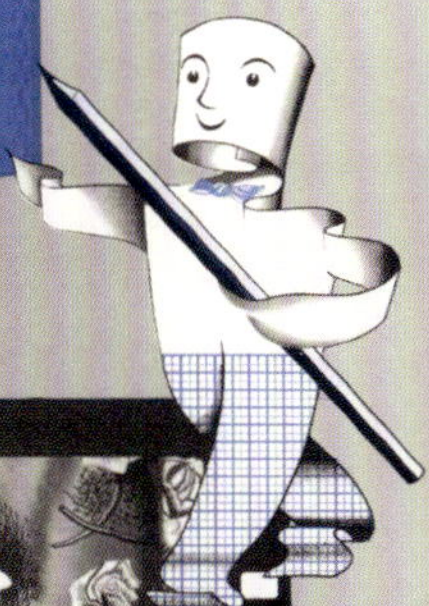

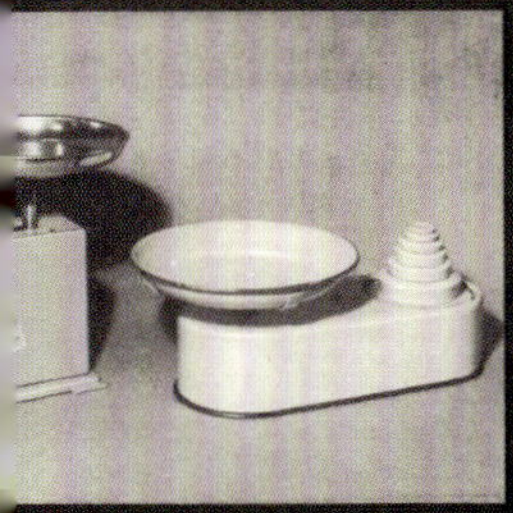

ient scales that are easy to clean.

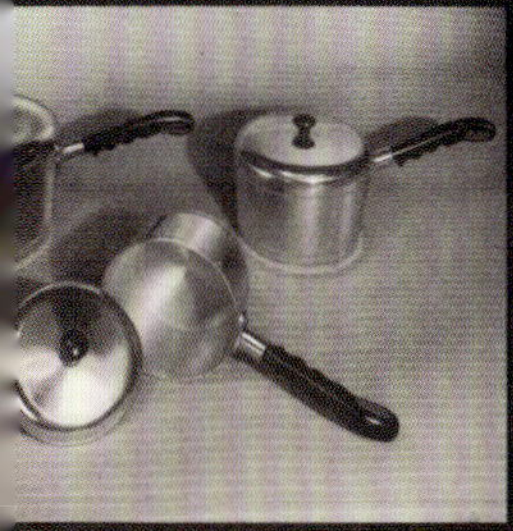

ucepans with insulated handles.

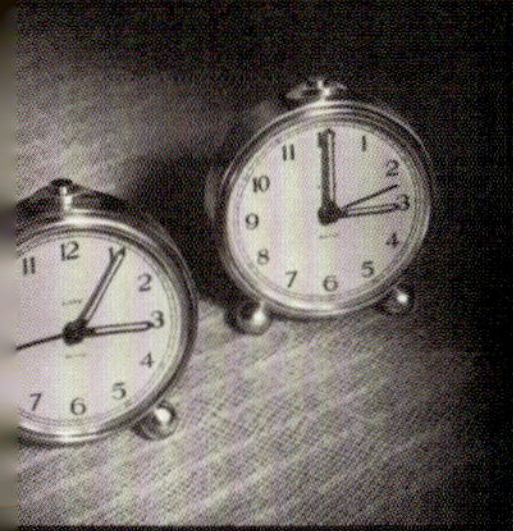

le, but a change from jazzy ornament.

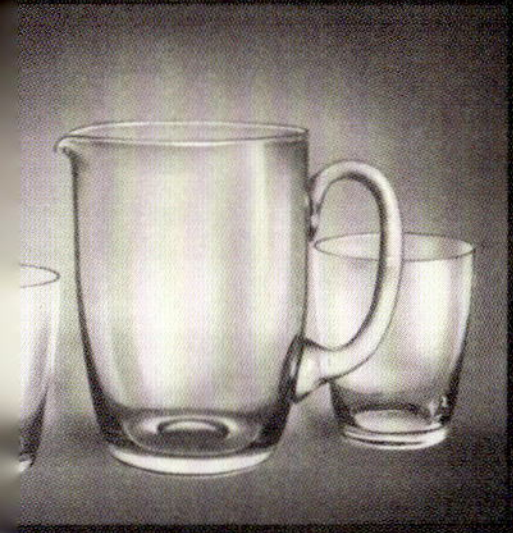

n glass of fine shape and quality.

t well made?

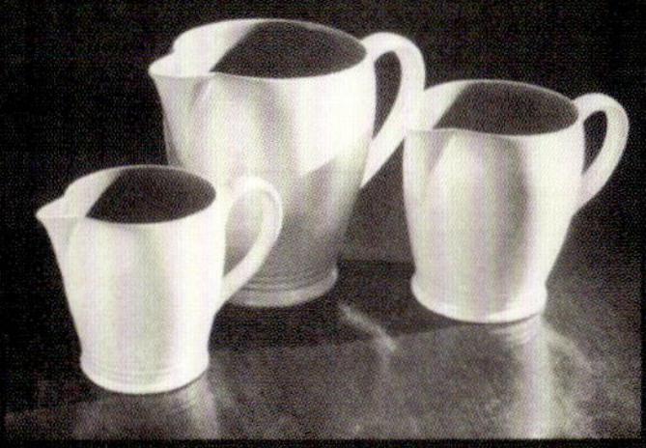

Strong, easy-pouring jugs in clear colours.

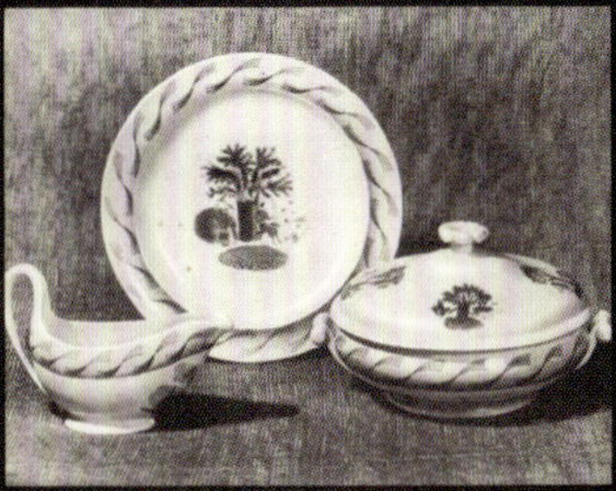

A gay pattern that really suits the shapes.

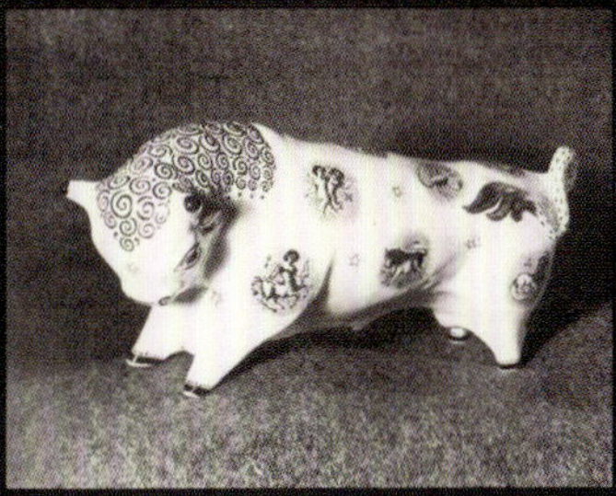

Merely ornamental; but what fun!

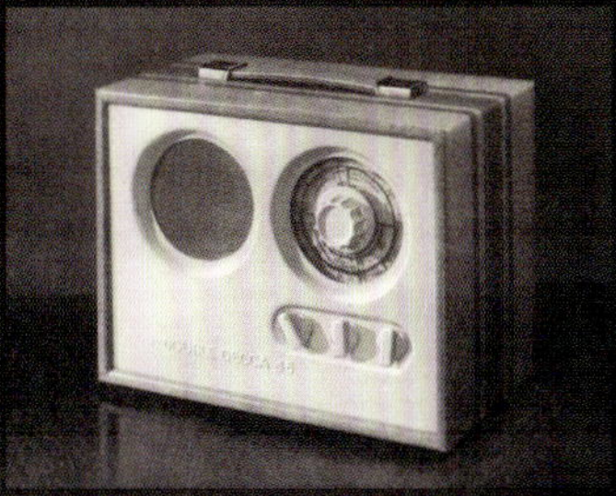

An honest-looking machine that fits a home background.

Does it look well?

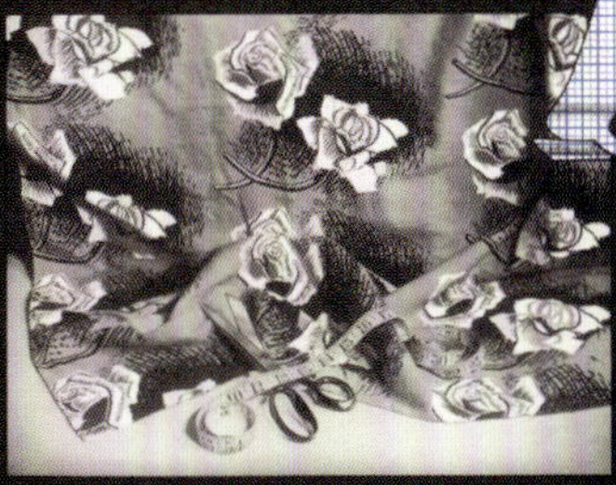

An old motif handled in a modern way.

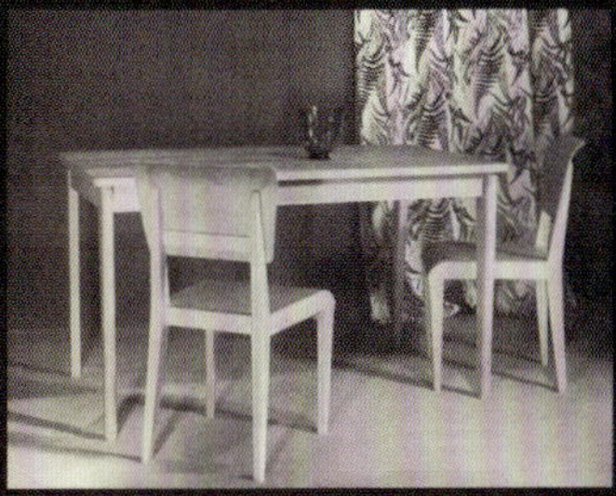

Plywood makes possible a new elegance and grace.

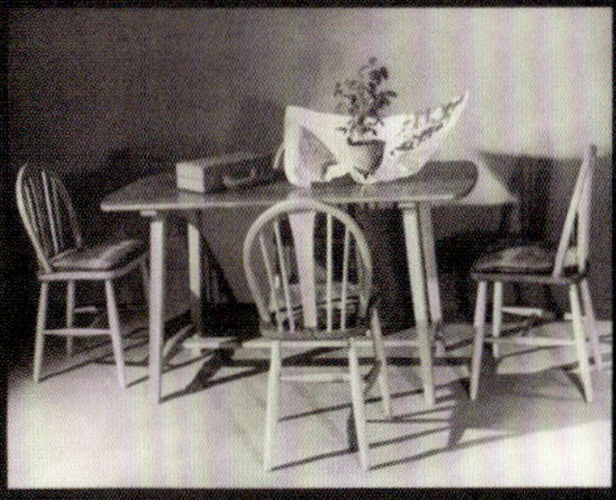

Traditional forms in furniture take a lot of beating.

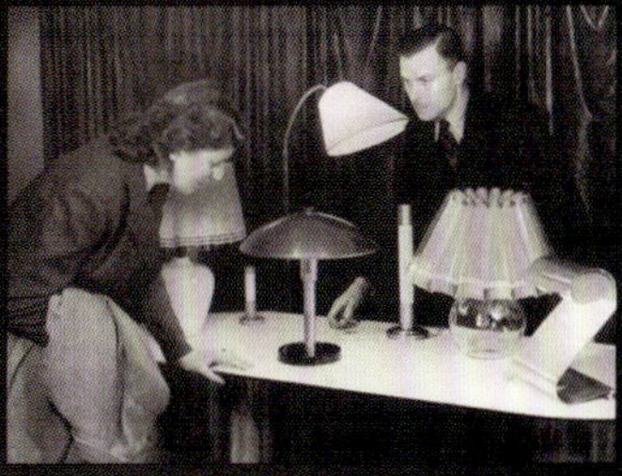

Some of these things are in the shops now; many others will be before long. Let's be sure we know how to choose when the shops are full again.

10 **Foreword** | Sir Terence Conran

12 **Recollections** | Raph Bonella

14 **Acknowledgements**

15 ***Britain Can Make It*** **at the University of Brighton Design Archives**
Lesley Whitworth

17 *BRITAIN CAN MAKE IT*: THE DREAM OF THE FUTURE
Diane Bilbey

29 BRITAIN IN 1946 | David Kynaston

33 THE V&A 1939–46 | Katrina Royall

39 NAMING THE EXHIBITION | Diane Bilbey

46 ASHLEY HAVINDEN AND THE *BRITAIN CAN MAKE IT* POSTER | Margaret Timmers

52 STATISTICS, STATISTICS | Diane Bilbey

60 FROM WAR TO PEACE | Diane Bilbey

69 WHAT THE GOODS ARE MADE OF | Diane Bilbey

76 SHOPWINDOW STREET | Diane Bilbey

82 SILVERWARE | Eric Turner

88 JEWELLERY | Clare Phillips

94 POTTERY AND GLASS | Alun Graves

104 DESIGN QUIZ | Diane Bilbey

116 DRESSING THE GOODS: PACKAGING & POST-WAR RECONSTRUCTION | Paul Rennie

122 WOMEN'S FASHIONS | Edwina Ehrman

130 TIME FOR TEA | Diane Bilbey

139 INTRODUCTORY STORY TO THE FURNISHED ROOMS | Diane Bilbey

145 THE FURNISHED ROOMS: ROOMS MADE TO FIT
Alexandra Goddard

153 DISPLAYING ART AT *BRITAIN CAN MAKE IT*
Diane Bilbey

165 THE MURALS | Diane Bilbey

172 WALLPAPER | Gill Saunders

178 THINGS FOR CHILDREN | Catherine Howell

185 RULED BY FASHION: ASHLEY HAVINDEN AND THE MENSWEAR DISPLAY | Paul Jobling

192 GREAT BRITISH DESIGNERS: MADE IN BRITAIN
Jonathan M. Woodham

198 BIRTH OF AN EGGCUP: WHAT INDUSTRIAL DESIGN MEANS | Lesley Whitworth

204 UTILITY AND OTHER FURNITURE
Alexandra Goddard

210 BOOKS AND PRINTING | Martin Andrews

219 DESIGNERS LOOK AHEAD: THE HALL OF THE FUTURE
Diane Bilbey

231 SURREALIST STRATEGIES IN EXHIBITION DESIGN
Andy Groarke

238 MASS OBSERVATION AND *BRITAIN CAN MAKE IT*
Simon Garfield

241 REFLECTIONS ON *BRITAIN CAN MAKE IT* | Diane Bilbey

---

250 Index
254 Contributor biographies
256 Select bibliography | Photographic credits

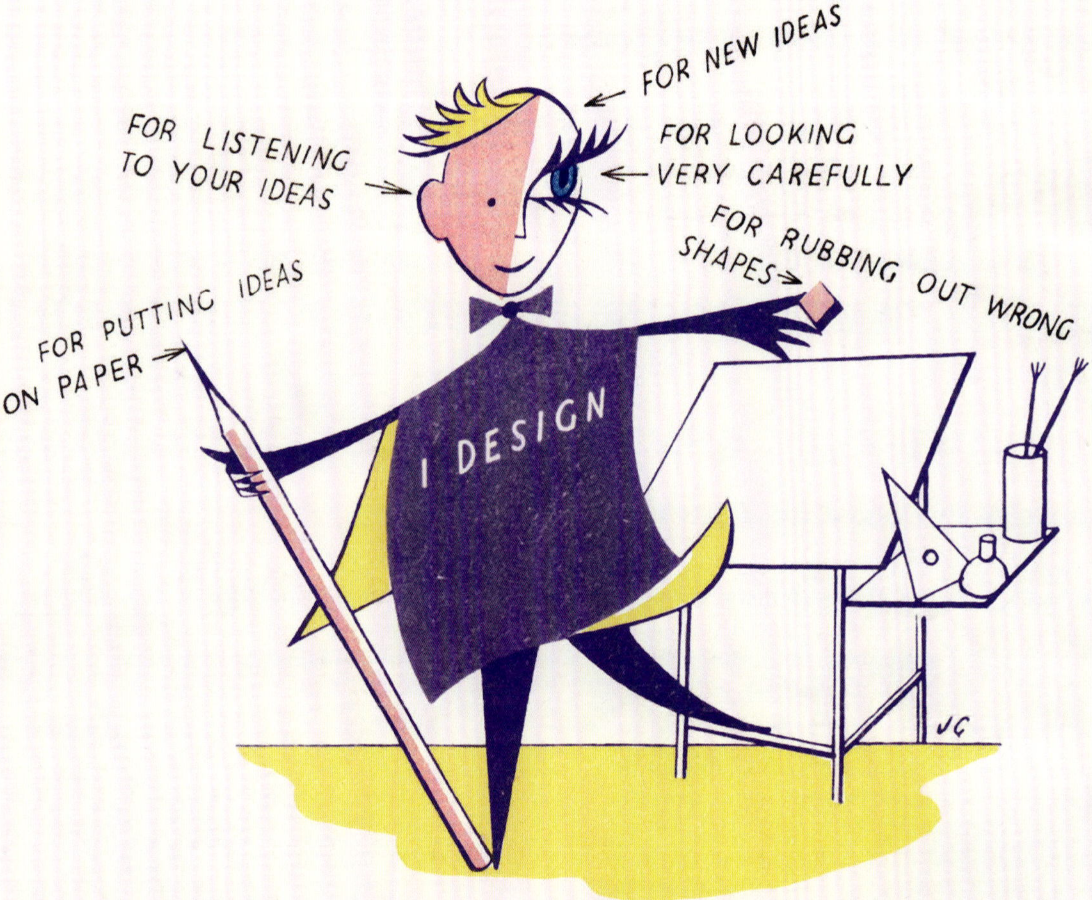

Here is the Designer. When *you* look on the other side you will see what he says about his work and how it affects you.

Foreword

SIR TERENCE CONRAN

I've always been passionate about making things. As a young child my favourite present was a large bag of wood off-cuts with all sorts of nails, a hammer and a gimlet. I was also somewhat of a fanatical potter in my youth. Throughout my career I have always believed that you shouldn't design something you wouldn't know how to make yourself, and concerned myself with the practical part of the design process.

The *Britain Can Make It* exhibition wasn't my initial introduction to design but, having spent most of my childhood in the countryside, it was the first industrial manifestation of design I'd seen face-to-face. Until that point I'd just enjoyed the simple act of making things, but visiting the exhibition turned my mind to something much more serious.

It is over seventy years since my visit but I remember it clearly and very fondly. We travelled up in a bus from Bryanston School near Blandford and we were all terrifically excited. For all of us it was our first visit to London since the end of the war and, having been sheltered from the worst of it, we were quite shocked as the bus drove through the suburbs and in to the city streets. London was at its lowest; bomb damaged, grey and weary of austerity. Curiously enough I remember buddleia seemed to be sprouting everywhere.

Then the V&A came in to view. I remember we gasped at the queues and felt slightly disappointed we couldn't just waltz straight through the doors, but the wait wasn't too long. Ashley Havinden's enormous poster loomed over the entrance welcoming you in with what felt like such a confident statement. If Britain was weary, the V&A and Council of Industrial Design weren't timidly asking if Britain could make it. They were boldly telling the 20,000 visitors a day that Britain Can Make It – it was a statement, not a question. Perhaps most crucially of all, the exhibition was forward thinking, cheerful and optimistic in showing the world as it might be tomorrow, rather than what it had been.

I remember being very impressed with the room sets, in particular those that provided a glimpse of the future. They felt so new at the time and I remember thinking 'that is a room I'd like to live in one day'. We all loved Gordon Russell's radio cabinets and wirelesses in laminated wood and later I tried to persuade my mother, unsuccessfully, that the family simply had

Exhibition designer James Gardner's pictorial explanation of the role of a designer, from one side of the *Britain Can Make It* guide for children to have one. I was also drawn to the textiles section as I loved pattern and colour even back then.

The way materials such as metals and timber could be used was also very thought-provoking and definitely influenced the way I thought about design, and the way I worked in the future. I liked how simple, practical objects and furniture sat alongside ingenious, innovative pieces that offered new ways of looking at problems. It was often surprising and delightful. It was also uplifting to see that even simple household objects could be made beautifully, an idea that had seemed to get slightly lost in the war.

The exhibition successfully captured just what Britain needed by doing something contemporary, and by recognizing just how important design and manufacturing would be to rebuilding our rather beleaguered nation. It was an important exhibition and reached out to British manufacturing very successfully, and anyone not involved certainly wanted to be part of this future the Council of Industrial Design were putting on display.

I think the messages from an exhibition over seventy years ago resonate with the Britain we live in today. We may not have endured a brutal war recently, and the country is certainly more aware of design than it was in 1946, but we are weary of austerity and as a nation we feel utterly divided and beleaguered by Brexit. We also seem to have lost the ability to make things here on an industrial scale. That is partly why I founded The Design Museum, to encourage and inspire young designers to use design and manufacturing techniques to create a better, more sustainable world.

We have the most amazing craftspeople in this country, and the UK's creative industries are the finest in the world. So we need to ensure we keep making things and utilize and value these skills, because they were a vital part of our past and we must find a way to make them part of our future – they must not be lost.

Perhaps today we do need a terrific, cheerful and forward thinking exhibition that focuses the world on Britain, and the idea that we *CAN* be a very innovative nation.

FEBRUARY 2019

Queue outside the V&A, stretching down Cromwell Road, of people waiting to get into BCMI

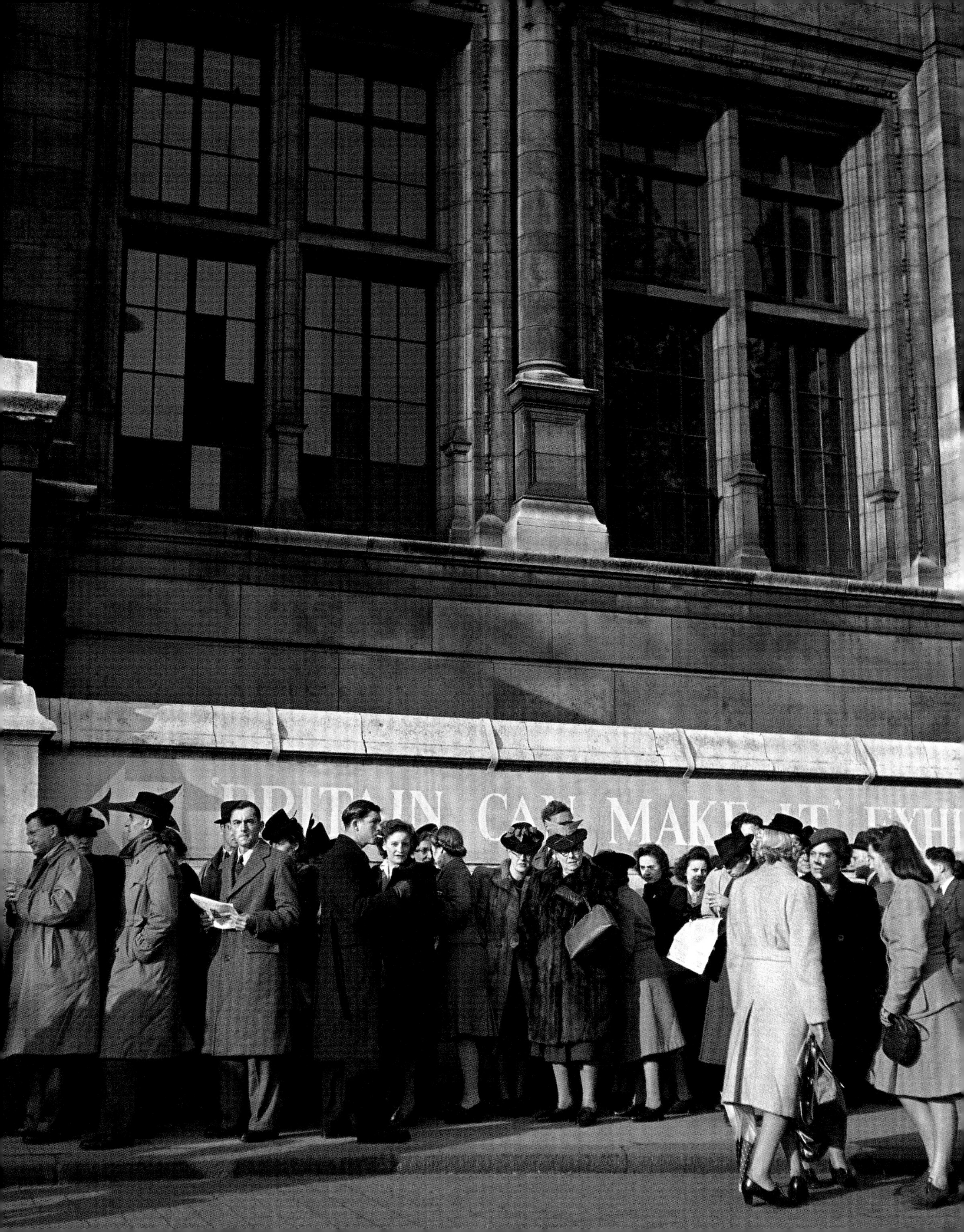
'BRITAIN CAN MAKE IT' EXHI

Recollections

RAPH BONELLA

Raph Bonella (1921–2016) worked on *Britain Can Make It* as an architectural assistant between 11 February and 16 September 1946, one of a team of twelve working under the supervision of Basil Spence for the Council of Industrial Design. These are his recollections of that time.

As an architectural assistant, newly 'demobbed' from the RAF, I joined a small team assembled by the Council of Industrial Design and headed by Basil Spence, as architect, and James Gardner, as industrial designer. Both men were first-rate designers from whom I was to learn a great deal. So the job was the opportunity of a lifetime, because I still had to submit a thesis before I could call myself an architect. The design team were not concerned with the actual exhibits – unless they were very large, like the Spitfire that had to be eased through a window with inches to spare. We had to correlate the work of the individual designers of the various Halls and to design the circulation spaces.

London was, at that time, a very dreary place. Though still the capital city of an Empire 'on which the sun never set', it had suffered not only enemy bombing but also five years of no painting or refurbishment. The exhibition was intended not only to showcase Britain's products but also to bring light and jollity! But there were some who thought the project a misuse of scarce resources, and questions were asked in the House of Commons. The Minister reassured the House that only two standards of timber would be used and all could be re-cycled. At that very moment, we were taking delivery of the tenth standard. As to the re-cycling, I doubt it – except perhaps as kindling.

Basil Spence laid great stress on using the natural colour and texture of the construction materials rather than applied decoration. Although commonplace today, this was a new concept at the time, so far as the public were concerned. One of Spence's favourite sayings was: 'If something is in an awkward place, don't hide it, make a feature of it'. A good example of his design was a central feature composed of electric flex, with copper strands covered with transparent plastic, intertwined with similar flex with steel strands. Magic! Until then, all flex had been covered in brown cloth.

So-called 'Modern' designs had not yet reached the High Street. Generally, furniture was still stained 'dark oak' and public buildings still had light-green walls with dark-green dados. That had to change. Painting and 'graining' were replaced by chrome plating and clear varnishing. Much attention was given to detail. The whole team spent a day drawing on the floor with chalk sticks, designing clouds for one of the halls!

When work started on site, I was sent to the V&A to liaise between the design team and the site team, including the V&A Clerk of Works; Mr Driver, the Ministry of Works architect; the formidable Mrs Whitaker and myself. Mrs Whitaker's principal interest was to stop anything being fixed to the walls of her beloved V&A. Most of the building was an empty shell, all the peace-time exhibits having been removed to escape the bombing. I had a small timber office (a garden shed really) in one of the Halls. It was a very relaxed atmosphere.

My job was in two parts – to design the links between the Halls and to deal with the disputes and complaints of the various artists designing the Halls. If only all the artists had been the same as the very affable F.R.S. Yorke, a leading architect at the time. The egos of some of the lesser designers were unbelievable.

Finally, the exhibits began to arrive and the glamorous boys and girls who were to man the displays replaced Beck & Pollitzer's workmen. The King opened the Exhibition and my job was over. There was a drinks party at the Rembrandt Hotel over the road, where everyone congratulated everyone else. But, it seemed that *Britain Can Make It* was soon forgotten, overshadowed by the very successful Festival of Britain that succeeded it. Perhaps our exhibition was just a practice run for the later one, since that was from the same design team, but I only read about it in the Swedish press, as having completed my thesis I went to live in Sweden during the Festival of Britain and so never saw it myself. Had we achieved the 'WOW factor' for which we had striven? I hope so.

JUNE 2016

Acknowledgements

This publication is dedicated to Alexandra Goddard (1972–2018), a much-loved friend and invaluable contributor to the book

This book has had a long life prior to its publication. The original proposal was submitted in 2013, so I must firstly extend my greatest thanks to all the contributors, not only for their excellent essays, but also for their patience in sticking with me and seeing this project through.

Heartfelt thanks must initially be extended to The Commissioners for the Exhibition of 1851 for providing a Special Award towards the publication costs. To Paul Holberton Publishing for happily agreeing to publish this book; to Laura Parker for her superb, perceptive design, and to Katherine Bogden Bayard for her generously given help.

The book would not have been possible without access to the material held at the University of Brighton Design Archives. I must sincerely thank Lesley Whitworth for providing this access and for her unrelenting help with this long-lived project. Images of the exhibition provided by the University of Brighton Design Archives (part of the Design Council Archive in their care) are a tremendous contribution. Thanks must also be extended to Design Council, formerly the Council of Industrial Design, whose records are greatly used throughout the book. The Victoria and Albert Museum have generously supplied images, and I am grateful for the continued help and advice provided by Tom Windross at V&A Publishing.

Colleagues in the National Art Library and at the Archive of Art & Design, particularly James Sutton, have assisted me hugely. Samantha Blake and Jessica Hogg at the BBC Written Archives Centre, for providing guidance and advice on the amazing scripts held there. Rose Lock and Caroline Marchant-Wallis, Special Collections, University of Sussex for help with accessing and citing the Mass Observation Archive. Fiona Orsini, RIBA Drawings & Archives Collection, and to Evelyn Watson for providing access to the archives of the Royal Society of Arts. Also to Gale Cengage whose online resources have been invaluable. I would also like to thank those who helped in the earlier stages of this project, including Catherine Moriarty, formerly of the University of Brighton Design Archives, and Mark Eastment, formerly of V&A Publishing.

Special thanks must go to: my husband Vince Bridgman for his unwavering support and positive response to countless requests for help and advice; to Margaret Timmers for her great enthusiasm and support; and to Michael Twyman for his encouragement and much appreciated comments on the text. My great friends who have given me invaluable advice and encouragement from the very start, and who have endured years of conversations about *Britain Can Make It*: Lucy Cullen, Alex Goddard, Louise Goodliffe, Charlotte Hubbard, Fiona Leslie, Charlene McGroarty, Emma Minns, Dolly Montague, Peta Motture, Ann Pillar and Katrina Royall.

Thanks to all those who have kindly helped with supplying images: Barbara Taylor and Sirpa Kutilainen, University of Brighton Design Archives; Karen Fick and Frankie Chappell, V&A; Greag Mac a' tSaoir and Laura Feliu Lloberas, National Galleries of Scotland; Mandy Williams, The Geffrye Museum of the Home; Lauren Alderton, RIBA Drawings & Archives Collection; Flora Smith, TopFoto; Sophia Tobin, The Goldsmiths' Company; Luke Hayes; Philip Brooks, Historic Environment Scotland; Carole Dumoulin, Condé Nast Publications; John McDonald, and Brian Webb, Webb & Webb Design Limited.

Those who have assisted with permissions queries include: Ann Berni, Barbican Centre; Oliver Black; Jenna Burlingham; Claire Donald, Scott Brownrigg; Nathalie Feitsch, University of Applied Arts Vienna; Malcolm Garrett; Kate Hutchins, Liberty; Neil Jennings; Sandy Kamaryc, O-I Glass Limited; Lucy Lead, The Wedgwood Museum; Martin Long, Ernest Race Furniture; Simon Martin, Pallant House Gallery; Elspeth Millar, University of Kent; Mary Mullin; Gillian Newberry; David Oelman, The Fry Art Gallery; Justine Sambrook, British Architectural Library, RIBA; Peyton Skipwith; Marco Verdoia, Shell Brands International AG; Kate Walter, Shell Heritage Art Collection; Laura Wood, Portmeirion Group. I am also grateful to the following for their help in various ways: Anne Crosbie; Verity Elson, Gordon Russell Design Museum; Nick Gilmour; Esther Giangrande, Charlotte Higson, Mariah Nielson, Amy Rowntree and Corinne Timmis, Carmody Groarke Ltd; Deborah Kellard, Design Council; Sara Kestelman; Stephen Walton, IWM; and Nigel Williams, Royal Commission for the Exhibition of 1851. Lord March for giving me the opportunity to work on a display about the *Britain Can Make It* exhibition at the Goodwood Revival in 2016. Mat Riches for his long-standing help. And a special thank you to Sir Terence Conran for the wonderful, evocative recollections in his foreword.

Finally, special thanks to all those who worked on the 1946 exhibition, named and unnamed in this book. I hope they will excuse the fact that it was not possible to include chapters on all sections of the exhibition. To the reader, I hope you enjoy this book – the sequence roughly follows the route around the exhibition devised by the exhibition's Chief Designer, James Gardner.

DIANE BILBEY
MARCH 2019

Britain Can Make It at the University of Brighton Design Archives

Mr Designer, the James Gardner creation who you will encounter throughout the book, is a favourite motif of colleagues here in the University of Brighton Design Archives (www.brighton.ac.uk/designarchives). Here he boldly holds our gaze, born out of wartime ingenuity, and blazoning the way into the future, just as his visual counterpart, Mr Therm, had done in the pre-war period. His shiny angularity and all-seeing eye humanizes the enterprise that came to fruition in the gloomy basement spaces of the Victoria and Albert Museum, and gives this austerity-era offering a heroic aspect that its later, and more popularly acclaimed successor, the 1951 Festival of Britain, arguably lacked. The Council of Industrial Design's *Britain Can Make It* (BCMI) records, both administrative and photographic, constitute a significant part of the Design Council Archive, which was our founding collection. And the relationships and discussions captured in these files remind researchers irresistibly of the immense challenges posed by that period, and of the scale of the organizers' achievement.

Exhibition design has proved to be a consistent feature in the 22 archives in our care – a thread of professional practice at the leading edge that runs through much of the material. The concerns of these skilled spatial practitioners and/or their commissioning bodies are writ large in the documentary record, which captures something of the tropes and preoccupations of the day. Readers of this book whose curiosity is piqued are welcome to further their enquiries by writing to designarchives@brighton.ac.uk.

LESLEY WHITWORTH
MARCH 2019

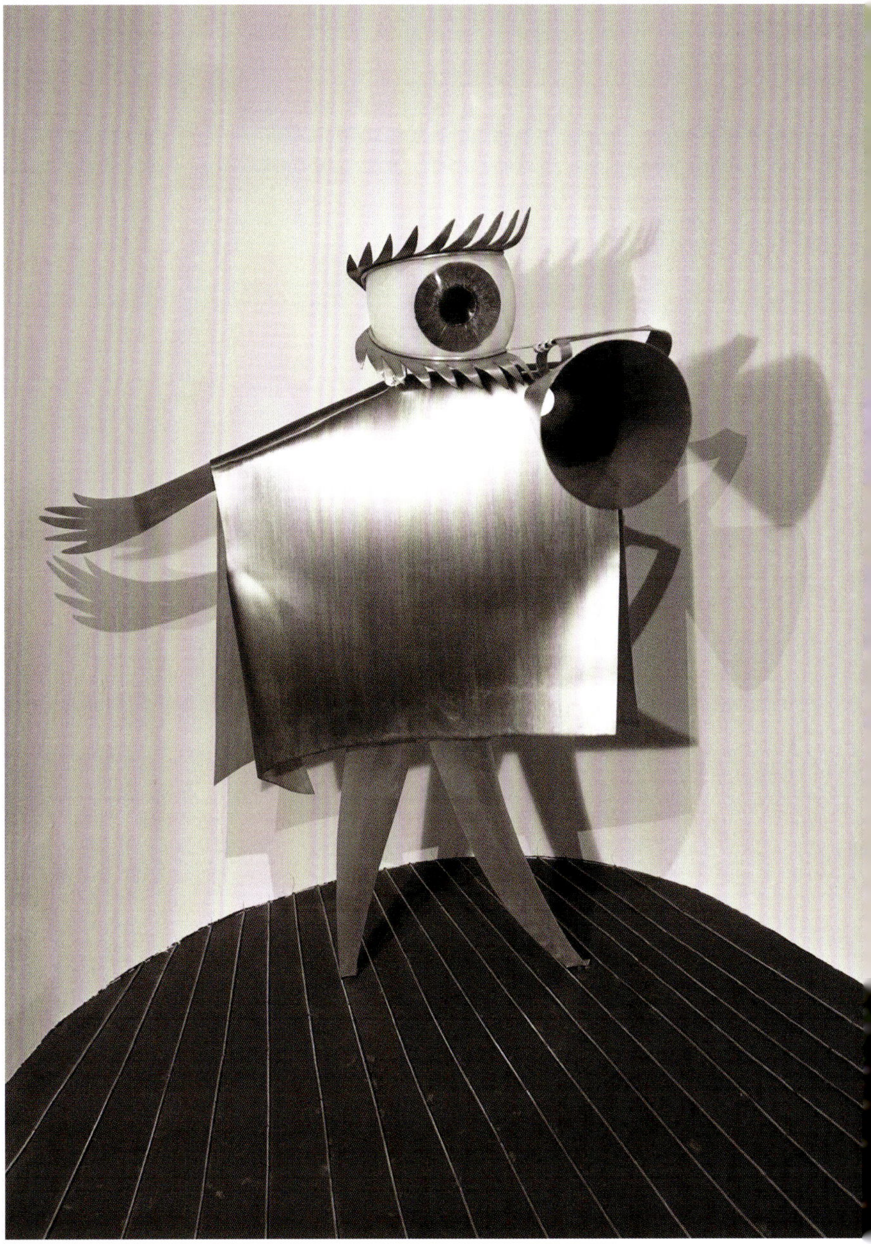

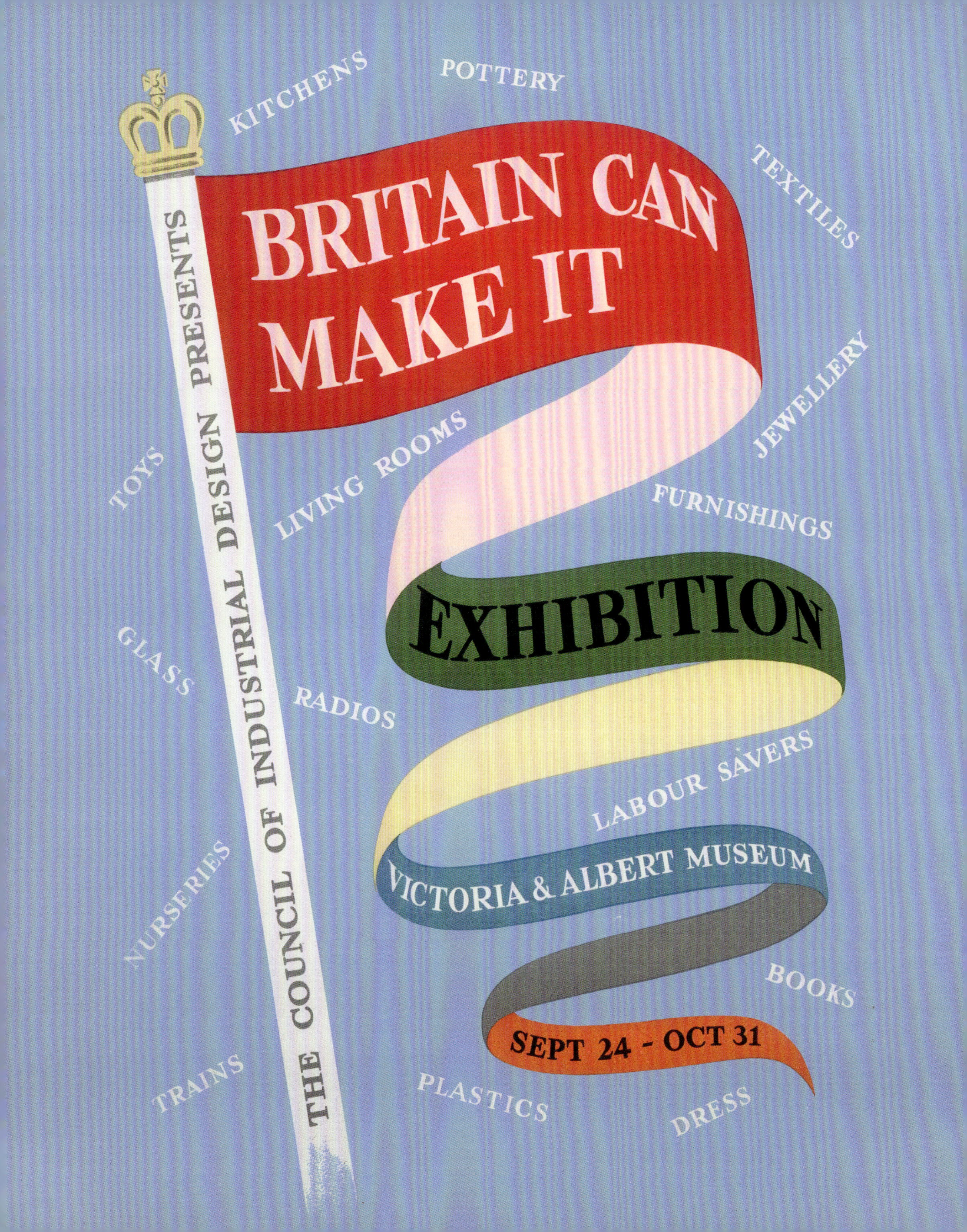
THE COUNCIL OF INDUSTRIAL DESIGN PRESENTS
BRITAIN CAN MAKE IT
EXHIBITION
VICTORIA & ALBERT MUSEUM
SEPT 24 - OCT 31
KITCHENS
POTTERY
TEXTILES
JEWELLERY
FURNISHINGS
TOYS
LIVING ROOMS
GLASS
RADIOS
LABOUR SAVERS
NURSERIES
BOOKS
TRAINS
PLASTICS
DRESS

DIANE BILBEY

Britain Can Make It: The Dream of the Future

> 'Britain Can Make It'. Like all famous slogans, this one thrives on practical demonstration. So, in the autumn of 1946, the 'Britain Can Make It' Exhibition was staged in one of London's great museums. The public was offered a glimpse of Today and Tomorrow instead of Yesterday. And the public poured in. War-weary, tired of rations, tired of austerity, they came to see what the future holds for them. They came at the rate of 20,000 a day.[1]

So began the rousing commentary to a Central Office of Information film devoted to the *Britain Can Make It* (BCMI) exhibition, whose aims were simple and yet ambitious – to increase exports, encourage well-designed manufactured goods, and to educate the British public in recognizing good design. Of course, it was also a propaganda tool, intended to boost the morale of a British public worn out by war and austerity. Organized by the Council of Industrial Design (CoID, later Design Council), on behalf of the government's Board of Trade, the exhibition was held at the Victoria and Albert Museum from September to December 1946.

Long before the end to Second World War hostilities, the government's Post War Export Trade Committee recognized the vital importance of promoting the country's manufacturing capabilities. In October 1942, with an eye on the post-war export trade, planning had begun for an exhibition of 'National Importance'. The establishment by the Board of Trade in December 1944 of a state organization for the improvement of industrial design – named the Council of Industrial Design – was a promising start to the road to recovery. 'For Britain must export or starve' was the stark and unsavoury prospect.[2]

The idea for an exhibition to promote British design and manufacturing was announced by Sir Stafford Cripps, then newly appointed President of the Board of Trade, within a few days of the armistice with Japan on 2 September 1945. The ousting of Churchill's Conservative government

1. CoID poster promoting the exhibition, designed by Ashley Havinden. V&A: Circ.459–1971 (Given by the artist)

in the General Election of 5 July 1945 had been an historic win for the Labour Party (and a surprise result to most):

> The 1945 voter was not so much casting his ballot in judgement of the past five years as in denunciation of the ten before that. The dole queue was more evocative than El Alamein, the lack of roofs at home more important than any 'national' non-party edifice, the peace that might be lost far more influential than the war that had been nearly won. And by refusing to come to grips with these problems at all the Conservatives signed and sealed their defeat.[3]

It was a Britain in debt to the USA, a Britain where rationing and austerity measures were as bad, if not worse, than during the war, and the proposed welfare state would need funding. Britain's need to promote, manufacture and export its goods was urgent. It would be a challenge for the newly elected Labour government.

By November 1945, it is clear that Cripps was impatient for progress. He called for 'a first-class exhibition of the best of British goods next year'. The suggestion of waiting until 1951 to stage an exhibition showcasing British manufacturing (with the centenary of the 1851 Great Exhibition in mind) was not an option for Cripps:

> In six years the pattern of world trade will already have begun to take shape; the British exporter will be well on the way to success, or will be fighting a rearguard action. We cannot wait six years to give our country a lead, or to offer British Industry the chance of showing its achievements to the world.[4]

Stafford Cripps's great personal contribution to the success of BCMI was later reflected upon by S.C. Leslie, Director of the CoID, as part of a tribute to Cripps in *Picture Post*: 'He [Cripps] wanted to get people away from the drabness of the war years and to show them that the post-war world could be colourful, and that the rewards would be worth working for. I have always felt that his concern with these matters of design, and more widely with all beautiful things, represented for him a large part of what politics is all about.'[5]

The positive impact that thriving, innovative British manufacturing industries might have on the spirit of the British public was summed up by R. Dudley Ryder, Chief Exhibitions Officer for the CoID: 'Empty shops are apt to demoralise the public. If we don't do something to inspire the manufacturer and build up public morale, the public will, in desperation, be forced to buy trash.' The National Savings Committee also tapped into the tangible desire of the British public to be able to look towards a prosperous future by commissioning a promotional film about saving from British Movietone News. In cooperation with the CoID, the film, called 'Design for Better Living', followed the visit of a couple, Mr and Mrs (or John and Joan) Britain, to BCMI. The introduction to the short ten-minute film highlights the hopes of this representative young British couple (both just out of the Forces) and the first glimpse, offered by the exhibition, 'to plan and build new homes in a new world'. After a tour round the exhibition, the couple leave to go home – passing a BCMI poster on a wall that fades out to become a National Savings poster, with the aspirational slogan 'I SAVE for my target'. John's dialogue reinforced the propaganda message that saving was something for *now* by comparing it to the lack of availability of products shown at BCMI, 'Most of that stuff's only available sooner or later, but I know one thing that's available right now!'. The shot pans to a National Savings Certificates booklet with an 'Available now' label similar in style to those used in the exhibition [2].[6]

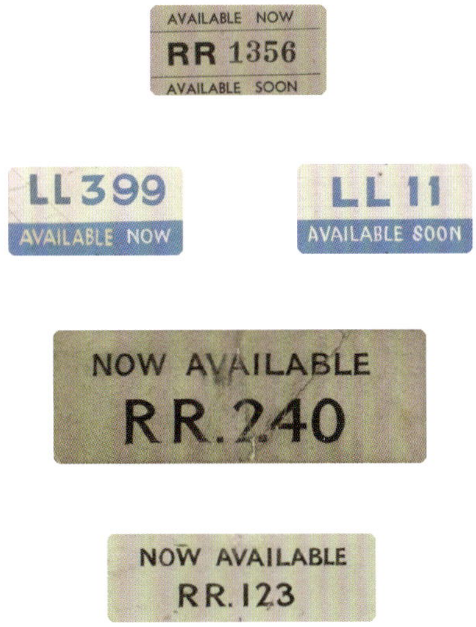

2. Some of the labels displayed with products at BCMI, indicating whether goods would be available 'soon', 'now' or 'later' for the home market

Part of the attraction of BCMI to visitors must have been the stark contrast it offered to everyday life, providing a sense of welcome unreality, so soon after the end of the war. *The Architect and Building News* noted:

> Outside is the world of queues, drab and inadequate homes (or no homes at all), shortages, substitutes, shoddy goods at high prices. Within, all is light, warmth, soft music, seemingly miles of beautiful, practical, desirable things. A shilling has paid for a passage through fairyland; now it is time to wend homeward through the shabby and bomb-scarred streets and warm up a tin for supper.[7]

As noted by Michael Sissons and Philip French in their book *Age of Austerity 1945–1951*, '... in many ways the atmosphere of the post-war years is incredibly remote. It is difficult to recall a time when so much idealism was in the air ... when TV was only a metropolitan toy, ball-point pens a source of wonder, and long-playing records a transatlantic rumour'.[8]

The genesis for BCMI can be traced back to the *Art in Industry* exhibition that took place at the Royal Academy in 1935. Organized by the Royal Society of Arts, this exhibition included many of the same themes that would later be seen at BCMI. Rooms full of ceramics, glass, silver and jewellery were set alongside others displaying domestic equipment or demonstrating book production and commercial printing, with further 'specimen rooms' much like the 'Furnished Rooms' at BCMI. John A. Milne, Chairman of the Council of the Royal Society of Arts, wrote that the 'Royal Society of Arts can rightly lay claim to some credit for the proposed 1946 Exhibition. It was the pioneer work done by the Society in promoting, with the collaboration of the Royal Academy, the Exhibition of Art and Industry ... which started the ball rolling.'[9]

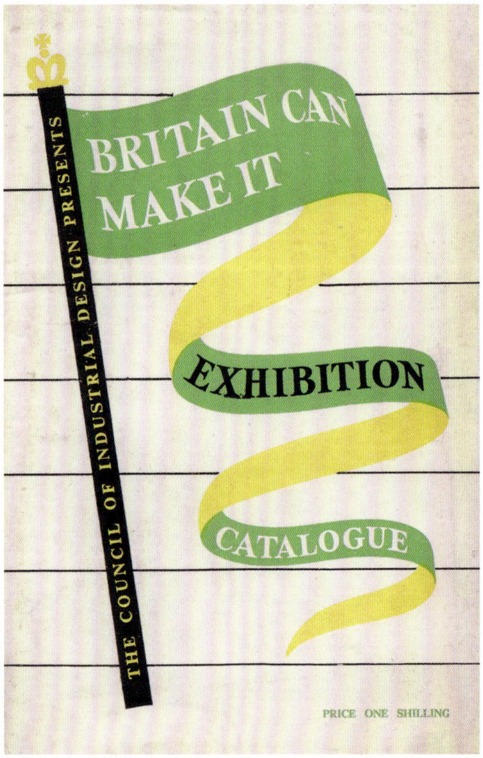

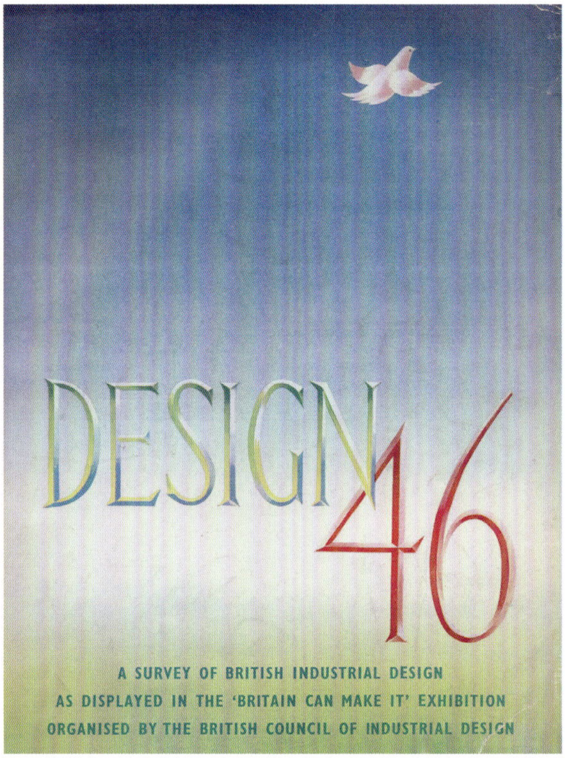

3. Front cover of the BCMI catalogue, designed by Ashley Havinden

4. *Design '46* was the illustrated book accompanying the exhibition and the nearest thing to a souvenir that was on offer to visitors. Intended as a handy reference guide to 'help you buy with discrimination', it consisted of 27 essays by expert commentators – mainly industry specialists – on selected goods shown in the exhibition. It also included a rather academic essay by Bernard Shaw on 'Aesthetic Science'. Press publicity for *Design '46* explained optimistically that it was showcasing a 'Preview of To-morrow's World', and that 'The new things you dreamed of in the days of austerity are taking shape'.

The BBC broadcast a number of features about the exhibition, reflecting the close relationship that existed between the BBC and the CoID at the time. This included a fifteen-minute Home Service radio programme devoted to BCMI on the day it opened, featuring the opening speech made by King George VI [see **72**, **130**], in which the vital importance and timing of the exhibition was emphasized:

> In the overseas markets upon which our prosperity, our solvency, and our standard of life must depend in years to come, and also in our own home markets, our manufacturers will have to compete with a steady improvement in the design of foreign productions. Let us today set out to make British design a hall-mark of pre-eminence in the eyes of the world. As British materials and workmanship have long been.[10]

From the outset, the focus of BCMI was to be on 'manufactured' goods. Artisan crafts or hand-made goods – 'precious stuff' – were excluded, at the insistence of Sir Stafford Cripps. But this was no trade fair, and display space was not sold to manufacturers. Instead products were put forward by manufacturers, and those deemed the best examples were scrutinized and selected by specialist committees.

An accompanying catalogue [3] detailing the manufacturers of products – and, significantly, wherever possible the names of their designers – could be bought by visitors from bookstalls dotted around the exhibition [5]. At the same bookstalls, visitors could also purchase a copy of *Design '46* [4]. Together with small labels that were placed alongside the products on display, the catalogue also indicated the availability of the products on show for the home market, 'now', 'soon' or, as was most often the case, 'later' [2].

A trade enquiries room was set up at the exhibition, where buyers could obtain further information on the products they had seen. The importance of this 'small back room behind the vast shopwindow' was highlighted in the Central Office of Information film devoted to the exhibition. The soundtrack noted that prospective buyers from sixty different countries came 'with a stream of very practical questions ... and they went out to the factories and placed orders worth millions of pounds .... More work, more goods and money coming in from abroad, a better standard of living – that is what this room meant and what the whole Exhibition meant to all those people who queued up for it; and to every worker and housewife in the country.'[11] Despite the success in bringing in orders, another key objective of BCMI was to demonstrate the quality of British products and manufacturing, and to act as 'a long-term national prestige advertisement'.[12] However, the period that immediately followed the BCMI exhibition was a desperate one, compounded by the severe winter of 1946–47 and crippling fuel shortages; it was the beginning of the end for the Labour government [see **105**].[13]

Devised and assembled by creative innovators, and displaying only committee-selected British products, BCMI was – and remains – the most well attended exhibition ever staged at the V&A. The reason for its mass appeal was touched upon by James Gardner, the exhibition's designer, when he was interviewed for a BBC documentary in 1984. In it he explained the post-war situation, 'We'd got to get British manufacturers to produce well-designed goods quickly and to cheer the British public up. They were so depressed. Give them something to look forward to. You know, this was the dream of the future, if you like.'[14] The sentiment was echoed by Sir Hugh Casson, who became Director of Architecture for the Festival of Britain in 1951, when he wrote that *Britain Can Make It* was a 'knockout for tired eyes'.[15]

Around eighty designers 'representing the best British display talent' (many of whom had previously worked for the Ministry of Information) were responsible for the decoration of the exhibition space, under the supervision of James Gardner and Basil Spence, 'two young men just out of the Services'. A press release noted among James Gardner's qualifications for the role as Chief Designer that he was a 'specialist in

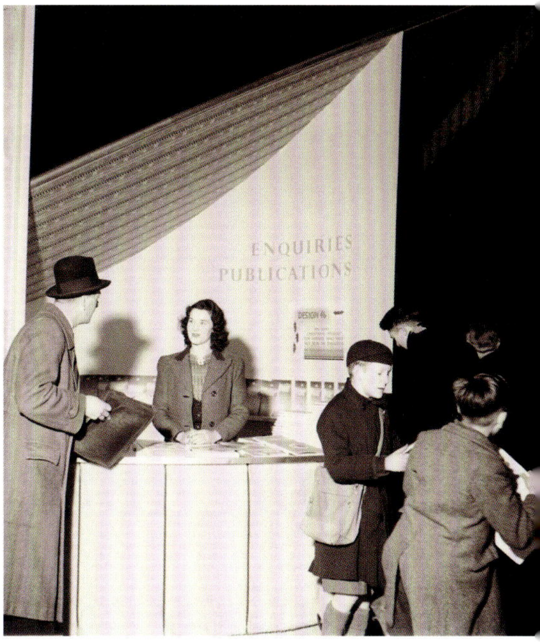

5. One of the bookstalls at BCMI

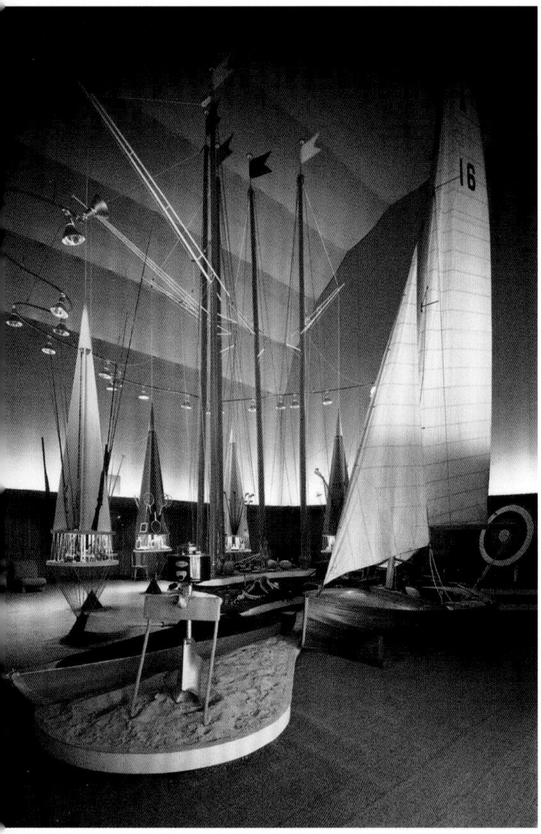

6. Robert Goodden's sports display, part of the 'Weekend' section, one of the highlights of the exhibition

certain aspects of visual deception'. During the war Gardner – known as 'G' to all who knew him – had been Chief Development Officer for Camouflage, and designed special military equipment and 'camouflage effects which were used so successfully in the invasion of Europe'.[16] He had already established a reputation for himself as an effective communicator, designing Puffin books for children during the war.[17] In the spring of 1939 Gardner had been asked to design an exhibition at Shell-Mex House, entitled *How They Fly*, on the history of aviation, with the artist Barnett Freedman. He wrote of this experience in his first autobiography, *Elephants in the Attic*, noting that 'This job happened to launch me into the world of exhibitions' [see **51**]. Others involved in BCMI had also worked in camouflage during the war, including Basil Spence, the exhibition's Chief Architect, John Hutton, Ashley Havinden [see **146**] and Robert Goodden.[18]

The originality that was brought to the exhibition, including the use of innovative new materials and display devices, such as lighting to convey mood, sound features and smell (perfume), were highlighted by Sir Stafford Cripps when he enthused:

> The display which has been devised by James Gardner with a team of young designers, takes your breath away. I don't believe there has been anything to compare with it. It gives you the impression, not of post-war tiredness and frustration, but of youth and of vigour and of freshness.[19]

The 'lavish use of colour and the bold layouts' that characterized BCMI were part of the plan from the start.[20] According to Raph Bonella, an architectural assistant for the CoID working on the exhibition, 'there was a general drive to achieve a "light and airy" atmosphere. This was very much a reflection of the time. Not only was there a desire to get away from the browns and dark greens of the inter-war years, but an attempt to counter-balance the generally dreary appearance of London after five years of no maintenance or repainting'.[21] Robert Goodden's design for the Sports and Leisure section, picked out by many as a highlight of the exhibition, heralded the creativity that was to feature at the Festival of Britain some five years later, 'Top marks for showmanship to Robert Goodden's brilliant composition of masts, cones, ropes and sea-gulls, the colour scheme of which catches the breath' [**6**].[22]

An elusive colour film of BCMI commissioned from Moviegram Ltd by the CoID has not yet been located, if indeed it still exists.[23] However, what is revealed by the comprehensive photographs commissioned by the CoID, now part of the Design Council Archive held at the University of Brighton Design Archives, is a thorough story of an immensely ambitious and inspiring exhibition. Gardner reflected on the transformative qualities and idealistic aims of exhibition design in a lecture he gave on the subject of 'Exhibition Display', just one year on from BCMI: 'Finally, I hope I have been able to substantiate my belief that in exhibition display we have a means of uplifting and knitting together the people in a common purpose'.[24]

The possibility of there being no goods or, just as bad, only poor-quality goods to display at BCMI was a real one. Sir Kenneth Clark, then Director of the National Gallery and member of the CoID, wrote to S.C. Leslie about the proposed exhibition, 'All I feel is that the exhibition should not be too large because there simply will not be the stuff to show, and it is better to tell the President [of the Board of Trade] so now than to find ourselves letting in second rate things of which we shall all be secretly ashamed when the time comes when we have to get enough first rate material to fill our rooms'. Having visited the exhibition, one contemporary critic agreed, commenting that the outcome of the exhibition was substantively to provide 'a glittering frame for a mediocre picture'.[25]

Clark's warning of the possibility of there being no goods was timely. Whether or not British industry would be up to providing the well-designed goods it was supposed to showcase required a leap of faith. With the luxury of hindsight, it can be argued that the value of the exhibition as an event was far greater than the quality of the goods displayed, as Dorothy Braddell (who designed one of the furnished rooms in BCMI) later wrote: 'The psychological effect, just after the war, of the gay and enchanting settings was, in itself, valuable and, since the quantity and quality of the first post-war goods were difficult to assess in advance, the exhibition had to be considered from that angle'.[26]

Selection committees responsible for choosing products for the exhibition were given guidance on judging products submitted by manufacturers. Although the emphasis was to be on good new designs, the best pre-war designs were not to be excluded, if they were to go back into production. Traditional designs were also not ruled out if they were 'fresh' and not merely bland 'reproductions'. Not every item had to be in production, however: 'Stunt exhibits must be avoided but prototypes will be included if they are the forerunners of intended production ranges'.[27]

Although it was appropriate in many ways that BCMI should be held at the V&A – that was itself built on the successes of the Great Exhibition of 1851 and whose mission was 'To promote the application of art to manufacture' – the use of the V&A as the venue proved to be a contentious one. Although the founding principles of the V&A were highlighted in the CoID's annual report, suggesting that 'The holding of the first post-war exhibition of design [at the V&A is] therefore, a suitable as well as a happy omen',[28] in reality both the CoID and the V&A had their reservations.

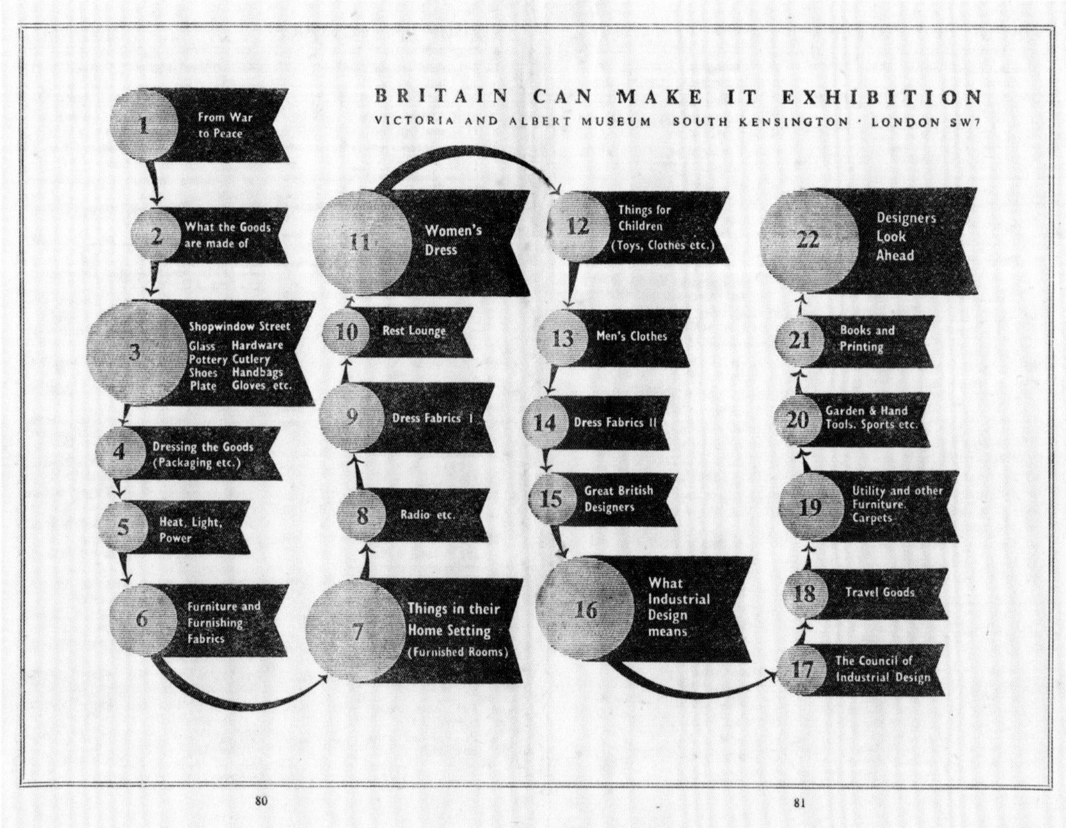

7. The route for the exhibition explained in the BCMI catalogue:

1. From War to Peace
2. What the Goods are made of
3. Shopwindow Street
4. Dressing the Goods
5. Heat, Light, Power
6. Furniture and Furnishing Fabrics
7. Things in their Home Setting (Furnished Rooms)
8. Radio etc.
9. Dress Fabrics I
10. Rest Lounge (Tea Lounge)
11. Women's dress
12. Things for Children
13. Men's Clothes
14. Dress Fabrics II
15. Great British Designers
16. What Industrial Design means
17. The Council of Industrial Design
18. Travel Goods
19. Utility and other Furniture
20. Garden & Hand Tools, Sports (Weekend)
21. Books and Printing
22. Designers Look Ahead: The Hall of the Future

V&A officials were wary of the anticipated criticism they might face in putting on the CoID's exhibition, when, surely, their priority should be in returning their own objects to display, following their removal for safekeeping during the war. When the formal decision to use the V&A as the venue for BCMI was about to be announced, Leigh Ashton, then Director of the V&A, sought to distance the museum from any responsibility, by requesting that it was made clear it was a government decision. Any criticism, thought Ashton, must be offset by explaining publicly that the exhibition had greater importance than the V&A's normal activities, and was 'part of the campaign for our vital overseas trade'.[29] The slightly testy relationship between the V&A and the CoID at the time of BCMI was touched upon in relation to a talk given in 1952 by Leigh Ashton at the Royal Society of Arts, on '100 Years of the Victoria and Albert Museum'. Following the talk, Gordon Russell (then Director of the CoID), commented:

> I think one of the great things about Sir Leigh is his interest not only in the historical things but in the things of to-day. He gave my Council an opportunity

> and backed us up in every possible way, at the end of the war over the big exhibition of 'Britain Can Make It'; and in my more irresponsible moments I argue with him as to whether the Victoria and Albert Museum put the Council of Industrial Design on the map, or the Council of Industrial Design showed Londoners where to find the Victoria and Albert Museum. We have never settled this one.

Gardner also reflected on the somewhat incongruous relationship between BCMI and the V&A in his second autobiography, *The ARTful designer*: 'Our plans shocked the Director of the museum. I was converting his dignified marble halls into theatre – marrying the new Mars-Group's "functional" mode with pageantry; a sort of Camelot, a romantic twilight of tented drapes. Too romantic? Anyway, a change from boarded-up windows, leaking sandbags and weed-covered bomb damage.'[30] Indeed, the connections to the Modern Architectural Research (MARS) group and also to the Surrealist exhibition of 1936 (which Gardner visited) were picked up in a review in *The Architects' Journal*: 'Presentation generally speaking excellent and gay – better than the objects presented. Display style – born out of Paris 1937 by MARS 1936 ...'.[31]

Gardner's fixed route design meant that visitors would be guided through a prescribed sequence of themes and products [**7**; see **187**]. A handout for teachers explained that 'The layout of the Exhibition is rather like a book with various chapters each written by a separate hand'.[32] The writer, Herbert Read was forceful in his distaste for Gardner's forced route, describing it as 'a labyrinth, a vast intestinal tract, convoluted, with here and there a bulbous expansion or a blind duct: traffic is peristaltic and progress inevitably in one-direction.'[33]

The introductory sections to the exhibition began with a display marking the end of war and the beginnings of peace, followed by another explaining the variety of materials from which the goods were to be made. Manufactured products representing key consumer groups, including clothing, leisure and household ware, were displayed throughout the rest of the exhibition. These were diverse and wide-ranging, from pottery and glass to cutlery, radios, women's and men's wear, furniture, fabrics, toys, jewellery, cookers, taps, buttons, sporting equipment, books and printing. 'Furnished Rooms', displayed in the basement area of the Museum, sought to show how a range of different people might live. Other sections in the exhibition explained the role of the CoID and developments in packaging design; and there was an historical survey of great British designers. One attempted to explain 'What Industrial Design Means'. The 'Hall of the Future', at the very end of the exhibition, aimed to leave visitors with something to think about.

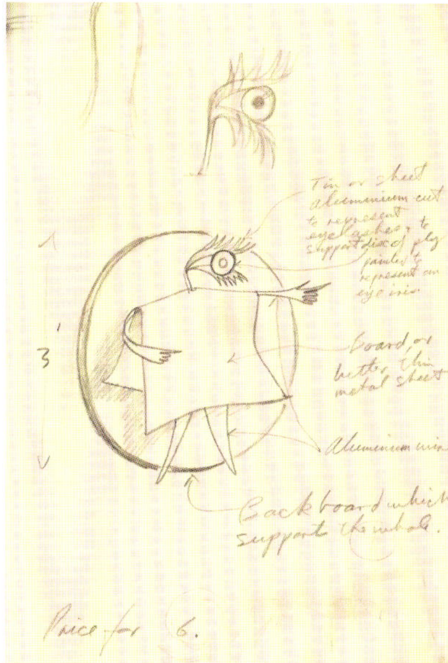

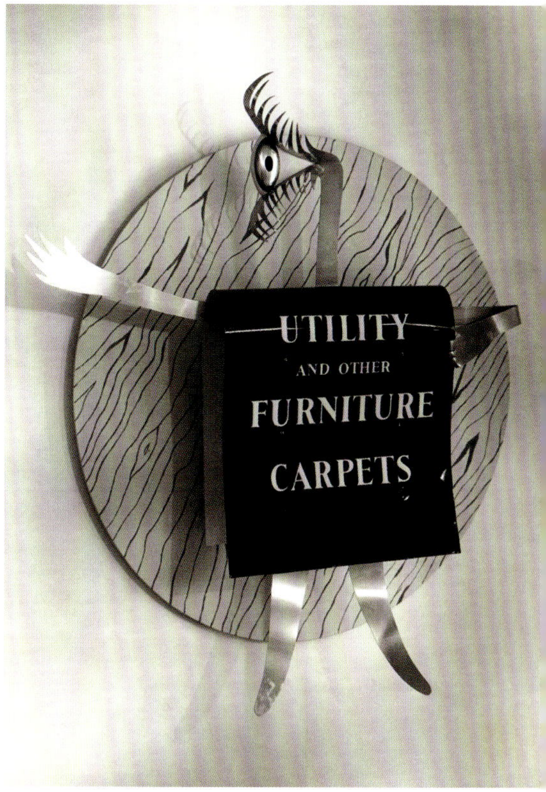

8. James Gardner's original sketch design for the 'Mr Designer' symbol used throughout the exhibition

9. One of the directional 'Mr Designer' figures

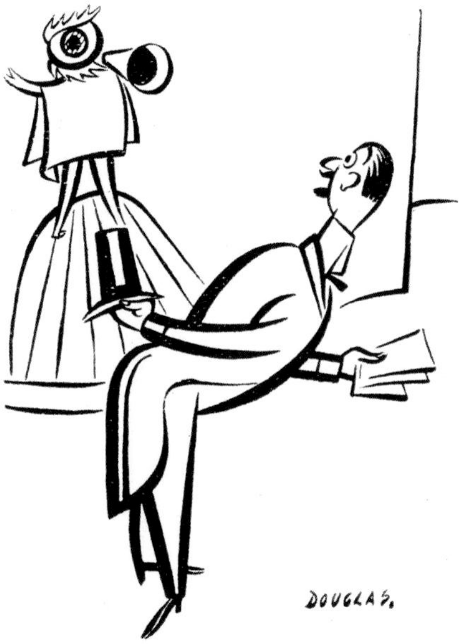

10. Thomas Douglas, a *Punch* cartoonist lampoons the speaking 'Mr Designer'. Published 30 October 1946

A key figure in BCMI was a quirky mannequin and 'symbolic sandwich board man' made of sheet aluminium with a giant bloodshot Perspex eye. This was the character 'Mr Designer', a motif of the whole show 'with an eye for design and a message'. Designed by James Gardner [8] and located throughout the exhibition at key points, 'Mr Designer' assisted the visitor with wayfinding to the various sections of the exhibition, as well as to the restaurant, rest areas and toilets [9]. He was also used to engage the public directly in displays by highlighting key themes for the visitor, pointing at signs, one of which said: 'Take a good look. Think over what you see. And judge for yourself.' Some 'Mr Designer' mannequins also 'spoke' to visitors through a hidden loudspeaker, providing recorded information about what they might see in the next display. *Punch* thought visitors would feel 'a faint uneasiness at the welcoming figure with the single enormous eye and the loud-speaker, but be reassured when its message ("Britain" – pause – "Can Make It!") carries the comfortable ring of B.B.C. English' [**10**; see **85**].[34] Further reinforcing the role of the designer, versions of the 'Mr Designer' figure displayed in the introductory section to the exhibition had the words 'I DESIGN' emblazoned across his aluminium chest in bold white hand-painted lettering.[35] The eye motif seen on the 'Mr Designer' figures was also used to dramatic effect by Basil Spence in the 'Hall of the Future', where the gigantic 'all seeing designer's eye' looked down on to the visitors and display [see **171–173, 188**].

Cutting it extremely fine, hundreds of staff were working overtime in the week of the opening to get the exhibition ready. This is perhaps unsurprising for a project that was only begun in January of the same year, with initial ideas sketched out in the autumn of 1945. One of the organizers is reported as saying on the eve of the royal opening, 'It will be a close scrape to finish it. We shall be pushing the painters out of one door as we usher in the King at the other.'[36] This was no exaggeration. The large 3-D version of Ashley Havinden's poster, erected above the entrance to BCMI on Exhibition Road, was being installed when King George VI and Queen Elizabeth arrived to open the exhibition around the corner at the Main Entrance on Cromwell Road [see **frontispiece**].[37]

Despite the queues and eager visitors, in reality things were a little different. The hurried nature of the preparations meant that there were complaints over the 'verminous condition of the premises [at the V&A]. In the hurry of preparation the place was opened before it was

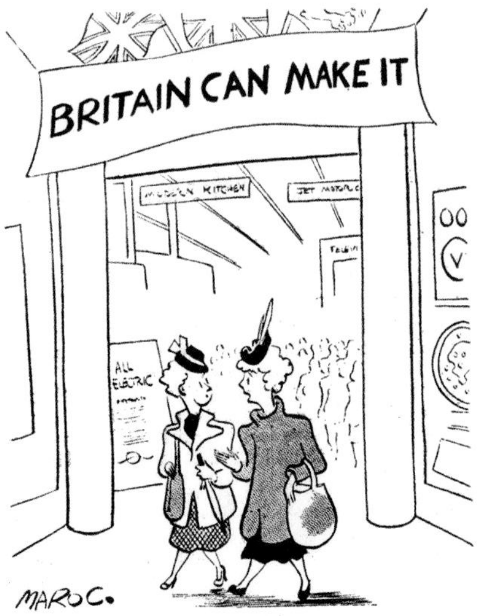

"And now I simply MUST *hunt round and try and find a second-hand mouse trap."*

11. The realities of life (compared to the idealism of BCMI) are commented upon by *Punch* cartoonist Robert Coram (Maroc). Published 6 November 1946

thoroughly cleansed. All day since it has been a seething mass of humanity including some who suffered from soap rationing.'[38]

The exhibition was criticized for tantalizingly showing the British public what it could not immediately have, or what it might be able to have at some unspecified time in the distant future. 'Britain Can't Have It', 'Britains can't buy it', 'Britain Can't Get It' were amongst the comments of its many detractors. A dream of the future perhaps, but one very much grounded in the cold realities of post-war Britain **[11]**.

NOTES

1 The National Archives (TNA) INF 6/595: Commentary Britain Can Make It issue no. 12. This was a Central Office of Information film made by Films of Fact Ltd. A copy of the film (Britain Can Make It No. 12) is held by the British Film Institute and can be accessed via BFI Player.

2 Noel Carrington, 'State and Industrial Design in Great Britain', *Graphis* (1947), vol. 3, no. 18, pp. 108–09.

3 Anthony Howard, 'We are the Masters Now'. The General Election of 5 July 1945', in Michael Sissons and Philip French (eds.), *Age of Austerity 1945–1951* (London 1964), p. 17.

4 The Council of Industrial Design, '"Britain can make it"'. The first post-war national exhibition. "Good design – and good business", reporting on Sir Stafford Cripps speech from 2 November 1945' (London 1945), 4 page unpaginated leaflet, copy in V&A Archive, Exhibition File, AAD/1977/4/12, Britain can make it. Also reported in *Display* (December 1945), vol. 27, no. 9, p. 43.

5 'The Man Cripps', *Picture Post* (10 May 1952), p. 25. The letter mentioned by S.C. Leslie in the *Picture Post* article referring to Cripps's comments about 'no special stuff', is held in the Design Council Archive at the University of Brighton, hereafter DCA: 516/Letter from Sir Stafford Cripps to S.C. Leslie, dated 10 August 1945.

6 R. Dudley Ryder, 'Inspire Industry: Build morale', *Display* (March 1946), p. 36. The Council of Industrial Design (CoID) was set up by Sir Stafford Cripps, President of the Board of Trade, in December 1944. Financed by the Government, its aim was to 'promote by all practicable means the improvement of design in the products of British industry'. Activities included the setting up of design centres and staging exhibitions. It also produced publications, films and photographs, in order to reach the public. The CoID sought to forge links with industry, make studies in the role and training of designers, and provide advice on design for the Departments of State. See *The Council of Industrial Design*, leaflet, held at the Mass Observation Archive, University of Sussex: SxMOA1/2/26/1/B; see also The Council of Industrial Design, *The Council of Industrial Design, First Annual Report 1945–1946* (London 1946). The National Savings Committee film produced by British Movietone News was filmed at BCMI on 10 October 1946 and is available to view online via the AP Archive (story number BN47601) (aparchive.com).

7 'The Box is Opened', *The Architect and Building News* (27 September 1946), p. 185.

8 Michael Sissons and Philip French, *Age of Austerity 1945–1951* (London 1964), p. 9.

9 'General Notes. The Britain Can Make It Exhibition', *Journal of the Royal Society of Arts* (17 January 1947), vol. 94, no. 4706, p. 149. *Royal Academy Exhibition of British Art in Industry*, January to March 1935 (exh. cat.), Burlington House, London; 'General Notes. Exhibition of British Industrial Design', *Journal of the Royal Society of Arts* (7 December 1945), vol. 94, no. 4706, pp. 41–42. I am grateful to Eve Watson for her help with providing me with information on the Royal Society of Arts. Designers involved in BCMI who already had the accolade of being given the Royal Society of Art's Royal Designers for Industry (RDI) award included Wells Coates, Milner Gray (awarded pre-BCMI); Basil Spence, Misha Black, George Him, James Gardner, Robert Goodden and Ashley Havinden were awarded the honour post-BCMI. Gardner, Goodden and Havinden in 1947. An online resource listing RDI's whose work was included in BCMI is available at the University of Brighton Design Archives webpages. The origins of the Council of Industrial Design were also considered to have been traced, 'through its predecessor, the Council for Art and Industry, to the impression made in official circles and on public opinion by the Exhibition of British Art in Industry, 1935'. An 'Exhibition of Exhibitions' organized by the Royal Society of Arts in 1951 illustrated five pioneer exhibitions held or initiated by the RSA, and included BCMI.

10 BBC Written Archives Centre Film no. T126: Opening of Britain Can Make It Exhibition included King George VI's speech. Pathé News also recorded the speech, 'Britain Has Made It – Shop Window', Film ID 1410.22. Another Pathé News film with the speech in full was never issued – 'King opens Britain Can Make It Exhibition 1946', Film ID 2122.01. For a discussion of the relationship between the CoID and the BBC, see Michelle Jones, 'Design and the Domestic Persuader: Television and the British Broadcasting Corporation's Promotion of Post-War "Good Design"', *Journal of Design History* (2003), vol. 16, no. 4, pp. 307–18.

11 The reference to 'no precious stuff' is cited above in note 5. Britain Can Make It film (cited note 1).

12 'BRITAIN CAN MAKE IT. Summary of Final Report on Exhibition', *British Plastics* (February 1947), vol. 19, no. 213, p. 61.

13 For a thorough examination of this period and impending crisis, see David Kynaston, 'Christ It's Bleeding Cold', in *Austerity Britain 1945–51* (London 2007), pp. 185–205.

14 BBC Written Archives Centre Film no. 57/58 Now the War is Over. 6. Britain Can Make It, first broadcast 1 November 1985 on BBC 2 Film no 57/58, page 2 of transcript. This eight-part series by the BBC was accompanied by a book: Paul Addison, *Now the War is Over. A Social History of Britain 1945–51* (London 1985).

15 Hugh Casson, '1946: knockout for tired eyes'. Hugh Casson on 'Britain Can Make It' at the V&A', *Design* (1970), pp. 52–55.

16 V&A Archive, Exhibition File: AAD/1977/4/12, Britain can make it: Note for the Information of the Press (summary), National Exhibition of Design "Britain Can Make It", Opening in September; dated February 1946. The Ministry of Information exhibition division was created in November 1942 and led by Milner Gray. Other designers, including Misha Black and F.H.K. Henrion, were also amongst those involved in the MOI exhibition division who went on to work on BCMI.

17 James Gardner, *Elephants in the Attic. The Autobiography of James Gardner* (London 1983), p. 33. The *How They Fly* exhibition included working models. It was open from 4 January to 6 April 1939 at Shell-Mex House, where the Ministry of Supply was based. See also "See How They Fly", *The Times* (5 January 1939), p. 9. V&A Archive, Exhibition File: AAD/1977/4/13, Britain can make it: Press Release "Britain Can Make It" Exhibition September 1946. Chief Designer – James Gardner. Puffin books designed by Gardner include *On the Farm* (1940), *War in the Air* (1940), *The Battle of Britain* (1941). He also produced *How they Fly* (1939) for *Country Life* to accompany the Shell-Mex exhibition of the same name. Gardner also designed the small booklet *An Air Ministry Account of the Great Days from 9th August–31st October 1940*. I am grateful to Michael Twyman for sharing his insights into James Gardner.

18 See Henrietta Goodden, *Camouflage and Art. Design for Deception in World War 2* (London 2007). CoID officials such as Jack Beddington (member of the Council's Policy Committee, and on the Selection Committee for garden tools and 'Designers Look Ahead'), Hugh Casson (judge in the 'Design Quiz'), and Sir Kenneth Clark member of the CoID and on the Selection Committee for the 'Furnished Rooms' and Art School Exhibits, were also linked with camouflage activities during the war.

19 BBC Written Archives Centre radio interview on BCMI by Sir Stafford Cripps broadcast on 27 September 1946. Home Service. Film ref. T101. The total maximum number of staff engaged on BCMI (excluding contractors), including architect's staff, administrative staff, guides, cashiers and ticket sellers, publications staff, warders, security men and turnstile attendants was 436. V&A Archive, Exhibition File, AAD/1977/4/14A, Britain can make it: Britain Can Make It. Maximum number of staff engaged on exhibition dated 27 February 1947.

20 '"Britain Can Make It" closed. Over 50,000 buyers', *The Times* (1 January 1947), p. 2. The importance of the use of colour is remarked upon in Notes on an Informal Conference held at Tilbury House, Petty France on 10 October 1945; one of the points suggested by Sir Charles Tennyson was that 'colour should be used as much as possible'. V&A Archive, Exhibition File: AAD/1977/4/12, Britain can make it.

21 Private correspondence with Raph Bonella February 2015. I am grateful to Raph Bonella, who worked as an architectural assistant on BCMI for sharing his recollections of the exhibition and for answering many questions. His personal account of his time working on BCMI is also an important contribution to this book.

22 Astragal, 'Notes & Topics. Private View', *The Architects' Journal* (3 October 1946), p. 239.

23 V&A Archive, Exhibition File: AAD/1977/4/14, Britain can make it: Memo from D. van Vlymen to Mr Jarvis dated 18 February 1947, listing all filming that took place at BCMI. Moviegram Ltd are noted as being commissioned by CoID to make a colour film of the exhibition at a cost of £1,000. Shooting took approximately 12 nights starting on 9 December 1946. The colour film is also noted in *The Council of Industrial Design*. Second *Annual Report 1946–1947 (London 1947)*, p. 15. Discussions on the possibility of obtaining three Moviegram Daylight Cine Machines, to be imported from the U.S., to use in the exhibition did not come to fruition. These are discussed in correspondence from March to June 1946 held in DCA/742; and in DCA/58: Action Points arising out of eleventh progress meeting 20 June 1946.

24 James Gardner, 'Exhibition Display', *Journal of the Royal Society of Arts* (30 January 1948), vol. 96, no. 4761, p. 165. The lecture was delivered by Gardner on 24 November 1947.

25 DCA/312/Letter to S.C. Leslie from Kenneth Clark, dated 18 August 1945. Penny Sparke (ed.), *Did Britain Make It. British Design in Context 1946–86* (London 1986), p. 3.

26 Darcy Braddell, 'General Notes. Designing an Exhibition' *Journal of the Royal Society of Arts* (1 October 1954), vol. 102, no. 4935, pp. 898–99. (Dorothy Braddell is cited as Darcy, as she sometimes used her husband's name as a pseudonym in her writings.)

27 DCA/352/Council of Industrial Design. Note of Guidance for Selection Committees.

28 The Council of Industrial Design, *The Council of Industrial Design. First Annual Report 1945–1946*, p. 25. Leigh Ashton noted the connection between the museum's founding principles and the venue for BCMI in a letter to Eric de Normann, Ministry of Works dated 4 January 1945, although probably this is a typo and it is 1946. V&A Archive, MA/28/75/1. The early collections of the South Kensington Museum (later V&A) were largely formed from purchases made at the Great Exhibition of 1851, whose profits were also used to obtain the site. See John Physick, *The Victoria and Albert Museum. The history of its building* (London 1982), p. 19.

29 V&A Archive, MA/28/75/1: cited note 28.

30 '100 Years of the Victoria & Albert Museum. A paper by Sir Leigh Ashton, F.S.A.', *Journal of the Royal Society of Arts*, 26 December 1952, vol. CI, no. 4889, p. 90. Anthony Burton also mentions the unease at holding the exhibition at the V&A in *Vision and Accident. The Story of the Victoria & Albert Museum* (London 1999), p. 204. James Gardner, *The ARTful designer. Ideas off the drawing board* (London 1993), p. 131.

31 Astragal, 'Notes & Topics', *The Architects' Journal* (3 October 1946), p. 239. The MARS (Modern Architectural Research) group were architects, engineers and town planners, 'united by a common realization of the necessity for a new conception of architecture in its relation to the structure of society today', as noted by Wells Coates, in 'Unit One', *The Listener* (12 July 1933), p. 68. Founding members included many who went on to work on BCMI, including Wells Coates, Maxwell Fry, Ralph Tubbs, Frederick Gibbert and Elizabeth Denby. The MARS Group organized exhibitions, one held in 1938 at the New Burlington Galleries, London, included room sets much like those at BCMI.

32 V&A Archive, Exhibition File: AAD/1977/4/14, Britain can make it: Notes on Britain Can Make It" for teachers.

33 Herbert Read, 'Britain Can Make It. Herbert Read on the exhibition at the Victoria and Albert Museum', *The Listener* (3 October 1946), p. 429.

34 John Basil Boothroyd, 'Britain Can Make It. A Late Look Round', *Punch* (30 October 1946), p. 374.

35 DCA/742/Letter from James Gardner to Messrs J. Starkie Gardner Ltd, dated 16 July 1946 detailing the specification for Mr Designer. Mr Everyman figures were meant to be the civilian counterparts for Mr Designer and were included in the introductory section.

36 'Sleep 'with the lid off' in the electric bed', *Daily Mail* (23 September 1946), p. 3.

37 Personal communication with Raph Bonella (see note 21).

38 V&A Archive: MA/28/75/1: Letter to Tom Driberg M.P. from the journalist Colin Cathcart dated 12 November 1946. A copy of the letter was passed to Sir Stafford Cripps, then to the CoID and thereafter to Leigh Ashton at the V&A.

DAVID KYNASTON

Britain in 1946

Two future novelists arrived in Britain in 1946. 'Everyone looked small and tired and white-faced and badly nourished,' recalled J.G. Ballard, who had spent most of the war in a Japanese civilian camp.[1] For Fay Weldon (then Franklin Birkinshaw), coming from New Zealand, it was her 15th birthday and a day of dashed expectations. 'Where were the green fields, rippling brooks and church towers?' she asked herself. 'Here was a grey harbour and a grey hillside, shrouded in a kind of murky, badly woven cloth, which as the day grew lighter proved to be a mass of tiny, dirty houses pressed up against one another, with holes gaping where bombs had fallen, as ragged as holes in the heels of lisle stockings. "It's just Tilbury," my mother said. "It's always like this." Just Tilbury? The greyness was so vast, as far as the eye could reach.'[2]

Welcome to austerity Britain. A year after the end of the war, it was above all the food rationing – as severe as it had been during the war – and the general shortages that made daily life a continuing ordeal. Take Judy Haines, a housewife living in Chingford. Her experience one Tuesday, as recorded in her very engaging diary, was almost certainly representative of that of many others:

> Got ahead with ironing and then felt I must go in quest of meat as that little chop left over from our Sunday joint will not make a very nourishing Shepherd's Pie. Dyson's very empty. I enquired tenderly if the van had called and they informed me 'no' and there should be some rabbits. I have had my hopes raised like this before, falsely. But I went home out of the cold, made myself a cup of cocoa and when halfway through it saw the van. Hastily finished my drink and set off again. Yes, there were some rabbits but they weren't ready just yet. O.K. I'll come back again. Met Mother H. [her mother-in-law] who told me List's had some nice Mince Tarts. I hardly liked leaving Dyson's, but she said she would wait there, which seemed a help though Dyson's would only let their registered customers have rabbits. Wondered if all the tarts would be gone, but I was lucky, and this will make a nice sweet

> with some custard. No sign of rabbits, so I went into the Post Office to draw my allowance, cash money order and buy National Savings Stamps. Crowds in there, but I thought Dyson's couldn't sell out of rabbits very quickly. When I returned to the shop there was a queue and only about three rabbits visible. However, I waited and more came up. I was lucky.

'This shopping!' she added. 'All housewives are fed up to the eyebrows with it.'[3] One should not exaggerate. Not only was the food situation in 1946 far less serious than on many parts of the Continent, but the war itself was still sufficiently recent for the principle of food rationing – fair shares and all that – to be widely accepted. This was true even after the Minister of Food, Sir Ben Smith, had announced in February cuts in the bacon, poultry and egg rations (the last cut made much worse by the simultaneous decision to end the importing of dried eggs), and indeed even after the controversial introduction in July of bread rationing. Revealingly, it was now that the 'spiv' – big knot in tie, hair parted in middle with wave on either side, pencil moustache – emerged as a demonized figure, despite in practice many people, including the most respectable middle class, finding themselves having recourse to the black market.

For most adults, whether housewives or not, these were intensely difficult times. Food shortages were compounded by a severe housing shortage (alleviated to a degree by the mushrooming of prefabs); men grumbled particularly about being unable to buy a new suit; the process of demobilization proved agonizingly slow, not being completed until 1947; as semi-strangers returned home, many marriages were put under huge strain, and the divorce rate rose sharply. 'No one feels well or happy just now,' the novelist Sylvia Townsend Warner, living in Dorset, wrote to a friend less than a year after VE Day.

> No one in wartime can quite escape the illusion that when the war ends things will snap back to where they were and that one will be the same age one was when it began, and able to go on from where one left off. But the temple of Janus has two doors, and the door for war and door for peace are equally marked in plain lettering, No Way Back.[4]

Inevitably, the overwhelming desire was to return as best and as soon as possible to some sort of normality. If someone remarked, 'Ah, it's just like it was before the war,' that would almost invariably be meant as praise, not criticism. Most of the married women who had been working in offices and factories now returned to being full-time homemakers, for the most part uncomplainingly; the familiar sporting rituals reappeared (with professional cricketers at last being given their

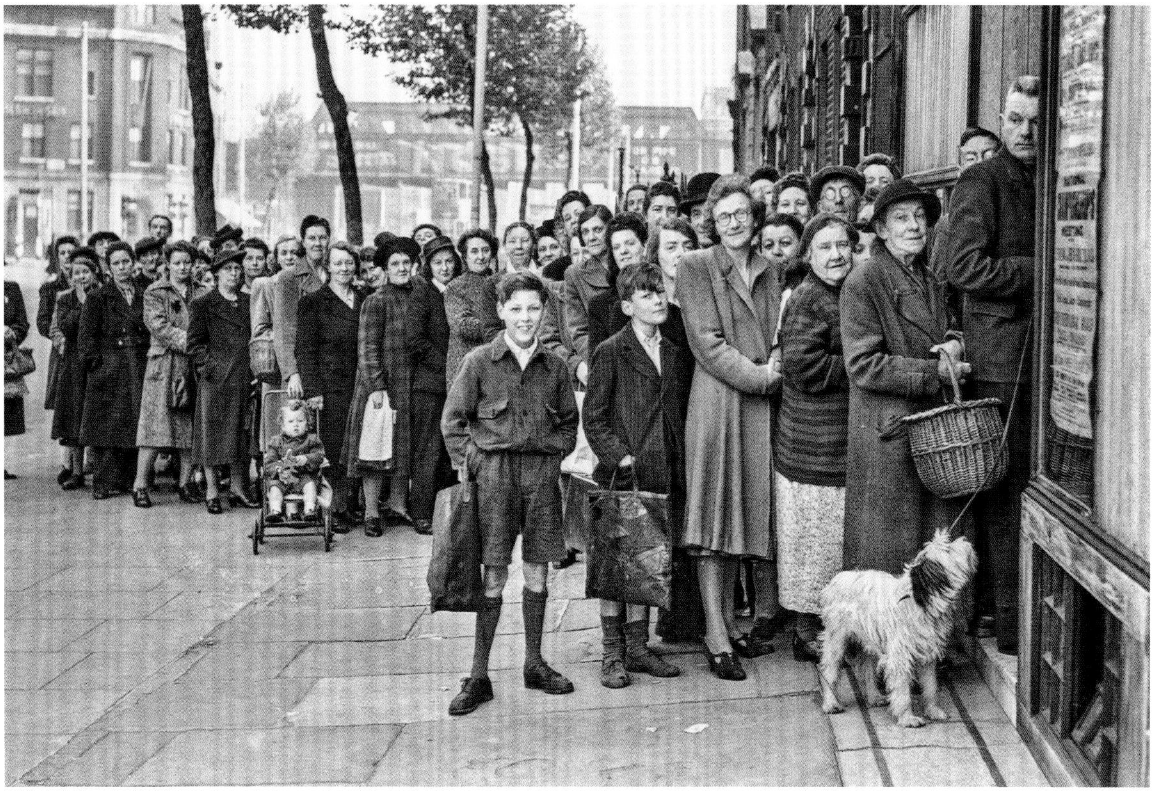

12. Queuing for bread in Blackfriars Road, London, in 1946

initials on scorecards at Lord's); and, in a last hurrah of the golden age of mass entertainment, the dance halls were heaving, the country's 4,709 cinemas were usually packed out (attracting during 1946 an all-time peak of 1,635 million admissions), and cherished programmes on the radio attracted vast listening figures, above all that great wartime favourite *ITMA (It's That Man Again)*, as the surreal yet warm Tommy Handley comedy vehicle successfully relocated to Tomtopia. In the course of the year, some 50 million tons of coal were consumed on open domestic grates: the consequence may have been appalling smoke pollution, but for most people there was no more supreme reward of peace than the quiet satisfaction of home and hearth.

Even so, amidst all this welcome normality, the signs were apparent of a rather different post-war world starting to take shape. The key political context was of course Labour's landslide election victory of July 1945 – a victory that may have owed a considerable amount to immediate circumstances (especially the housing and demobilization issues) but undeniably was also a reflection of a distinct wartime mood that looked ahead to peace as a unique opportunity to build a new Britain. By 1946 the process was under way of a widespread extension of public ownership, starting with the Bank of England and about to include the coal mines; the foundations were being laid of the modern welfare state, with that year's National Insurance Act providing universal protection from the financial perils of sickness, unemployment and old age, while Aneurin Bevan in March put forward the groundbreaking NHS Bill that would become more or less actuality just over two years later. The New Towns Act sought to create a healthier, more satisfying way of life in new, planned environments far away from the squalor of the Victorian slums; and a systematic attempt began, in large part through the recently established Arts Council and the still-Reithian BBC, to diffuse the best of culture to the population as a whole.

This great collective push was arguably vitiated by two key flaws. The first was its top-down character: too often the politicians and the Whitehall mandarins – and that whole cluster of economists, planners, architects, commentators, editors, etc., whom I would call 'activators' – were insufficiently willing to listen to the views, often more conservative and traditional than their own, of those they were honourably but paternalistically seeking to benefit. Most people did indeed want a material upgrade, for example in their living conditions; but, at the same time, they much preferred that upgrade to happen within a familiar, reassuring setting. The other flaw concerned a failure of nerve by Clement Attlee and his colleagues. 'In the social set-up there is no symptom by which one could infer that we are not living under a Conservative government,' wrote a disappointed George Orwell in 1946 in his progress report. 'Allowing for the general impoverishment,' he added, 'the upper classes are still living their accustomed life.'[5] In short, whether in relation to the House of Lords, or the class system as a whole, or private education, or gender inequality, or much else, there was no real attempt to bring about a social revolution. Probably the popular appetite was lacking for such a revolution; but given that the '1945' moment was a once-in-a-century phenomenon, it is a shame that the attempt was not made.

Of course, all this happened – or failed to happen – some seventy years ago. We no longer live in a world of shops and pubs on every corner, of heavy coins, of steam trains and smog, of Austin Sevens and Ford Eights, of white faces everywhere, of Bakelite wirelesses, of clothes put through a mangle, of make do and mend. But quite apart from its high intrinsic interest, there are still some lessons to be learned – positive and inspiriting as well as negative and cautionary – from that very distinctive world.

NOTES

1 J.G. Ballard, interview with Danny Danziger, 'The Worst of Times', *The Independent* (16 December 1991), p. 12.
2 Fay Weldon, *Auto da Fay* (London 2002), pp. 154–55.
3 David Kynaston, *Austerity Britain, 1945–51* (London 2007), pp. 105–06.
4 William Maxwell (ed.), *Sylvia Townsend Warner, Letters* (London 1982), p. 91.
5 S. Orwell and I. Angus (eds.), *The Collected Essays, Journalism and Letters of George Orwell*, vol. 4 (London 1968), pp. 220–21.

KATRINA ROYALL

The V&A 1939–46

> The damage to these walls is the result of enemy bombing during the blitz of the Second World War 1939–1945 and is left as a memorial to the enduring values of this great museum in the time of conflict.[1]

Although the Victoria and Albert Museum (V&A) initially closed at the outbreak of the war, it re-opened just two months later, on 13 November 1939, and remained open throughout the rest of the war (unlike many other museums and galleries). Staff continued to catalogue, acquire objects, plan and put on exhibitions. Understandably, this was not on the scale it had been in the preceding decades, and activities were severely restricted by the unique difficulties and circumstances of war. During the conflict, the Museum was hit many times, often severely, though very few objects were lost. Its officials also had to deal with a reduction by about half of its workforce, as staff were called to both the armed and non-combatant services, or lost their lives [**13**].

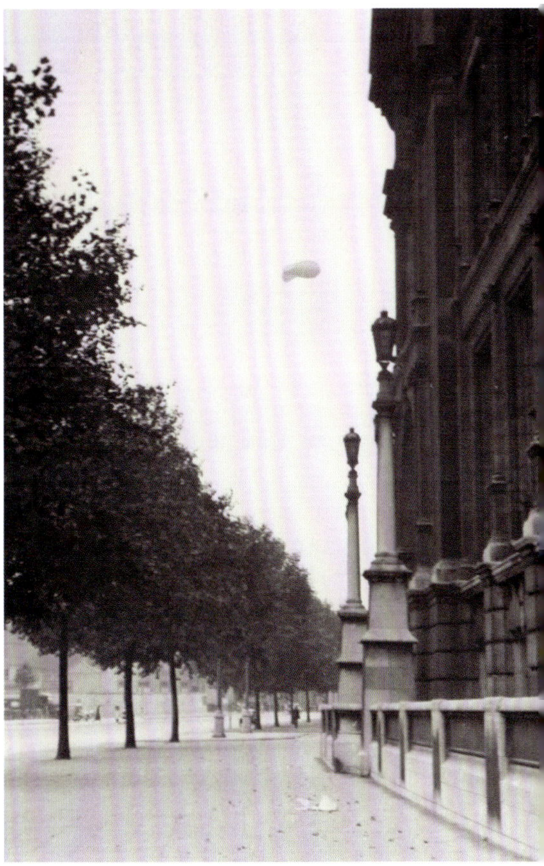

13. Barrage balloon in Exhibition Road, c.1939, with the V&A building on the right

The building and its objects

By 1939 the V&A, like other museums and galleries in London, was well prepared for the imminent outbreak of war. As far back as 1934 meetings had taken place to decide what should happen to the collections in the event of war.[2] A plan was put in place to ensure the safe removal of objects, and a rehearsal of these procedures had taken place in 1938 as a response to the Munich crisis that September, when it was feared that Nazi Germany's annexation of parts of Czechoslovakia made war imminent.[3] When war did break out, many of the smaller, portable objects were placed in bomb-proof stores in the basement of the Museum. Larger objects, such as the sixteenth-century fresco by Floriano Ferramola, remained in situ, sandbagged for protection [**14**]. A large proportion of the collections, including illuminated manuscripts, carpets, textiles, prints, drawings and furniture, were packed up within the first few months of the war and sent by lorry to Montacute House, a National

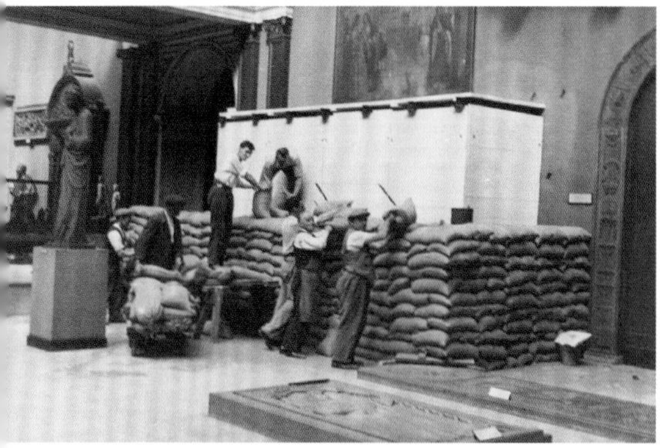

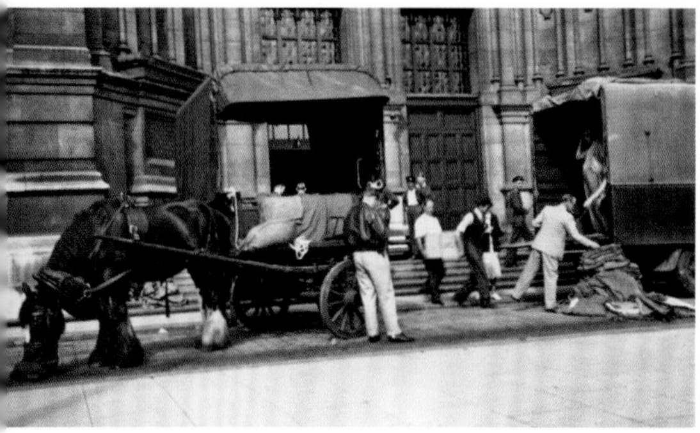

14. Preparations for war included the sandbagging of objects that could not be moved. Here, the fresco by Floriano Ferramola (inv.no. 8878–1861), c.1511, then displayed in gallery 50, is being protected by a wall of sandbags, c.1939.

15. Objects being packed up outside the Main Entrance to the V&A on Cromwell Road for removal to safe storage, c.1939

Trust property in Somerset, or the 'house in the west of England', as it was referred to for security reasons.[4] Other objects, including those held by the Departments of Architecture & Sculpture, Ceramics & Glass, Indian, and Circulation, were removed to the disused passages in Aldwych tube station, where they joined many items from the British Museum, including the Elgin Marbles [**15**].[5]

Air raid precautions (ARP) were also well in hand at the outbreak of war, with the Secretariat (west side of the V&A) shored up and reinforced to accommodate over four hundred people. These included Royal College of Art students, members of the public using the National Art library, as well as Museum staff. When the siren was sounded, visitors were quickly ushered out of the Museum and into the public air-raid shelter on Exhibition Road. As part of the ARP arrangements members of staff were asked to volunteer as firewatchers, and most signed up, working on a rota system to fit in alongside their work. Many bombs, mostly incendiary, dropped on the V&A, starting fires which were extinguished by the Museum's firemen. Their services were clearly much needed, as noted in a report by Eric Maclagan, the Museum's Director:

> the whole of one end of the roof of the Cast Courts was ablaze and burning beams of wood kept falling with loud crashes on the floor. I expected at least a quarter of all the casts would be destroyed ... but in fact once the big fire in the roof of the Cast Courts had been got under control, the water became much more alarming than the flames. We were wading sometimes ankle deep in the passages and a cascade of water mixed with lumps of charred wood was pouring down the staircase in the Secretariat.'[6]

In November 1940, at the height of the Blitz, the Museum was hit badly. Maclagan noted, 'We had our first high explosive bomb last night. The results as usual are surprising. One very heavy packing case was picked up by the bomb and thrown on the roof about 50–60 feet above the ground, where it is now peeping over the cornice'.[7]

The greatest damage to the Museum, however, took place on 19 April 1941, when the Geological Museum (part of the Natural History Museum), on the opposite side of Exhibition Road, was hit by two high explosive bombs: 'It has practically wrecked the west side of the museum,' described a report on the damage. 'The surface of the masonry was badly knocked about and the Exhibition Road doors were blown in. Practically all windows, frames and iron grilles were destroyed and we have lost most of the glass roofing on that side of the museum.'

Maclagan's first priority was however, 'the large mass of material still stored in the galleries and exposed to the wind and the weather'.[8] Despite the damage, the Museum was closed only briefly.

Whilst staff dealt with these pressing issues, the Museum was also being put to other uses. By early summer 1941 it had become a school for evacuated children from Gibraltar, much to the consternation of Maclagan, who saw it as diminishing the space for the remaining collections that still remained on site. In addition, not insignificant changes to the museum's infrastructure were required to accommodate the 350 school children and staff, including extra sanitation and space for exercise.[9]

Further worries for Maclagan surfaced in 1942, when the RAF effectively requisitioned the North and South Courts of the Museum. They were needed to house the canteen for Radio Signals Station Number 7, based next door in the Science Museum, which catered for around 1300 signal men and women a day [**16**]. Maclagan was again worried, this time about steam from the catering damaging the Leighton frescoes and the Raphael Cartoons. The Cartoons, on loan from the Royal Collection, had been a real headache for the Director from the outbreak of the war – they were considered priceless, but almost impossible to move without being damaged. When war broke out they were located on the upper floor of the Museum and initially carpets of 'lesser quality', also from the Museum's collections, were put in front of them as protection. By September 1940, when the Blitz bombing began in earnest, it was decided to take the Cartoons down to ground level. The seven Cartoons, each weighing three-quarters of a ton, were gingerly manoeuvred using slings and ropes, and brought down to gallery 14 on the ground floor (the current restaurant) and bricked up in their own shelter. Although they would not have survived a direct hit, this brick shelter was in one of the safest and most structurally sound places in the building (except of course the basement, which had been bomb-proofed).[10]

During 1941, concerns that Montacute House was no longer a safe refuge for the collections emerged. The increased range of Nazi bombers, and close proximity of perceived targets, including aircraft factories and an American base, led to plans to relocate the collections to a more impenetrable repository. Along with their counterparts at the British Museum, V&A officials decided on Westwood, an underground Bath stone quarry at Corsham, near Bradford-on-Avon, Wiltshire. By 1942 many of the collections were on the move again to their new home. Many of them would not to return to the Museum until the late 1940s.[11]

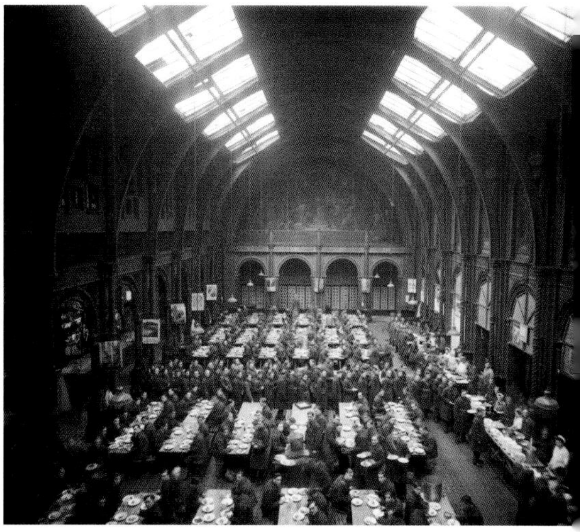

16. RAF canteen in the South Court of the Museum, c.1939

The V&A staff

Fourteen members of V&A staff died during the Second World War: nine were in the armed forces and five were civilians; twelve were men and two women. They are commemorated on a simple slate plaque in the Main Entrance of the Museum.[12] Some of those killed on active duty,

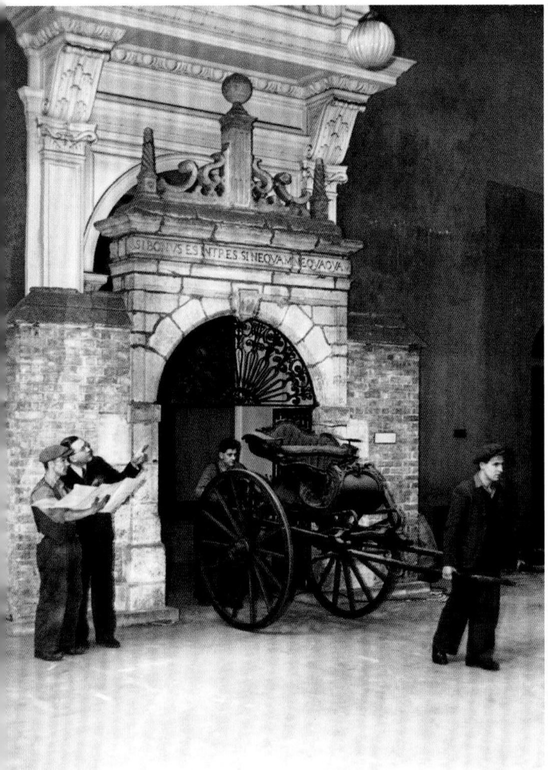

17. Sir Leigh Ashton, Director of the V&A, directing the movement of objects for BCMI, 1946. An eighteenth-century carriage (inv.no. 368–1886) is being wheeled through the sixteenth-century Ascott House gateway (inv. no. A.5–1925).

such as George Morton (Warder), a Gunner, died in the early days of the D-Day landings; Lance Sergeant John Melless (Warder) was killed in the Battle for Monte Cassino in Italy; John Roberts (Assistant Keeper), a Lieutenant in the Life Guards, was killed in action during the Moro River Campaign in Italy; Charles Cheverton (Attendant) was a Flying Officer and Test Pilot based in India and died in an accident while testing a Spitfire's readiness for combat, and Able Seaman Herbert Mooring (Labourer) was drowned when H.M.S. Avenger was sunk off Gibraltar.

The staff who were left at the Museum were also dwindling, often seconded to non-combatant war work or other departments of Government. Leigh Ashton, then Keeper of Architecture and Sculpture (later Director), went to the Ministry of Information (MOI); Stanley Hutchinson, Museum Photographer, to the Air Ministry, and Basil (Robbie) Robinson to the Ministry of Home Security (he would later join the Army and take part in the war in Burma). Ashton's work for the MOI took him to the Middle East, where, during the 1920s, he had travelled on behalf of the Museum. In addition to his war work, Ashton kept his eyes open for objects for the Museum. A flurry of communications between Ashton and Maclagan reveals that he bought about 200 objects whilst in the Middle East. Running out of money, he wired Maclagan for help. During the war the Museum had no acquisition fund, so Maclagan contacted the Art Collections Fund, who provided money. A relieved Ashton replied that it was not possible to ship the objects back under the present circumstances: 'I'm afraid we will have to wait now till after the war as we can't really use the shipping space. I hope Rommel doesn't get them.'[13]

Everyday curatorial jobs, including cataloguing of new acquisitions, mainly bequests, continued. Retired curators, such as Martin Hardie, former Keeper of Engravings, Illustration, Design and Painting, came in to help out. An example of one of the bequests in 1943 was from the artist Philip Wilson Steer of his paintings and sketchbooks, which were put on display in the Museum in Room 74, then part of the National Art Library.[14] Although much of the Museum was closed to the public, the National Art Library would remain open throughout the war. Even though much of the book collection was moved to Montacute House or the Bodleian Library for safekeeping, many books and periodicals remained on site. It proved to be a much used resource, as the Library ledgers show, allowing people to continue to research, read and reflect in a quiet space, despite the war.

Exhibitions

A programme of exhibitions continued at the V&A throughout the war, partly to raise public morale and relieve the boredom of life on the home front. Visitors looking for escape often included troops on leave. A letter to the Editor in *The Times* recorded:

> I was pleasantly surprised to find out, by accident, that the Victoria and Albert Museum is open, or sufficiently open to give a visitor a show well worth seeing ... there are many thousands of soldiers, sailors, and airmen here from many parts of the world who would probably enjoy seeing our many artistic treasures housed there.[15]

With the National Art library remaining open, displays had to be installed in galleries that were en route to the library from the Exhibition Road entrance, which acted as the main entrance to the Museum during the war. To lessen the possible damage to the museum's collections the format of many of the exhibitions at this time was to use perceived lesser material, such as prints and photographs of objects, rather than 'originals' themselves. Significantly, V&A officials were prepared to risk 'sacrificing' one of the sculptures by Eric Gill – *St Sebastian* in Hoptonwood stone – by including it in a exhibition devoted to his work in 1941.[16]

Further exhibitions took place on a wide range of topics, including Van Dyck (1941); watercolour drawings and sculpture (1942); British landscape in oils (1942); Old Master drawings from the Witt Collection (1943) and Holbein (1943). Other exhibitions included *Recording Britain*, showing drawings and paintings made during the war (1944); *How to Understand Modern Painting* (1945); and a memorial exhibition on the calligrapher Edward Johnston (1945).[17] As the war was coming to an end plans by the British Council to shore up Anglo-French relations resulted in a much talked-about exhibition of the works of Picasso and Matisse in 1945.[18]

Britain Can Make It

Although the Museum had shown great resolution in continuing to put on exhibitions during the war, when the idea of an exhibition to promote British manufacturing and exports was mooted the V&A was not seen as the best venue, and more suitable exhibition areas including Earl's Court were suggested. The plans did not initially win the support of Leigh Ashton, who had become the V&A's Director in 1945 [**17**]. He worried that it would prevent the return of objects to display and expose the Museum to criticism. Months before *Britain Can Make It* (BCMI), in a letter to Eric de Normann (Ministry of Works), Ashton wrote of his concerns:

> ... the impression seems to be abroad that the V&A is going to gain a great deal by having the Council of Industrial Design exhibition here. My own view and that of my Advisory Council is that it will in fact completely dislocate our reconstruction work for over a year, and that, while we shall get some advantages, these will certainly be offset by the condition in which the galleries will be left. So I hope that when it is all over I shall not be told in return for its gallant gesture, that the V&A has had all that it is entitled to, and that we cannot ask for any repairs or restoration.[19]

The final decision, however, was not in Ashton's hands, and his antipathy was tempered by the concession that labour and materials were to be immediately sent to the Museum for essential repairs.[20]

The use of the V&A as a venue for BCMI was derided by some, who felt it totally inappropriate. Others were amazed at the transformation the exhibition brought to some of the Museum's galleries: 'The basement of the Victoria And Albert has been temporarily transformed from a display case of the past into a shop window of the future'.[21]

That the V&A continued to open almost every day throughout the war is remarkable enough. That it was able to host an exhibition of the size and ambition of BCMI just 12 months after V.E. Day, despite post-war shortages and problems, is extraordinary.

NOTES

1 The inscription is incised into part of the bomb-damaged Portland stone on the exterior wall of the Aston Webb façade near to the Exhibition Road entrance. It was produced by the letter-cutter David Kindersley in 1987 (see Diane Bilbey with Marjorie Trusted, *British Sculpture 1470 to 2000. A Concise Catalogue of the Collection at the Victoria & Albert Museum* (London 2002), p. 458). For further information on the history of the Museum during the Second World War and just after, see John Physick, *The Victoria and Albert Museum. The history of its building* (London 1982), pp. 269–71, and Anthony Burton, *Vision and Accident. The Story of the Victoria & Albert Museum* (London 1999), pp. 193–97.

2 V&A Archive, ED/84/264.

3 Ibid.

4 V&A Archive, ED/84/267. The use of Montecute House in Somerset had been agreed with the National Trust before the outbreak of war at an annual rent of £450 per year.

5 Transport for London information on disused underground stations https//www.tfl.gov.uk/cdn/static/cms/documents/disused-underground-stations.pdf.

6 V&A Archive, ED/84/264: Report by Eric Maclagan dated 10 March 1941. The bombing took place on 8 March 1941.

7 V&A Archive, MA/2/R: Letter to Sir Maurice Holmes, Board of Education from Eric Maclagan, dated 8 November 1940.

8 V&A Archive, ED/84/264: Letter to Sir Maurice Holmes, Board of Education from Eric Maclagan, dated 21 April 1941.

9 V&A Archive, ED/84/263: Papers on provision of space for Gibraltar children etc.

10 The Raphael Cartoons are full-size designs for tapestries. They are approximately 10ft by 13ft and illustrate passages from the Bible, lives of Saints Peter and Paul. They have been on loan to the Museum from the Crown, initially from Queen Victoria, since 1865. When the Cartoons were removed to their brick shelter in gallery 14, Maclagan was concerned about their proximity to the RAF restaurant, and noted: 'I have no wish to be unnecessarily alarmist, but after thinking over the whole business again I still feel that it ought not to be allowed'. V&A Archive, ED/84/263: Papers on provision of space for Gibraltar children etc. Letter to Sir Kenneth Clark from Eric Maclagan dated 11 September 1941.

11 The quarry was made ready to house the collections at a cost of £20,000. V&A Archive, ED/84/269, and cited in David M. Wilson, *The British Museum. A History* (London 2002), p. 251.

12 See Bilbey with Trusted (cited note 1), cat. 755, p. 465.

13 V&A Archive, MA/1/A985: Letter to Sir Robert Witt from Leigh Ashton dated 17 November 1941. The objects, including Egyptian textiles, an agate bowl and a crystal bottle, remained in the British Embassy in Ankara until peace was declared.

14 Wilson Steer's bequest resulted in 45 sketchbooks being given to the Museum, inv. nos. E.268 to 312–1943.

15 Henry J. Wagg, letter to the editor, 'Victoria and Albert Museum', *The Times* (1 March 1942), p. 5.

16 The Eric Gill memorial display began in January 1941. The *Nude Figure* by Gill was given to the Museum in 1934, V&A inv. no. A.10–1934. It was transferred to the Tate in 1983 along with further British twentieth-century sculpture, to accord with the agreement between the V&A and Tate that twentieth-century sculpture would be their remit. Tate inv. no. T04745.

17 The *Watercolour drawings and sculpture 1919–1939* exhibition, held in 1942, was considered a great success. The artist William Russell Flint wrote to Maclagan, 'please accept a word of hearty compliment on the excellence of the watercolour exhibition at the V&A, I thought it admirable.' Letter from Russell Flint to Eric Maclagan dated 10 August 1942. V&A Archive, VX.1942.001. This exhibition was organized by the Department of Paintings at the V&A and took place between March and August 1942 in Room 74. A typescript catalogue is held by the National Art Library. For a complete list of exhibitions that took place at the V&A see Elizabeth James, *The Victoria and Albert Museum. A Bibliography and Exhibition Chronology, 1852–1996* (London & Chicago 1998), pp. 534–36.

18 *Picasso, Matisse* ran from 1 November 1945 to 11 January 1946. It was organized by l'Association française d'action artistique and The British Council. A catalogue from this exhibition is held in the National Art Library.

19 V&A Archive, MA/28/75/1: extract from letter to Mr De Normann dated 21 February 1946 from Mr Leigh Ashton.

20 V&A Archive, MA/28/75/1. The decision to hold the BCMI exhibition was made by 1 January 1946.

21 'Britain Can Make It. Victoria & Albert Museum', *The Architect and Building News* (20 September 1946), p. 177.

DIANE BILBEY

Naming the Exhibition

The Council has advised me that it wishes to rename the Exhibition, and to call it 'Britain Can Make It' – and I think it is a good title, signifying as it does that we have passed from the years of endurance to the years of achievement, from destruction to creation, and that Britain can not only 'make the goods' but 'make the grade' too.
—*Sir Stafford Cripps, President of the Board of Trade, 2 November 1945*[1]

London Can Take It

'Britain Can Make It' was not the Council of Industrial Design's (CoID) first idea for naming what would become its blockbuster post-war exhibition. The title did not appear out of thin air, either; wartime slogans and film titles reflect the same stoic tone – that of Britain 'making' or 'taking' it in the face of adversity.

The capitalization of all the initial capitals in the title of *Britain Can Make It* appears to be a wartime trend. The Ministry of Information film of 1940, *London Can Take It*, was a poignant and moving documentary showing the resilience of Londoners during one night of the Blitz. Narrated by the American writer and war commentator Quentin Reynolds, the film was produced for an American audience to rally emotional as well as tangible support for Britain by showing the effects of war on its civilian population, in its iconic capital city. Sponsored by Warner Brothers, *London Can Take It* was shown in 12,000 theatres throughout the United States, and a shortened version, renamed for a British audience to the more inclusive *Britain Can Take It*, was also shown in British cinemas.[2]

Britain Can Take It

During 1941 Her Majesty's Stationery Office (HMSO) used the 'Britain Can Take It' slogan for a propagandist pamphlet, adding, defiantly, '... and give it back'.[3] The artist Ruskin Spear was also asked to produce a painting with similar terminology in a commission by the War Artists'

Advisory Committee in 1942. Spear painted a sombre scene of three small children, survivors of the Blitz, standing on a bombsite. Behind them, the remains of a wall are daubed with the graffiti slogan of *WE CAN TAKE IT*.[4]

The rousing wartime speeches of Herbert Morrison, Minister of Supply and later Home Secretary in the Labour government, have also been cited as a source for the 'Britain Can Make It' slogan. Writing on the conference held in conjunction with *Britain Can Make It* (BCMI), the design and architecture commentator John Gloag remarked on the naming of the exhibition:

> This conference is held in what may prove to be the most significant week in the history of British industrial design, for on Tuesday the King opened the 'Britain Can Make It' Exhibition. The title, a memorable variation of one of Mr Herbert Morrison's wartime slogans, not only reminds us and other nations of the fact that we were once the workshop of the world, but underlines the fact that we can still maintain our reputation as machine craftsmen .... Had an exhibition of British industrial design been held in 1920, the title might well have been 'Britain Can Fake It'; for at that time we were still in the disguise phase of design in industry and mechanically-produced goods were still contemptuously stigmatized as 'machine made'. To say that a thing was 'machine made' was to imply a basic inferiority of conception, execution and material.[5]

Morrison had used the slogans 'Britain Can Take It', 'Britain Can Make It' and 'Britain is Making It' in a debate on the increased production of goods in the House of Commons on 28 February 1946:

> Nobody would suggest that we have yet gone any distance in meeting the vast accumulated demands of the war years, but we are moving, and the signs of progress are encouraging and material. On the other hand, it would be defeatist not to recognise the beginnings of some reward for all the nation's labour over these years. The simple fact is that the way to have plenty, is to make plenty. We have been through the blitz when our motto was, 'Britain can take it.' We have kept on telling our overseas customers, 'Britain can make it.' Now let us tell ourselves and the whole world, 'Britain is making it.' That is a true statement. We are making the things, and before this year ends, we shall have more men and women making goods than at any previous moment in our history.[6]

The *Daily Mail* reported on Morrison's House of Commons speech, and noted: 'He [Morrison] coined a new slogan especially for experts. "Britain Can Take It" he said, saw us through the blitz. "Britain Can Make It" and "Britain is Making It" should be the new cries.' Another *Daily Mail* article, entitled 'The best news for seven years about things to buy summed up in – Four little words', wrote about the title of the exhibition: '"Britain Can Make It", a paraphrase of the "Britain Can Take It" war-time slogan. Not very original, but it tells the tale in four little words'.[7]

The long list

Discussions amongst officials at the CoID on a possible title for the exhibition began with fairly unimaginative placeholder ideas, including 'Summer Exhibition 1946', 'The British Exhibition of 1946', and 'National Exhibition of Design'. Gordon Russell, designer, furniture manufacturer and member of the Council, suggested in a letter to S.C. Leslie, Director of the CoID, in the summer of 1945 that the exhibition should contain a patriotic and uplifting message. With reference to those who had come before in the history of British design, he suggested something along the lines of 'Let's go on being pioneers'.[8]

Suggestions for a title appear to have gained a more patriotic flavour as discussions with Council officials intensified. They culminated in a draft title reflecting the changeover from war to peace, 'Swords into Ploughshares: British goods for the new age', which was revealed to the public in a press release from the CoID on 3 October 1945. F.A. Mercer, editor of the journal *Art & Industry*, was one who was openly critical of this title, suggesting that it sounded like an agricultural show.[9] But the Council had still not settled definitely on a title. By 10 October 1945, when Sir Charles Tennyson, Chairman of the Exhibitions Committee at the CoID, wrote on the scope of the proposed exhibition, he noted that 'We have not yet found a title to our satisfaction. One of the visitors to the Conference thought that the note of the Exhibition must be something to appeal to the imagination and make the visitor think to himself "Something has happened to England". I would like something like this introduced into the title – something new and unexpected – something of the future.'[10]

A long list of possible titles (not yet including 'Britain Can Make It') was written in a memorandum of 17 October 1945 [18]. Out of a list of 31 suggestions, there were only two that C.A. Mesling (Industrial Liaison Officer at the CoID) thought might be suitable – 'British Made' and 'Power for Peace' – and these were put forward to S.C. Leslie. Further possibilities using war-time terminology included 'Britain's Victory Show', 'Passport to Peace' and 'Victory in Industry'. Others emphasized a perceived British superiority in design and industry – 'The Lion on the Label', 'Hallmark – British' and 'Good British Goods'.[11]

By 25 October 1945, however, with the announcement by Sir Stafford Cripps of the Council's decision to rename the exhibition, the title

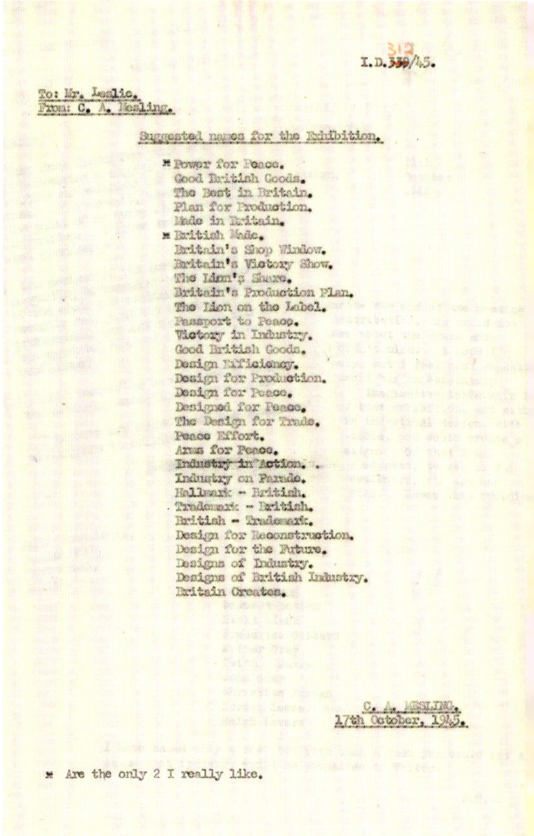
I.D.312/45.

To: Mr. Leslie.
From: C. A. Mesling.

Suggested names for the Exhibition.

* Power for Peace.
Good British Goods.
The Best in Britain.
Plan for Production.
Made in Britain.
* British Made.
Britain's Shop Window.
Britain's Victory Show.
The Lion's Share.
Britain's Production Plan.
The Lion on the Label.
Passport to Peace.
Victory in Industry.
Good British Goods.
Design Efficiency.
Design for Production.
Design for Peace.
Designed for Peace.
The Design for Trade.
Peace Effort.
Arms for Peace.
~~Industry in Action.~~
Industry on Parade.
Hallmark – British.
Trademark – British.
British – Trademark.
Design for Reconstruction.
Design for the Future.
Designs of Industry.
Designs of British Industry.
Britain Creates.

C. A. MESLING.
17th October, 1945.

* Are the only 2 I really like.

18. Memorandum from C.A. Mesling, member of the Exhibition Policy Committee to S.C. Leslie, Director of the CoID, dated 17 October 1945 showing list of possible names suggested for the exhibition. At this stage, 'Britain Can Make It' is not identified.

'Britain Can Make It' had somehow been decided upon. S.C. Leslie credited the naming of the exhibition to the Council's Librarian Miss Goodwin but, frustratingly, no reasons for her suggesting the name, nor for the Council deciding on it, are recorded.[12] Whilst endorsing the aims of the exhibition, the designer (and writer on design) Noel Carrington was immediately critical of the title: 'In spite of its somewhat meaningless title the exhibition's main theme is the function of the designer'.[13]

The 'Britain Can Make It' films

At roughly the same time, a series of twelve Central Office of Information films was produced, rather confusingly entitled 'Britain Can Make It'. These films were issued monthly and were described as a 'one-reel film magazine'. Beginning in January 1946, the series covered diverse aspects of British life, such as factory time-and-motion studies; the War Artists exhibition; women's shoes made from new plastics; the use of aluminium in pre-fabricated homes; and advice as to how the British people should adjust to the changeover from war to peace. The final film in the series (no. 12) was the only one directly connected with the BCMI exhibition, and was entirely dedicated to it. This was for non-theatrical distribution and was screened after the exhibition had closed in February 1947. Produced by Paul Rotha Productions (Films of Fact) on behalf of Crown Films Ltd, it was intended for distribution at home to factories and clubs, and also for overseas. This beautifully produced short film is now available for all to see via the BFI Player.[14]

Variations on 'Britain Can Make It'

Evidence of the catchiness of the name is reflected by the fact that, quite quickly, it was corrupted, with jokey variations being widely used. In January 1946, Ashley Havinden (who was to be appointed both designer of the men's wear section and of the poster publicizing the exhibition) commented, in an article on industrial design, that 'Britain Can Make It' 'was an "inspiring" title but no less important to establish that "Britain Can Sell It"'.[15] It appears that even Queen Elizabeth is reported to have said wryly (and privately) to one of her confidants following her official visit to the opening of the exhibition, 'But no one can get it'.[16]

Perhaps the most widely used corruption of the name was 'Britain Can't Have It'. Most regularly used in the popular press, it lampooned the stark contradictions between the aims of the exhibition and the realities of British life. During the opening week of the exhibition the *Evening Standard* used the headline 'Britons can't buy it', and explained exactly why it thought that this was the case:

But what Britain can make, the Government can take away. Exhibit after exhibit is catalogued 'available now for export', but only available 'soon' or 'later' for the home market .... Britain Can Make It is the most frustrating show on earth.[17]

Walter Bernays, a Swiss journalist sent to London to cover the exhibition, reported that derogatory and sneering remarks were widespread:

In my hotel and on the street I heard sarcastic remarks such as 'Britain Can't Get It'. Is it really so bad as that? I found in many London shop windows a remarkable number of goods which not so long ago British tourists in Switzerland told me were impossible to get .... Although I have come from a land of plenty and abundance, I have seen several things – goods and materials, not only unavailable, but so far unknown in Switzerland. No doubt, now Britain Can Make It – Switzerland will Buy it too![18]

The designer and organizer of the *Daily Mail* Ideal Home exhibitions, Ian Jeffcott, wrote that the title of the exhibition was misguided:

Sir Stafford Cripps invented, one must assume, the phrase: 'Britain Can Make It'. As a Briton, may I say to him quite personally, 'Really!' Did not Britain make it in the high days of British craftsmanship; in the 15th and 16th centuries; in the murky days of the Industrial Revolution, afterwards in the Victorian artistic twilight, and lately when we needed tools during the war? Cannot – has not – Britain always made it? Sir Stafford's slogan should have been, what in fact his exhibition has proved: 'Britain Can Design It'.[19]

Others focused their attention on different aspects of British life, including art and sport, to subvert the 'Britain Can Make It' title. Cartoons in the magazine *Punch* included 'Britain Shan't Break It' to describe a contemporary modern sculpture (of the Henry Moore reclining figure type) and 'Can Britain Make It?' relating to the 1946 Ashes series showing the Labour politician Emanuel (Manny) Shinwell as Umpire.[20]

The journal *Display* wrote enthusiastically:

'Britain Can Make It' proves the promise of its title. Whether its exhibits are a truly representative collection of the best that Britain can offer it is not for us to say. But the display which the vehicle for presenting them is so buoyant that it all but swamps them. The technique of presentation has been carried to new heights; display reaches new peaks of virtuosity![21]

Questions on BCMI's success and possible legacy were raised in the Houses of Parliament on 28 October 1946. In addition to being asked why the exhibition could not be sent to other parts of the country, Sir Stafford Cripps was quizzed by Major Tufton Beamish, then Conservative MP for Lewes, about the value of cultivating the home market:

> Is the right hon. and learned Gentleman aware that this Exhibition is described as the 'Britain Can't Get It' Exhibition, and that much greater incentive to production would be provided if he would make more goods available to the home market?'

Cripps replied, rather waspishly, that 'I think it is only so described by the very ignorant'.[22]

The impact and brand value of the 'Britain Can Make It' name overseas was just as significant. The Council's *Annual Report* noted that 'The phrase "Britain Can Make It", in English, appeared in the headlines of dozens of newspapers overseas, and it seems that, incidentally, the Council may have created a useful commercial slogan for the future'.[23] Coverage following the close of the exhibition in the all-important overseas press was also reported in the journal *British Plastics*:

> A very large number of overseas newspapers did not translate the title 'Britain Can Make It' into their own languages. The phrase has gained an international currency. The Exhibition has thus created a new and a valuable slogan for British industry in the markets of the world.[24]

Unlike the items on display, the exhibition's nay-sayers and sniping commentators were not in short supply. And that the title 'Britain Can Make It' was so divisive, lampooned and even derided may, in retrospect, be taken as a strong indication of its success in capturing the popular imagination and of it being so memorable. From today's perspective, 'Britain Can Make It' seems to be the perfect title for an exhibition seeking to inspire visitors, motivate manufacturers and designers, and present a confident, even defiant, face to the world.

NOTES

1 The Council of Industrial Design, '"Britain can make it". The first post-war national exhibition 'Good design - and good business' reporting on Sir Stafford Cripps speech from 2 November 1945' (London 1945), 4-page unpaginated leaflet, copy in V&A Archive, Exhibition File: AAD/1977/4/12, Britain can make it. The Council lunch meeting was also reported in 'Design as key to export trade. Government help for exhibition', *The Times* (3 November 1945), p. 6.

2 Douglas Williams and Quentin Reynolds, *Britain Can Take It; The Book of the Film; A photographic record in aid of the homeless* (London 1941). A. Aldgate and J. Richards, *Britain Can Take It: The British cinema in the Second World War* (London 2007), pp. 120–21.

3 A copy is held in the Imperial War Museum K.7300. I am grateful to the Trustees of the Imperial War Museums for allowing access to the collections.

4 Imperial War Museums. The Papers of the Second World War Artists Archive/Ruskin Spear GP/55/280. The subject and site of Spear's 'We can take it' and 'Blackout' paintings were decided by Sir Geoffrey Burton, Ministry of Supply. Ruskin Spear, *We can take it.* Watercolour, 1942. Imperial War Museums Art. IWM. ART LD 2580. Exhibited in *Blood Sweat & Tears – 1940 and all that*, Parkin Gallery, London, 15 February–17 March 1973, cat. 78. Spear's painting is included in Brian Foss, *Warpaint: art, war, state and identity in Britain 1939–1945* (New Haven and London 2007), p. 71. I am grateful to Claire Brenard for drawing my attention to the book by Brian Foss.

5 John Gloag, 'Design for Machinery', *Journal of the Royal Society of Arts* (25 October 1946), p. 696.

6 Hansard, HC Deb (28 February 1946), vol. 419, cc2125, reported in *The Times* (1 March 1946), p. 4.

7 Percy Cater, '''Patience'' pleads Cripps', *Daily Mail* (1 March 1946), p. 1. Graham Stanford, 'Four little words', *Daily Mail* (27 August 1946), p. 2.

8 Design Council Archive at the University of Brighton Design Archive, hereafter DCA: 312/'Summer Exhibition'/Letter from Gordon Russell with thoughts on an exhibition to S.C. Leslie Summer 1945, titled 'Cripps Exhibition. Let's go on being pioneers'.

9 DCA/312/The British Exhibition of 1946. The Council of Industrial Design. Proposed Exhibition – Summer 1946 Draft plan 2. A Proposal to the President of the Board of Trade from the Council of Industrial Design 22 August 1945; National Exhibition of Design, 3.10.45 no. 6. National Exhibition of Design. Press release 3/10/45 no. 6, 'British goods in the future. National display next year', *The Times* (4 October 1945), p. 8. V&A Archive, Exhibition File: AAD/1977/4/12: Summer Exhibition 1946. Notes on an Informal Conference held at Tilbury House, Petty France, SW1, 10 October 1945.

10 DCA/312/Summer Exhibition, Appendix to Council Minutes 12.10.45 CM (45) 9. Scope of the Exhibition. Report by Sir Charles Tennyson.

11 Patrick J. Maguire and Jonathan M. Woodham (eds.), *Design and Cultural Politics in Postwar Britain. The Britain Can Make It Exhibition of 1946* (London and Washington 1997), p. 145. ID312 'Suggested names for the Exhibition' memo from C.A. Mesling to S.C. Leslie 17 October 1945.

12 Maguire and Woodham (eds.) (cited note 11), p. 157. DCA/361/'The Council of Industrial Design. The First Meeting of the Exhibition Policy Committee, Tilbury House, Petty France, on Thursday 25 October 1946.

13 Noel Carrington, 'General Notes, Britain Can Make It', *Journal of the Royal Society of Arts* (27 September 1946), vol. XCIV, no. 4727, p. 664.

14 V&A Archive, Exhibition File: AAD/1977/4/14: 'Britain Can Make It' Films. Memo from Mr Jarvis to D. van Vlymen, 18 February 1947 detailing all filming that took place at the exhibition. The film Britain Can Make It no. 12 – a tour of the exhibition – took three days to make, starting 9 November 1946. A copy of the film is in the BFI and is available via BFI Player. 'A New Film Magazine, "Britain Can Make It"', *The Times* (17 January 1946), p. 6.

15 'No Nineteenth Century Eccentricity. Sir Stafford Cripps says designers must meet industry half-way', *Display* (January 1946), vol. 27, no. 10, p. 38.

16 William Shawcross, *Queen Elizabeth. The Queen Mother. The Official Biography* (Basingstoke and Oxford 2009), p. 606, p. 1026 note 53.

17 'Britons can't buy it', *Evening Standard* (28 September 1946), p. 2.

18 Walter Bernays, 'Britain Can Make It', *Design Digest* (October 1946), no page number.

19 'Britain Can Make It', *Display, Design & Presentation* (November 1946), p. 9.

20 Punch Almanack for 1947, 'Britain Shan't Break It', cartoon by George Morrow, p. 31. *Punch*, CCX, no. 5518 (2 October 1946), 'Can Britain Make It', cartoon by Leslie Illingworth, p. 255

21 'A Great Exhibition', *Display, Design & Presentation* (October 1946), p. 7.

22 Hansard, HC Deb (28 October 1946), vol. 428, cc262–64.

23 The Council of Industrial Design, *The Council of Industrial Design. Second Annual Report 1946–1947* (London 1947), p. 5.

24 '"Britain Can Make It". Summary of Final Report on Exhibition', *British Plastics* (February 1947), p. 61.

MARGARET TIMMERS

Ashley Havinden and the Britain Can Make It Poster

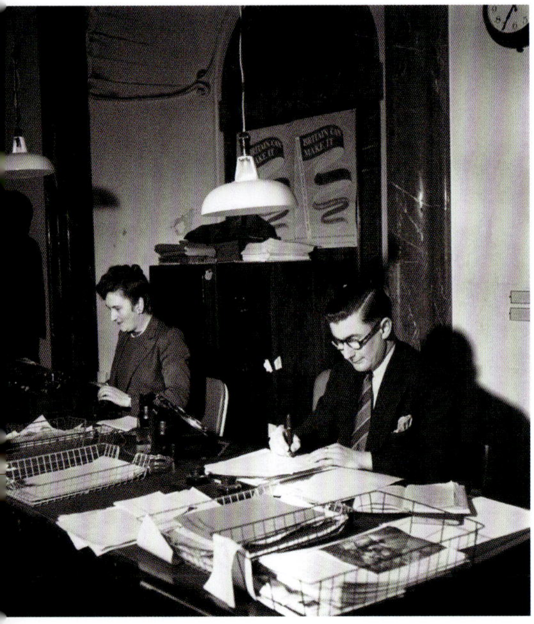

19. Mr D. van Vlymen of the Photographic Department in the Press Information Room at the *Britain Can Make It* exhibition, 1946. Two double crown versions of the exhibition poster are affixed to the wall of the BCMI Press Office, which was situated on the first floor of the V&A. Photograph. 15 x 20 cm. University of Brighton Design Archives, DCA, BCMI Photographic Album, book 5, print 385

> I believe the only ideal a designer should work to is complete clarity of expression. Not only clarity of words, or clarity of picture, but 'idea' clarity
> —*Ashley Havinden, OBE, RDI, FSIA (1903–1973)*[1]

The idea communicated in Ashley Havinden's *Britain Can Make It* (BCMI) poster is one of patriotic celebration [20]. The upbeat design features a crown-topped flagstaff lettered 'The Council of Industrial Design Presents' (meeting the Council's wish to bring its existence to wider public attention), from which flutters a multi-coloured pennant on whose surfaces the exhibition title, venue and date are proclaimed. Dispersed across a sky-blue background the names of the types of exhibits form a lively asymmetric pattern. The lettering, originally hand-drawn, is based on Times New Roman (bold weight), launched by Monotype Corporation for *The Times* newspaper in 1932, a fashionable font that would have been perceived as modern, authoritative and decidedly British.

Incorrectly, as it turned out, the Council of Industrial Design (CoID) viewed the V&A as a challenging venue for its exhibition in terms of public appeal:

> As the Victoria and Albert Museum is not a recognised exhibition centre and has probably never been visited for any purpose by a large proportion of the population of London (especially of the Eastern half) it is considered necessary to increase accordingly the estimate for expenditure on press and advertising and publicity on the hoardings and underground. The success of the Exhibition may turn upon our ability to induce the mass of people to accept the V&A as an appropriate place of resort for this purpose.[2]

A Public Relations Department was formed for general publicity and public relations, and press work [19], while advertising the exhibition was assigned to a group of agencies comprising the London Press Exchange, S.H. Benson, Everett's Advertising, and W.S. Crawford Ltd.

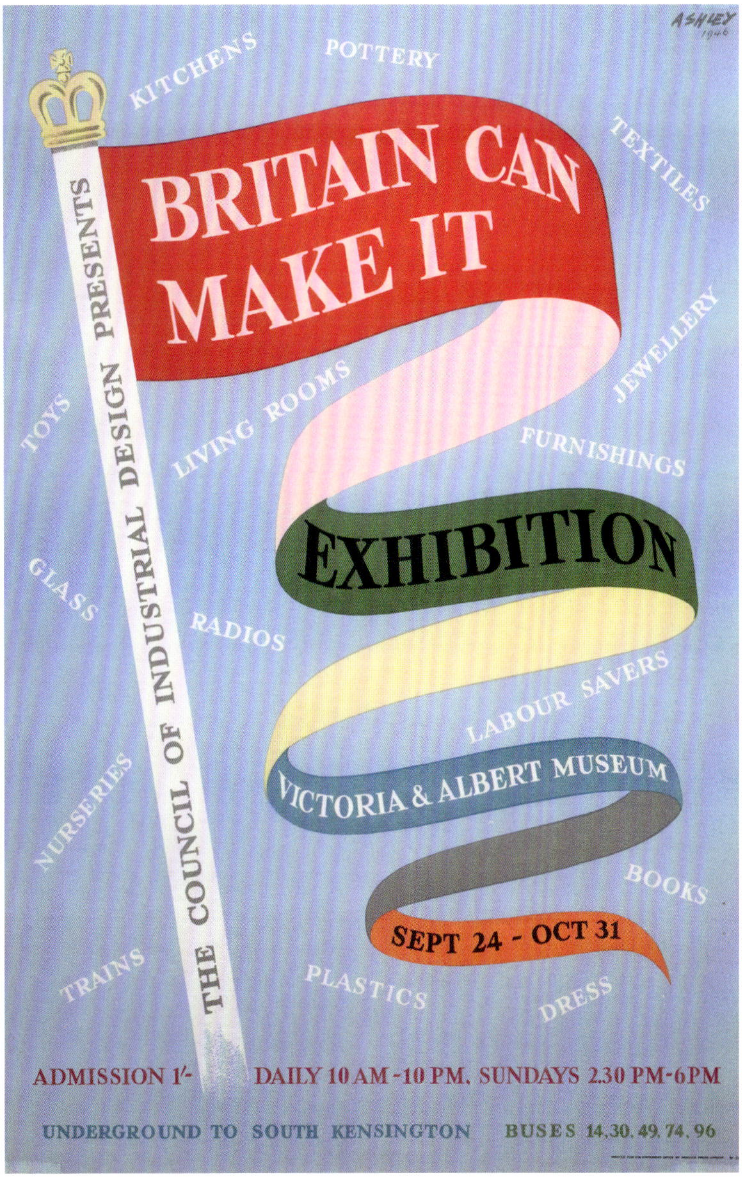

20. *Britain Can Make It* poster advertising the Council of Industrial Design Exhibition at the V&A, 24 September – 31 October 1946. Designed by Ashley Havinden. Printed for H.M. Stationery Office by Haycock Press, London. Signed Ashley and dated in ink 1946. Colour lithograph, 73.5 x 48.9 cm. V&A: Circ.459–1971 (Given by the artist)

21. The main Exhibition Road entrance to *Britain Can Make It* at the V&A, 1946. Captioned 'The Entrance to the Exhibition at 8am'. The bomb-scarred masonry of the Museum walls bear silent witness to the recent war, for which the exhibition was intended to provide a morale-boosting antidote. Photograph. 20 x 15 cm. University of Brighton Design Archives, DCA, BCMI Photographic Album, book 5, print 310

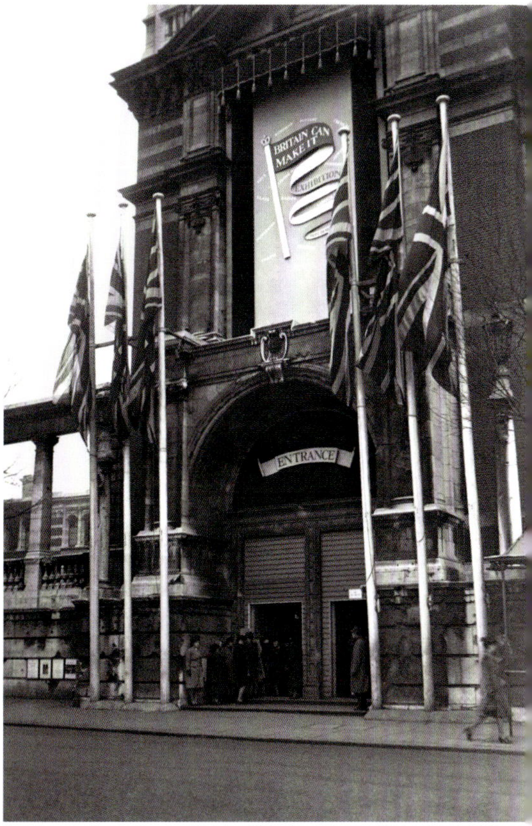

(where Havinden was Art Director). A committee of members of these four was responsible for the choice of a poster and the banner symbol.[3] Under them, Poster Services Limited looked after distribution. The committee was also responsible for display advertising, banner strips on buses, linen banners over roadways, etc. Three-dimensional poster displays were placed in all Odeon and Gaumont-British cinemas.[4] In spite of a printers' overtime ban (which did affect production of the exhibition catalogue) and post-war paper shortages, large numbers of the exhibition

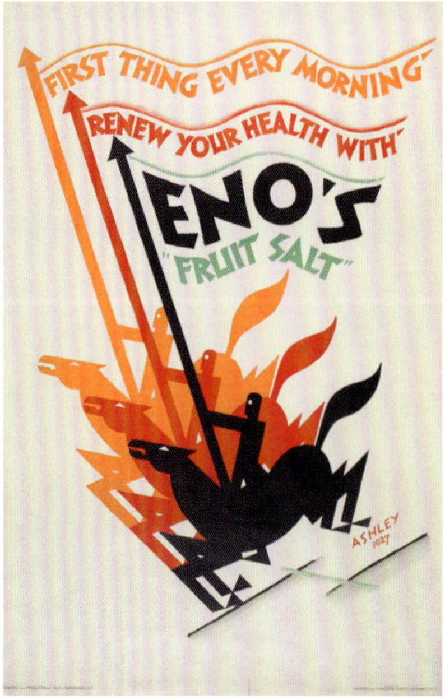

22. Eno's Fruit Salt poster, designed by Ashley Havinden and produced by W.S. Crawford. This dynamic design shows an early use by Havinden of the staff and pennant motif later employed in his BCMI poster. It also shows the influence of the poster designer Edward McKnight Kauffer, who also produced posters for Eno's and whose style Havinden admired. Signed and dated Ashley 1927. Colour lithograph, 76.2 x 50.5 cm. V&A: Circ.456–1971 (Given by the artist)

poster were produced in various sizes.[5] A relief-effect version, adapted to fit the available space, was skied above the Exhibition Road entrance to the Museum, where, colourfully flanked by six Union flags and above a banner-like 'Entrance' sign, it made a morale-boosting statement in defiance of post-war greyness and austerity [21].[6]

By the time Havinden (who signed his work, and was always known as, 'Ashley') was invited to design this poster he was an influential graphic artist and a leading figure in his profession, acclaimed for his sparkling designs for posters and advertisements, typography, packaging and corporate identity (which he called 'company handwriting')[7]. He had joined the advertising agents W.S. Crawford Ltd. (founded by William Crawford) in 1922 as a nineteen-year-old trainee layout artist. Early outside influences were the distinguished typographer Stanley Morison, an advocate for the improvement of design and typography in advertising, and Edward McKnight Kauffer, the American designer whose dramatic posters he revered. While travelling with Crawford to Germany in 1926 to visit type designers and foundries, Havinden became aware of Bauhaus graphic design, admiring a practice that favoured asymmetrical typography, the elimination of extraneous detail, and the use of white space as part of a dynamic composition. The following year he was sent to Berlin to found a branch of Crawford's to handle the Chrysler Motors European account.

Within Crawford's he rose to become Art Director and Board Member by 1929, and Vice-Chairman in 1960. During his early years there, Havinden formed a team with the young account director Margaret Sangster (whom he later married) and the copywriter G.H. ('Bingy') Saxon Mills, a master of the pithy slogan. Much of the success and reputation of Crawford's as one of the UK's most progressive agencies was attributable to Havinden's innovative work with this formidable team, including campaigns for Chrysler Cars, Eno's Fruit Salt [22], the Milk Marketing Board and Simpsons of Piccadilly. Furthering his interest in typography and layout design, in 1930 he created the lively 'Ashley-Crawford' typeface, based on display lettering evolved for Chrysler advertising, and in 1955 the 'Ashley Script', immortalizing the brush lettering style conceived in 1936 for his 'Milk' posters. In 1933 he published *Line Drawing for Reproduction*, a guide for commercial artists at the start of their career.[8]

Living in Hampstead during the 1930s, Havinden was part of an international group of artists which included Henry Moore (from whom he took private lessons), Barbara Hepworth, Ben Nicholson, Piet Mondrian and László Moholy-Nagy. Through these friendships he formed his own *avant-garde* art collection, which he displayed at home to elegant effect. During this decade Havinden exhibited posters and advertising design at the Union des Artistes Modernes, Paris, 1933, and at Lund Humphries Gallery, London, 1937, and was a member of the display committee for the British Pavilion at the Paris Exhibition of 1937, for which he designed the 'Books and Printing' section. In other fields, he showed abstract paintings at the London Gallery, 1937, and at the Reid and Lefevre and the Leicester Galleries, London, 1939, as well as rugs and textile designs at the Duncan Miller Gallery, London, 1937. After the war, during which he served as an officer in Camouflage, Havinden returned to Crawford's as Art Director, where he took on quasi-official work such as the BCMI commission, as well as devising new campaigns for commercial clients. Post-war successes included styling for Liberty [**23**], re-branding for KLM, packaging for Richard Shops, and modernist graphics for the Daks trade name. As a poster designer his work may be viewed in the context of other mid-century moderns, such as Hans Schleger ('Zero'), Abram Games, F.H.K. Henrion and Tom Eckersley, all of whom believed in the simplification of form for optimum clarity, and in the design of text and imagery as an integral whole.

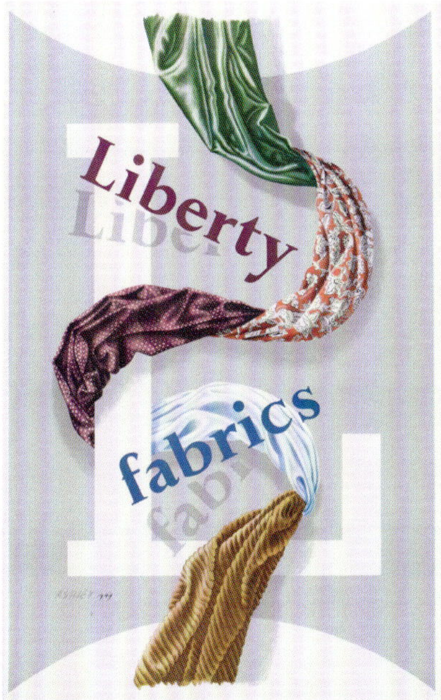

23. Liberty Fabrics poster, designed by Ashley Havinden and produced by W.S. Crawford. After the war, Havinden created unified styling for Liberty, including advertising, packaging, labelling, stationery and catalogues. In this poster he plays with his favourite pennant motif, which is transformed into a multi-patterned helix wound round the 'L' of Liberty to proclaim the quality and pattern of their textiles. Signed Ashley and dated in ink 1949. Colour lithograph, 75.9 x 50.6 cm. V&A: Circ.464–1971 (Given by the artist)

In the Ashley Havinden Archive held by the Scottish National Gallery of Modern Art in Edinburgh is a fascinating group of original sketches and artworks showing the genesis of the artist's thought as he tried out various concepts for the poster, catalogue cover and guide for the BCMI commission [**24–27**; see **3**].[9] In addition to the flagstaff device, he explored motifs based on a set of craftsman's tools, a radiant eye and a pointing finger, as well as alternative lettering designs. It may be that some of the ideas arose out of discussion with the Exhibition's other contributors, since 'pointing hands' indicated Misha Black's 'Birth of an Eggcup' section, and an 'all-seeing designer's eye' appears throughout: a huge Perspex eye was hung in the 'Designers Look Ahead' display in the Hall of the Future, designed by Basil Spence, and another 'Mr Designer' emblem with large eye, designed by James Gardner, appeared frequently in the exhibition [see **85**, **171–173**, **188**, **85**]. Possibly Havinden's design was an early thought to tie in with these ideas, but a flag-waving emblem was ultimately considered a more patriotic approach. Havinden would later recycle his own 'eye' image as a striking advertisement for the services provided by Crawford's agency.[10]

The chosen flagstaff emblem was a favourite in Havinden's repertoire – used to dramatic effect in his 'Eno's Fruit Salt' poster of 1927, in which three horsemen charge diagonally under the banner of Eno's. It appeared again in his poster for Liberty Fabrics of 1949, the pennant evolving into a multi-patterned helix. It may be thought that Havinden's device for the BCMI poster was unlikely to cause controversy, but the decision to depict a schematised crown on top of the flagstaff caused a flurry of correspondence between the Council of Industrial Design

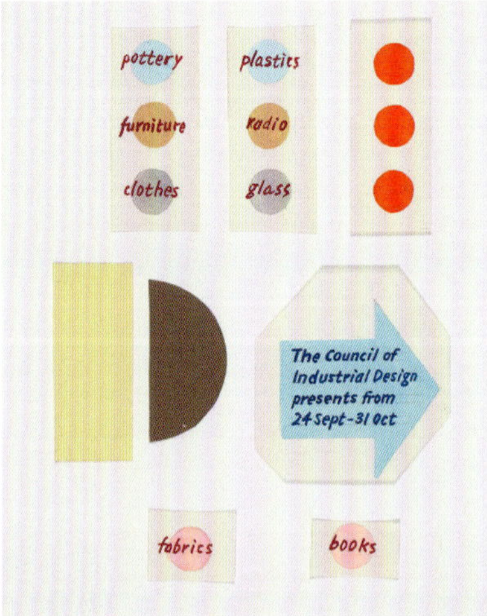

24

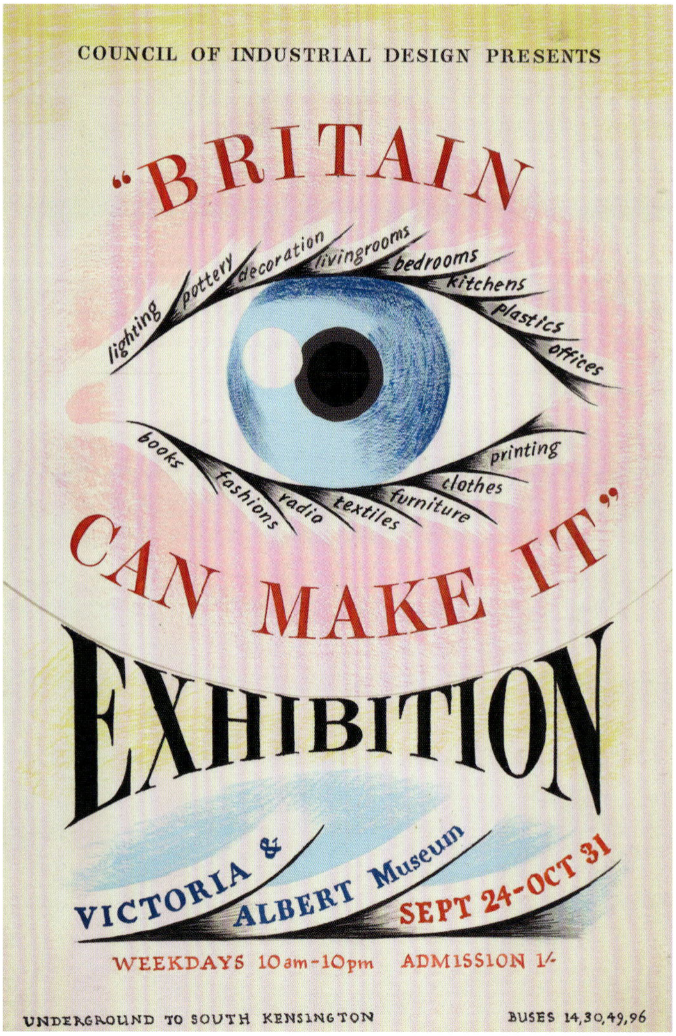

25

Ashley Havinden's trial designs. Ashley Havinden Archive, Scottish National Gallery of Modern Art Archive, GMA A39/1/153

24. Trial designs, patterns and colours for the *Britain Can Make It* commission. Body colour, various sizes

25. Trial design for the *Britain Can Make It* poster, overlaid with sheet showing alternative lettering design. Although Havinden's 'radiant eye' concept was not used for the poster, he later adapted the image as a striking press advertisement for the services provided by Crawford's agency. Ink and body colour, 25.3 x 17.1 cm

26. Preliminary design for the *Britain Can Make It* exhibition catalogue or guide. The text of this design includes the words 'Of Design' after 'Exhibition', and the venue and date 'London 1946', neither of which appears in the final versions. The use of variant colour-ways in the preliminary and final designs shows Havinden's inventive approach. The line drawing and cross-hatching style were carried forward, in adapted form, into the final version of the penny guide. Body colour, ink and collage, 27.9 x 22.6 cm

27. Trial design for the *Britain Can Make It* poster. There is a visual link between the floating abstract shapes of this unused design and the bands of lettering that appear on the final poster, giving an insight into the artist's thought process. The 'discs' on which the exhibit types are listed were a favourite in Havinden's design repertoire. Metal discs were used in a BCMI design quiz, and discs were incorporated into the cover illustration to the CoID's domestic advice book *Furnishing to fit the family* (1946). Body colour and pencil, 19.3 x 12.8 cm

and the Home Office, which had to adjudicate over the protocol for use of a royal emblem. The latter took the view that 'To show it over a pennon carrying wording (which is itself rather questionable as a matter of heraldic practice) does ... raise special considerations.'[11] Finally it was agreed that 'having regard to the form which they have adopted for the crown, we can safely say that the Home Secretary would not wish to take exception to its use for the purpose in mind'.[12] This was fortunate in light of the poster's popular appeal.

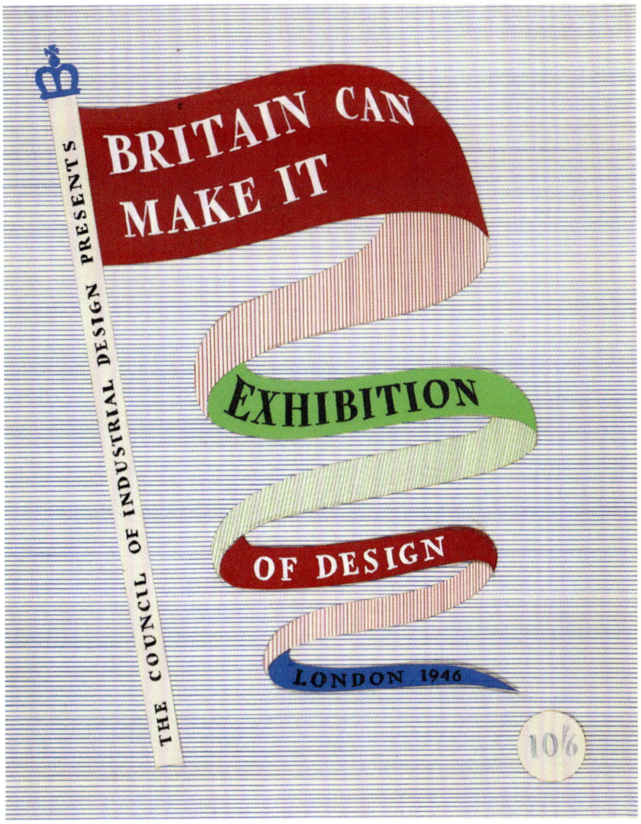

26

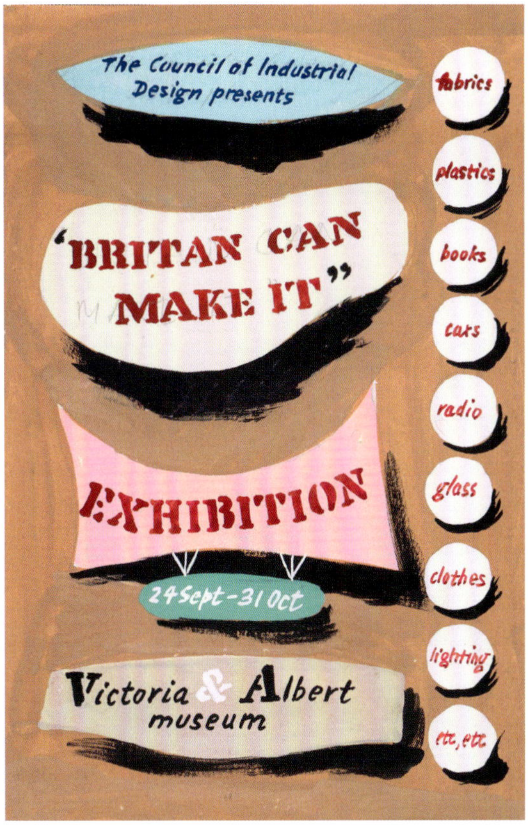

27

NOTES

With thanks to the following for their advice and expertise: Diane Bilbey; Eve Watson, Head of Archive, RSA; Kirstie Meehan and Neil Ogg, Archivists, Scottish National Gallery of Modern Art; Dr Lesley Whitworth, Deputy Curator, University of Brighton Design Archives; and Christopher Marsden and colleagues at V&A Archive.

1 RSA, London: Ashley Havinden, 'Visual Expression', a paper read at the 63rd dinner of the Double Crown Club at the Café Royal on 18 November 1937, p. 7.

2 V&A Archive, Exhibition File: AAD/1977/4/14, Britain can make it: In 'Preliminary Estimate of the Expenditure...', 16 January 1946, p. 5, note 7 'Publicity'.

3 V&A Archive, Exhibition File: AAD/1977/4/14, Britain can make it: In report to W. [H.] Newman [Editor of Exhibition catalogue and of *Design '46*] from A. Jarvis [Public Relations Officer], 27 March 1947.

4 Ibid.

5 V&A Archive, Exhibition File: AAD/1977/4/14, Britain can make it: Posters were produced in the following quantities and formats: 3,500 16-sheet; 400 4-sheet; 22,000 double crown; and 15,000 crown folio.

6 V&A Archive, Exhibition File: AAD/1977/4/13, Britain can make it: Flags were purchased from the Supplies Division of the Ministry of Works, and flagpoles borrowed from the same source, for display outside the Museum; see document 'Outside Display and Construction', 27 February 1947.

7 Ashley Havinden, 'The Importance of Company Handwriting', *Penrose Annual*, 1955, vol. 49, pp. 58–61.

8 Ashley Havinden, *Line Drawing for Reproduction* (London and New York 1933).

9 Ashley Havinden Archive, Scottish National Gallery of Modern Art: GMA A39/1/153. Contains over thirty designs for the BCMI commission in media including pencil and crayon, pen and ink, colour wash, water- and body-colour and collage.

10 Reproduced in P. Ray (ed.), *Designers in Britain*, Society of Industrial Artists (London 1947), pp. 282–83.

11 The National Archives (TNA) HO 144/22743: Letter from R. Wells to S.C. Leslie, 17 April 1946.

12 Ibid., Letter from R. Wells to S.C. Leslie, 26 April 1946.

DIANE BILBEY

Statistics, Statistics

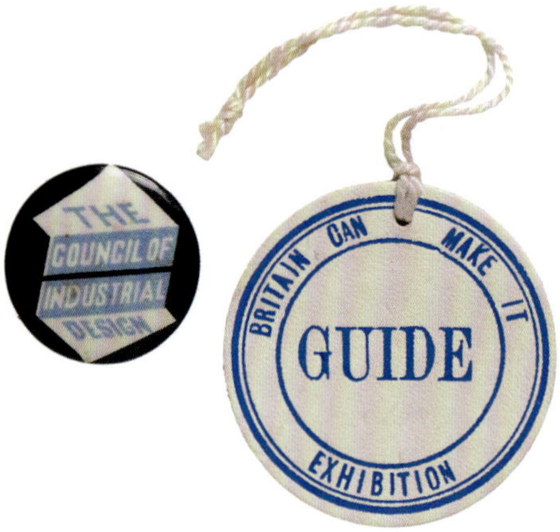

28. Badges worn by Guides at BCMI, designed by James Gardner

- 1,432,369 visitors
- 3/5 of visitors were under the age of 40
- 161,000 children visited
- Average daily attendance 21,000
- 2,000 visitors passed through turnstiles every hour
- Open for 14 weeks between 24 September and 31 December 1946 (it was planned to close earlier on 31 October, but extended to the end of the year)
- Opening hours 10 am–10 pm weekdays and Saturday; 2.30–6 pm Sundays
- Mile-long queues
- 90,000 square feet of exhibition space
- Over 50,000 trade buyers – 7,106 from 67 overseas countries
  - 420 trade buyers from the United States
  - 435 from Belgium, the highest number of trade enquiries from one country
  - 401 trade buyers from South America
  - 43,300 from Great Britain
- 18,000 products put forward by 3,582 firms of which 6,000 were chosen as exhibits from more than 1,300 firms
- Maximum number of staff working on the exhibition totalled 436, including: 142 Guides; 11 staff assigned to James Gardner as Chief Designer; 13 staff assigned to Basil Spence as Exhibition Architect. Administrative staff amounted to 90
- Total cost for the exhibition was £330,493 16s 4d (equivalent today to approximately £11.7 million)
- Total revenue from ticket sales £71,840 11s 6d (equivalent today to nearly £2.6 million)[1]

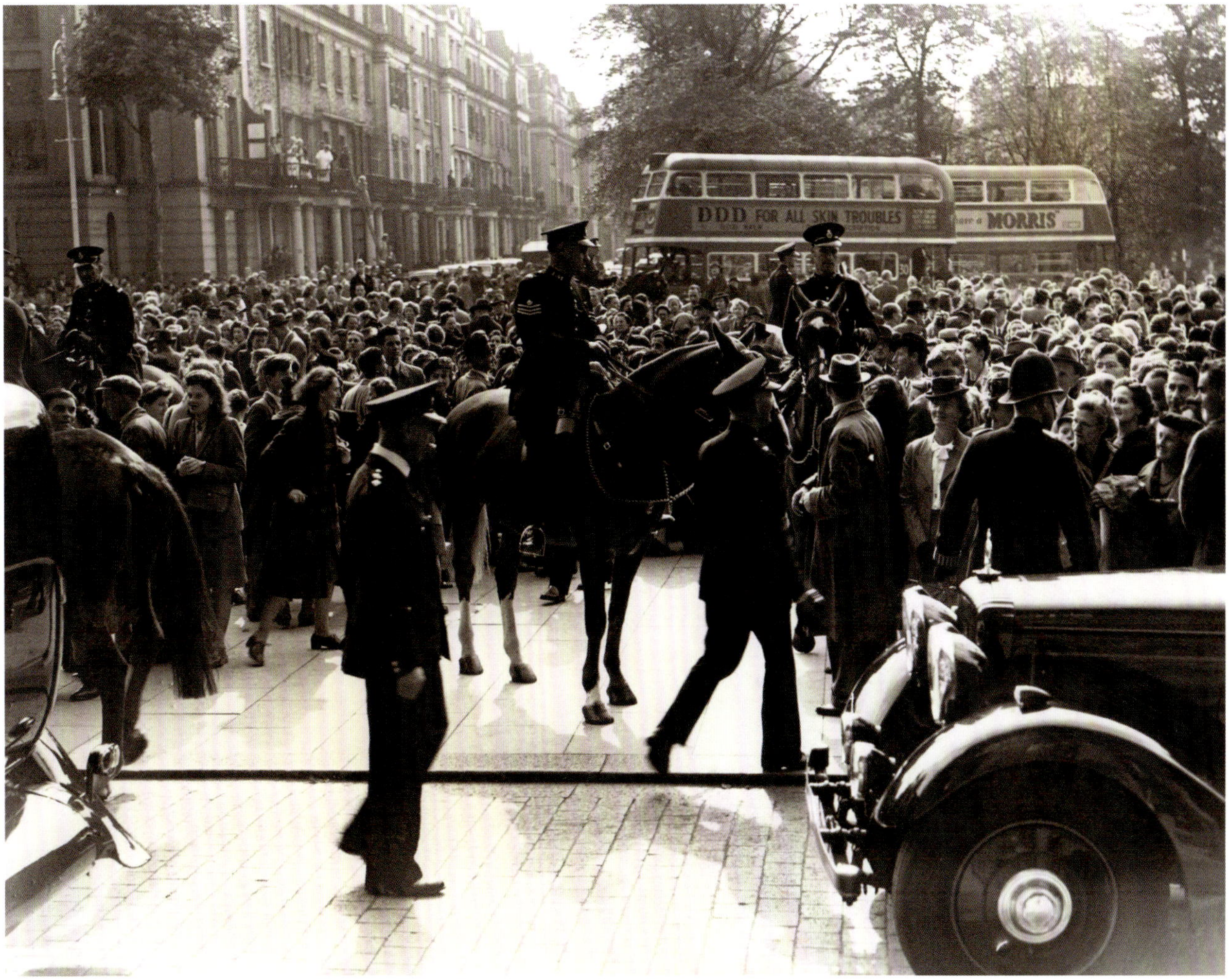

For over two months the longest queue in Britain was the one which stretched away from the doors of the Victoria and Albert Museum.[2]

29. Expectant crowds await the arrival of King George VI and Queen Elizabeth to open the exhibition on 24 September 1946. The royal party entered through the Main Entrance of the V&A on Cromwell Road.

King George VI and Queen Elizabeth arrived at the Victoria and Albert Museum at 11.15 am on 24 September 1946 to open the exhibition officially, having returned to London from Balmoral by royal train the night before. They entered the Museum via the Main Entrance on Cromwell Road, not the actual entrance to *Britain Can Make It* (BCMI), which was located around the corner in Exhibition Road. Officials had decided that the Main Entrance was grander and more suitable for such an arrival. They eventually left the exhibition at 1.40 pm, extending their scheduled tour by half an hour, such was their apparent interest in what was on display.

Outside the Museum a mile-long queue had formed, and at 2 pm the first eager visitors shuffled through the turnstiles. The vast number of visitors that followed over the next three months had not been expected. According to an article in *The Architect and Building News*, entitled rather sarcastically, 'Queue here for Fairyland', it appeared that 'The only people seemingly surprised by the queueing and congestion at South Kensington are the Council of Industrial Design themselves'.[3] But, it seems, the queues, which were widely reported and often ridiculed in the popular press, became an integral part of the BCMI experience. This was described as the 'Hollywood Approach' by Mass Observation (M-O), whose investigators reported that queuing was no deterrent. In fact, those who came 'treated it as a form of entertainment, especially in the days when the buskers were in full swing. Indeed observation on queue behaviour shows that, for many, the queue was something of a social occasion whose bustle and anticipation is enjoyable'.[4]

The exhibition preview should have given the Council of Industrial Design (CoID) some idea of the public's overwhelming interest in the exhibition. Three thousand invitations were sent out, each admitting two people. Out of a possible 6,000 visitors, a total of 5,011 were recorded passing through the turnstiles – a near 85% turn-out. The Council's records show that in the first seventeen days, from its official opening on 24 September to 10 October 1946, a total of 318,121 people had already visited.[5]

Reporting success

The CoID marked the unprecedented and unforeseen success of the exhibition with regular updates on visitor numbers, and these were reported by the popular press, the BBC and British Pathé. People were then, as now, fascinated by such statistics.

The Times pondered the public's interest in statistics in an article about one Private Mitchell, the 500,000th soldier to be de-mobbed, and went on to comment:

> In other walks of life these round numbers of great magnitude have inspired similar favours. The millionth visitor to the 'Britain Can Make It' Exhibition was welcomed by a film actress and presented with a soft leather travel bag because she was the millionth visitor. But, unlike the soldier at York, she soon had a rival in the 'last' visitor, who was presented with a portable radio set because she was the last visitor – not, be it noted, because she was the 1,432,369th visitor, which she also was.[6]

The first visitor

The first paying visitor to walk through the turnstile, after the King and Queen had left to return to Balmoral, was Mrs Rita Mostyn of Derwent Road, Twickenham.[7] Little did she, or anyone else (most notably the exhibition organizers), imagine that more than 1.4 million visitors would follow in her footsteps before the exhibition closed some 14 weeks later. On that first day, between 2 and 4 pm, more than 5,000 people had paid their shilling (6d for children) and been admitted to the exhibition, and by the end of the day the estimate was that 15,000 people had visited.[8]

The half-millionth visitor

Less than a month later, John Kenneth Nunn of Sheffield queued for three hours with his fiancé Miss M.L. Ardern, and unknowingly became the 500,000th visitor [30].He was rather unceremoniously picked out by CoID staff as he walked through the turnstiles on the afternoon of 20 October 1946. The BBC interviewed a somewhat startled Mr Nunn, after he and his fiancé had been treated to a guided tour of the exhibition, followed by a substantial tea (with sandwiches), which seemed to Mr Nunn the highlight of the event. The occasion was rounded off with the presentation to Mr Nunn of a radio (a Cossor Model 464, manufactured that year) worth £40, according to the *Daily Mail*.[9] The BBC interview reflected on Mr Nunn's experience of being identified as the half-millionth visitor, and the resulting interview was broadcast that same evening on the Home Service. In it Mr Nunn appeared a little uneasy at his moment of fame:

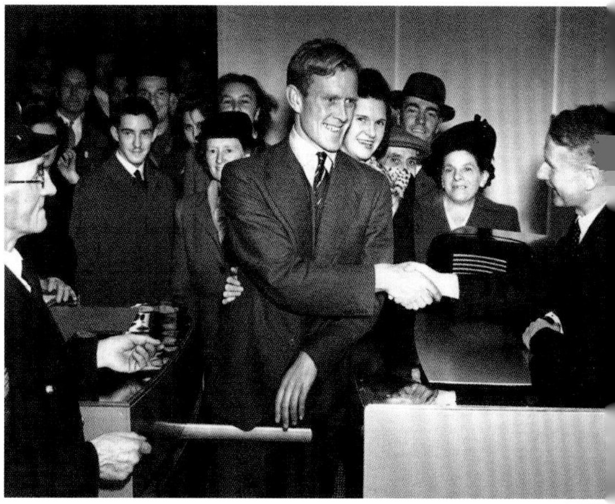

30. The 500,000th visitor to BCMI, Mr John Kenneth Nunn is greeted by Mr N.E. Kearley, Deputy Director of the CoID as he enters through the turnstiles, and is presented with a radio (a Cossor 464). His fiancée, Miss Ardern, peeks in behind.

> Well, I thought I'd done something wrong. I came down on Friday to London, to see my fiance [sic], and we intended to come to the Exhibition – we thought we'd get some ideas for our home-to-be and we came along yesterday afternoon, that is Saturday afternoon, and there was such a terrific queue we immediately went away and said we'd come early on Sunday. And of course we did that. And as I was going through the turnstile I saw that quite a lot of people seemed interested in the number by the turnstile, and then suddenly when I was there I was grabbed hold of. I looked round and I thought – well, I didn't realise I'd done anything wrong. They rushed hold of me and, of course, presented me with this set. I was very delighted, of course, and we had a young lady to take us round the exhibition, which made it much better. And, were given quite a nice tea. The only thing is the press gave us some beautiful sandwiches and then asked us questions and we couldn't very well eat them – well I couldn't anyway!

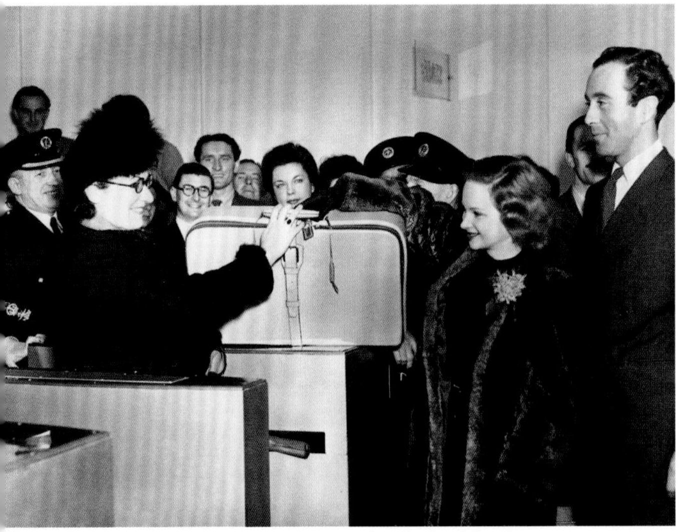

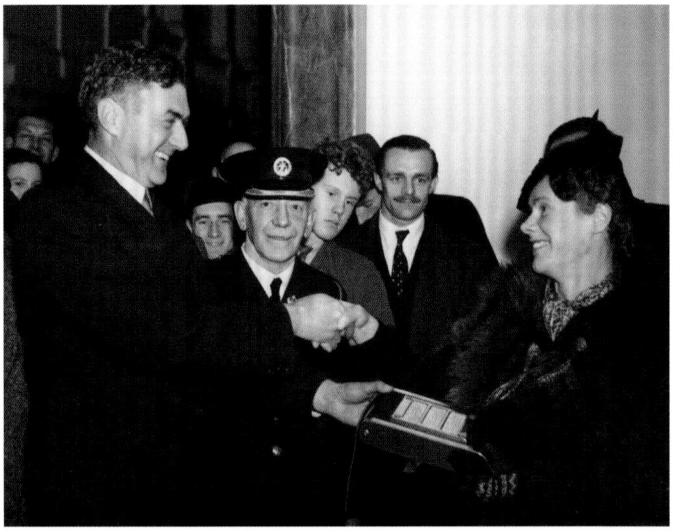

31. Mrs Irene Kelly, the millionth visitor to BCMI, being presented with a pig-skin travelling case by film stars Joan Greenwood and Dennis Price

32. The final visitor to BCMI, Mrs Eileen Stanhope, being presented with a portable Romac radio by S.C. Leslie, Director of the CoID

Later in the interview Mr Nunn was asked what had impressed him most about the exhibition and, in tune with many other visitors, he said that he thought the miner's cottage kitchen in the 'Furnished Rooms' section was the display that appealed to him most: 'I think it did this because it was more severely [sic] practical and not 'up-in-the-air' as a lot of the things tended to be' [see **111**].[10]

The millionth visitor

The millionth visitor was Austrian-born Mrs Irene Kelly, of Earls Court Gardens, London, whose visit on a rainy 19 November 1946 was captured on film by British Pathé [**31**].[11] The millionth visitor was naturally a landmark moment for the exhibition's organisers and Mrs Kelly (wife of Mr Douglas Kelly, a sanitary engineer) was treated like a film star attending an awards ceremony. After being counted in through the turnstile by Sergeant-Major R.A. Cooksey, the chief turnstile commissionaire, Mrs Kelly was picked out as the millionth visitor and greeted by British film stars Dennis Price and Joan Greenwood, who presented her with a modern soft pig-skin, zip-fastened, leather travelling case of a type shown in the exhibition.[12] The CoID added a nice touch to the event (and publicity masterstroke) by inviting Mrs Rita Mostyn, the first visitor to the exhibition, to be there to meet her.[13]

The final visitors

The Christmas and New Year holiday period did not see a reduction in opening hours or those who wanted to visit, and the exhibition remained open Christmas Eve, Boxing Day and New Year's Eve from 10 to 6 pm to accommodate last-minute visitors, eager not to miss out. They filed in through the 'clicking chromium plated turnstiles' at a rate of 1,800 an hour. The BBC Home Service sent Audrey Russell to report on the final day: she interviewed some of the final visitors, and exhibition staff, who expressed sadness that the exhibition was coming to an end. Mrs Haley was one of those who had left it to the last minute to visit. She was asked what she was most interested in, 'Well, household things. Naturally, being a housewife, I naturally am. And also children's toys – not that I've got any of my family little, they've grown up. Still, I like to see what their children are going to have when they have any.'[14]

The very last visitors, a husband and wife on their wedding anniversary, were Eric and Eileen Stanhope of Bicknoller Road, Enfield. Their impromptu visit, en route to an evening at the theatre on 31 December 1946, was included in the same BBC broadcast and also reported in an article by the *Daily Mail*. They became the last visitors to enter the exhibition on New Year's Eve 1946, passing through the turnstiles just before 6 pm, when 'the great "Britain Can Make It" exhibition, one of high significance in the history of British trade, closed its doors after three months'.[15]

These final visitors to what had become a record-breaking exhibition were presented with a portable camera-style radio (worth £18) 'developed from the type of wireless communication instrument used by underground agents during the war' by S.C. Leslie, Director of the Council of Industrial Design [**32, 33**].[16] An example of the radio was displayed in the 'From War to Peace' section of the exhibition, as one of the many products demonstrating how wartime developments could be adapted for the manufacture of peacetime products [see **40**]. The radio produced by Romac Industries Ltd, London, was from their Olympic range. About the same size as a woman's handbag – and worn like one over the shoulder – the strap also functioned as its aerial. The up-beat caption to a photograph issued by the Acme press agency, showing the ubiquitous 'British housewife' carrying the Romac radio, suggested that the portability of the radio was 'ideal for the British housewife who sometimes spends long hours in lines to obtain food and other necessities'. The significance of the Romac radio should not be underestimated. Its development marked a significant technological advance, and has been described as a 'precursor of what was to become a dominant theme in post-war radio design, personalization through miniaturization and portability'.[17]

A team of around 200 staff volunteers began dismantling the exhibition just hours after the last visitors had passed through the first sections of the exhibition. BCMI was cleared out in their wake. There was no shortage of volunteers, paid double the hourly rates, and

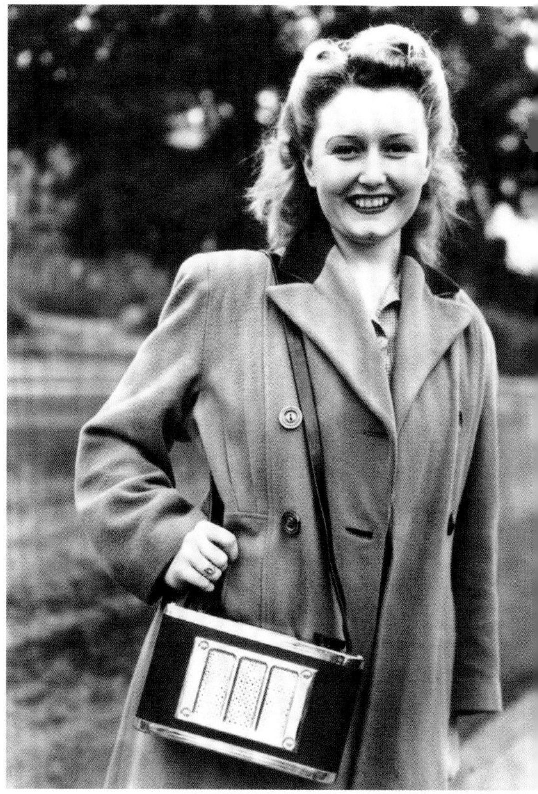

33. The Romac portable radio being modelled by a typical British housewife. The photograph was issued by Acme and intended for the American press.

'only really reliable guides were selected, with very gratifying results'.[18] The BBC reported,

> As far as the staff were concerned, it was a day of regret. There wasn't a single guide or information officer (and they'd been standing answering thousands of questions seven hours a day for three months) who did not say how sorry they were it was all coming to an end.[19]

Indeed, the *Daily Mail* reported that 'hundreds of people were being turned away, having arrived too late. Until nearly 10 p.m. there were "disappointed" visitors'.[20] But by 10.30 pm that night all of the smaller items in the exhibition had been packed up, and the removal of the furnished room contents and outside decorations began the next day, on 1 January 1947.[21] The hard-working staff who worked dismantling the exhibition were treated to an impromptu New Year's Eve party. Those who had worked as guides also had their own party held opposite the V&A at the Rembrandt Hotel on 3rd January, with another 'Smoking' party, organised on 7th January for the manual workers (men), warders, electricians etc.

With 1,432,369 visitors in just over three months, BCMI ranks as one of the most well attended exhibitions in British history. Although such figures appear high to us now, they are nowhere near the attendance for the two most well-known public events ever to take place in Britain – The Great Exhibition in 1851 and the Festival of Britain in 1951, where both events attracted several millions of visitors. The Great Exhibition, housed in the Crystal Palace at Hyde Park, took place from 1 May to 15 October 1851 and received over 6 million visitors. Records show that the Festival of Britain South Bank site attracted 8,455,863 visitors between 3 May and 30 September 1951. In 1863, the wedding presents for the Prince of Wales and Princess Alexandra were displayed for a period of just eight days, at what was then known as the South Kensington Museum (later renamed the V&A), attracting 127,560 visitors.

More recent V&A exhibitions have achieved what seems more modest attendance figures in comparison. Of course, these are in confined exhibition spaces, often with timed ticketing and with health and safety regulations restricting numbers. At the V&A, *Art Deco 1910–1939* (a paying exhibition in 2003) had 359,499 visitors over a four-month period, whilst *Disobedient Objects* (a free exhibition from July 2014 to February 2015) attracted 416,867 visitors over six months. The most successful V&A exhibition according to attendance records was *Alexander McQueen: Savage Beauty* (a paying exhibition in 2015) with 493,043 visitors. The *David Bowie is* exhibition had 310,760 visitors in London and reached 1,752,406 on its world-wide tour that took it to

11 further locations, totalling over 2 million visitors. Touring exhibitions, therefore, do get closer to those of BCMI*: Art Deco 1910–1939* attracted 1,358,761 visitors; *Vivienne Westwood* around 844,000; and *Surreal Things: Surrealism & Design* 881,000 visitors.[22] However, it is hard to imagine any exhibition – particularly one concerned with design and manufacturing – that could ever attract such an optimistic and eager audience as those who attended BCMI at the V&A in the autumn/winter of 1946.

NOTES

1 'Royal visit to exhibition. Mile-long queue', *Daily Telegraph*, 25 September 1946, p. 1; 'Fruits of "Britain Can Make It". Orders and prestige for industry', *The Times* (9 January 1947), p. 2; '"Britain Can Make It". Achievements in Art and Industry', *The Times* (24 September 1946), p. 2; V&A Archive, Exhibition File: AAD/1977/4/14A, Britain can make it: "Britain Can Make It". Income and Expenditure Account for the period 1/4/46 to 31/3/48; "Britain Can Make It" Maximum Number of Staff engaged on Exhibition dated 27.2.47. For historical currency values, see The National Archives (TNA) online Currency converter: 1270–2017.

2 'Britain Can Make It. Summary of Final Report on Exhibition', *British Plastics* (February 1947), vol. 19, no. 213, p. 61.

3 'Queue here for Fairyland', Events and Comments, *The Architect and Building News* (4 October 1946), CLXXXVIII, no. 4059, p. 2.

4 Design Council Archive at the University of Brighton Design Archives, hereafter DCA: 903/ Mass-Observation report entitled 'Those who came'.

5 V&A Archive, Exhibition File: AAD/1977/4/13, Britain can make it: 'Britain Can Make It' (preview and royal opening).

6 'Hundreds and Thousands', *The Times* (28 February 1947), p. 5.

7 Mrs Mostyn is cited as the first visitor in an article about the millionth visitor to the exhibition, Mrs Irene Kelly. 'Millionth Visitor. Ceremony at "Britain Can Make It" exhibition', *The Times* (20 November 1946), p. 2.

8 Home News, "Britain Can Make It" Exhibition opened by the King', *The Times* (25 September 1946), p. 2; 'The show 15,000 went to see', *Daily Mail* (25 September 1946), p. 4. The actual number, recorded in attendance figures collated by the Council, was a little less, at 14,065. V&A Archive, Exhibition file: AAD/1977/4/14, Britain can make it: BCMI Exhibition attendance figures.

9 'Queued 3 hrs – £40 present', *Daily Mail* (21 October 1946), p. 3.

10 'The Half millionth visitor to the "Britain Can Make It" Exhibition', BBC Written Archives Centre broadcast on the Home Service, Sunday 20 October 18.08 GMT, Film T382.

11 '1,000,000th Visitor to Exhibition', *British Pathe*, Film ID 2355.12. The footage has no sound and was never issued. Her visit was also reported in the *Daily Mail*. 'The Millionth. She gets a prize', *Daily Mail* (20 November 1946), p. 3.

12 The manufacturers S. Clarke & Co. Ltd of Bowling Green Lane, London, exhibited twelve items of luggage in the Travel Goods Section of the exhibition, see *Britain Can Make It Exhibition Catalogue* (London 1946), p. 113, cat. nos. 49–60; John W. Waterer, 'Luggage', *Design '46* (London 1946), p. 116 illus.

13 *The Times* (cited note 7).

14 Audrey Russell, 'Recordings at the Britain Can Make It Exhibition', BBC Written Archives Centre 31 December 1946, Film T450.

15 "Britain Can Make It" closed. Over 50,000 buyers', *The Times* (1 January 1947), p. 2; 'But 200 worked all night', *Daily Mail* (1 January 1947), p. 3. There is a slight discrepancy for the total attendance figures for the exhibition. 1,432,369 was widely reported in the press and official CoID documents, although a CoID's final attendance report, and the CoID Annual report for 1946–47, puts the final actual visitor numbers at 1,432,546. There is further reported confusion as the *Daily Mail* article noted here suggests the last visitors, Mr and Mrs Stanhope, were numbers 1,432,393 and 4.

16 'New Domestic Radio Receivers', *The Times* (21 August 1946), p. 8.

17 R. Chamberlain and G. Rayner, 'Domestic Equipment and Product Design', *Austerity to Affluence. British art & design 1945–62*, pp. 92, 95, PD18.

18 V&A Archive, Exhibition File: AAD/1977/4/14A, Britain can make it: Report on Staffing and General Arrangements "Britain Can Make It" exhibition.

19 Audrey Russell, 'Recordings at the Britain Can Make It Exhibition', (cited note 14).

20 *Daily Mail* (cited note 15).

21 V&A Archive, Exhibition File: AAD/1977/4/14A, Report on Staffing and General Arrangements, "Britain Can Make It " exhibition (cited note 18). Dismantling of the exhibition was completed on 5th February 1947.

22 The display of wedding presents for the Prince of Wales and Princess Alexandra took place between 15 April and 4 May 1863 at the then South Kensington Museum (later renamed Victoria and Albert Museum). V&A Archive BR/ B14/068: table 1. I am grateful to James Sutton for this reference. 'End to Festival', *The Times* (1 October 1951), p. 4. I am grateful to Daniel Feeney, Gordana Buck and Anna Fletcher for supplying me with the attendance figures for recent V&A exhibitions.

DIANE BILBEY

From War to Peace

> The whole object of the display is to show the public two things: One – that even out of the horrors and ruin of war, material good can accrue and that the lessons of war can be and are adapted to the needs of peace; two – that British factories have switched from war to peace even quicker than they did from peace to war, and that our scientists and technical staffs are in every way as good as their opposite numbers in other countries; also that we have reason to feel that British Industry can and will lead the world.[1]

'From War to Peace' was the first main space visitors experienced as they entered the *Britain Can Make It* (BCMI) exhibition. It was a powerful and unapologetically patriotic display, showing the transition from wartime production to peacetime consumerism. With a bold conviction that set the tone for the whole exhibition, this section optimistically sought to show Britain's potential pre-eminence in design and manufacturing, following an anticipated, essential post-war recovery. An early draft plan for the exhibition indicates that the Council of Industrial Design (CoID) envisaged that the 'From War to Peace' section would be 'the big "talking point" of the Exhibition'.[2] Original (rejected) suggestions for the title of the actual exhibition, including 'Spitfires to Saucepans', 'War to Peace', and 'Swords into Ploughshares', also sought to anticipate the transition towards peace.

Not surprisingly, themes connected with 'war to peace' were common in immediate post-war exhibition displays. Ministry of Information (MOI) so-called 'documentary exhibitions' showcased military achievements made during the war, including the *Victory over Japan* exhibition that took place in a temporary building on a bombed-out site in Oxford Street in 1945.

The emphasis on Britishness was an important concept to impress upon visitors in the 'From War to Peace' section, and it was undertaken with propagandist fervour. Although BCMI aimed at attracting foreign buyers, who were to be reminded of the quality of British manufacturing,

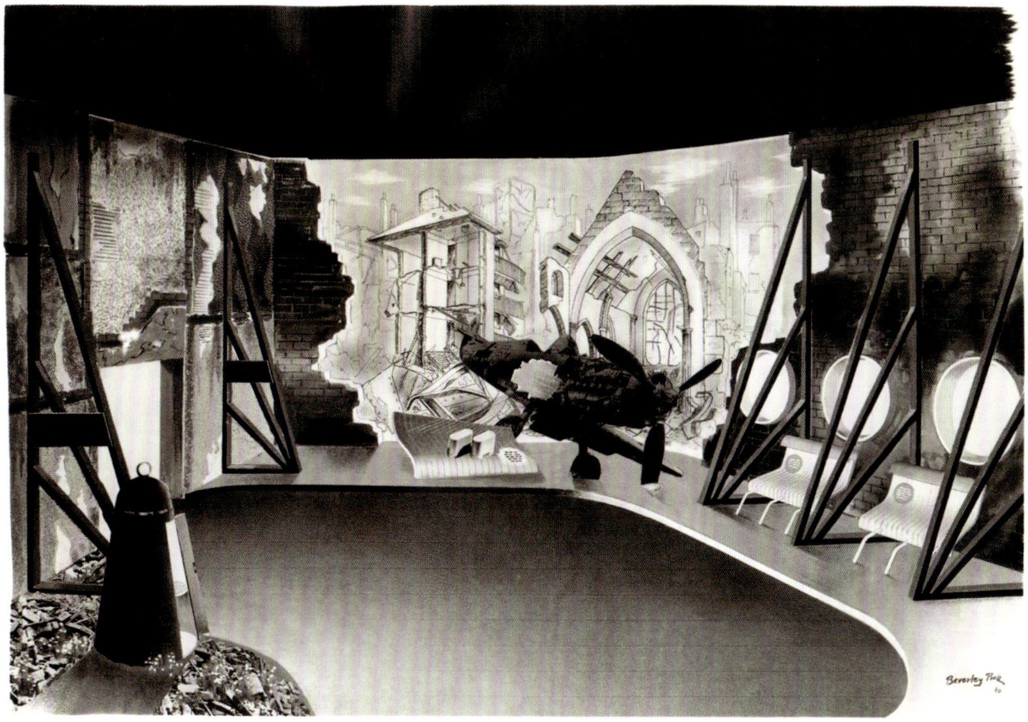

34. One of the preliminary sketches by Beverley Pick for the 'From War to Peace' section

it was also important to engage the British public in this same belief. Information provided to the guides working in the 'From War to Peace' section reinforced this ethos: 'Many of the war-time examples of inventions have been and are claimed as American but in every case in this Section items shown are entirely British, and are built by British labour and British brains'.[3] The display sought to convey hope for the future, and an insight into the possibilities of what that might hold, by demonstrating how post-war domestic products could be made from the new materials and processes pioneered during the war. Exhibits included in this section covered many aspects of domestic life, with products that included radios, clothing, toys and taps, and demonstrated diverse uses of materials including Perspex, rubber, glass, aluminium and wood laminate. Fabric with waterproofing incorporated into the weave, developed during the war for waterproof suits to keep aircrew alive if they ditched in the sea, was used in peacetime to produce a stylish raincoat. The *Daily Mail* reported,

> Textile industries learned a lot in war that they are now using in peace. For six years their technical and chemical Back-room Boys worked to protect the fighting men from the extremes of heat and cold; rain and sea water: Pacific typhoons, Middle East sandstorms, and Arctic blizzards .... Buyers of 1946

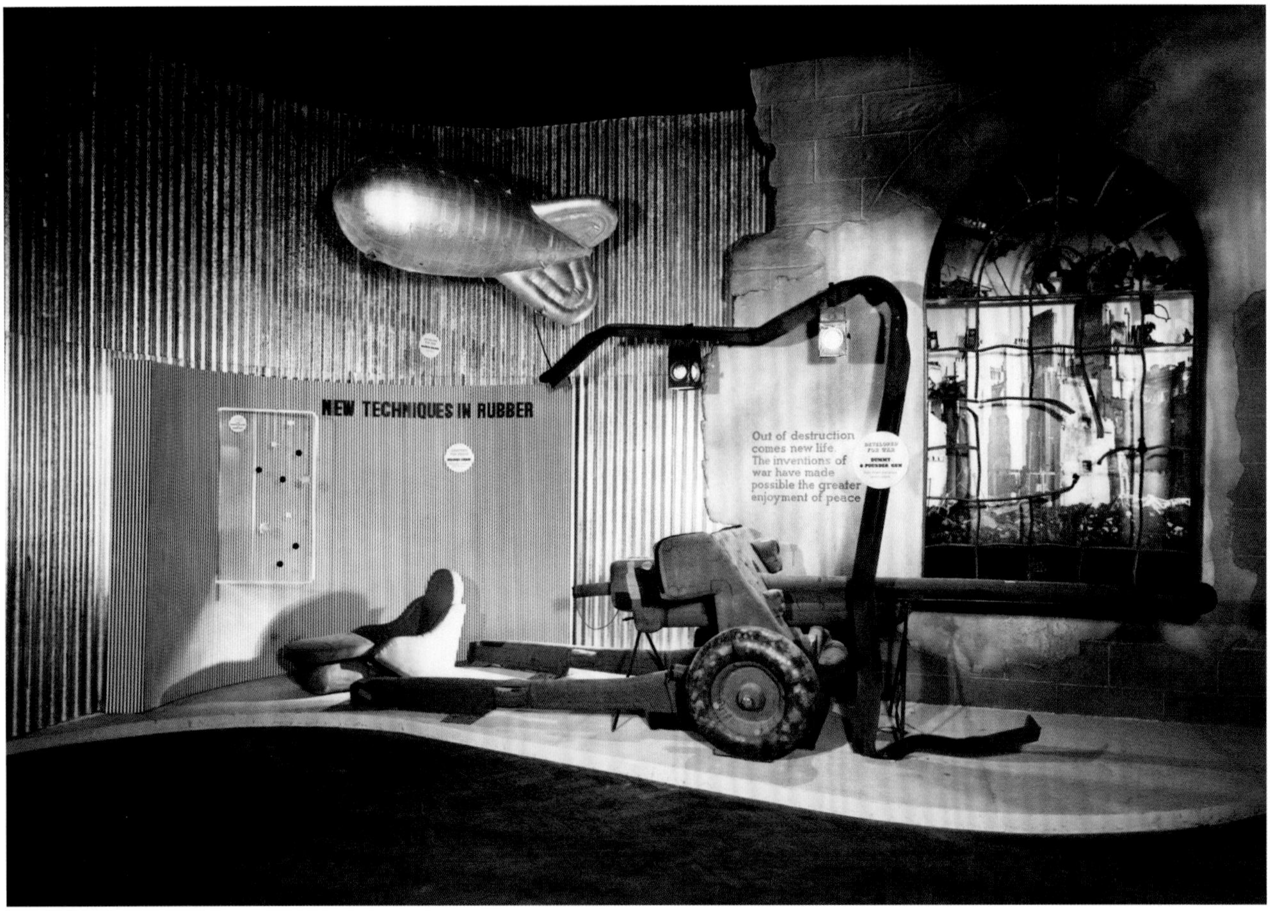

35. Part of the 'From War to Peace' section showing new technologies in rubber demonstrated by a barrage balloon and inflatable chair manufactured by RFD Co. Ltd of Guildford [36], developed from technology that produced the six-pounder inflatable decoy guns during the war (an example of which is also illustrated). Made entirely of rubber, these guns could be packed into a small case, and so were easily transportable. James Gardner was involved in their development and wrote in his biography, *The ARTful designer. Ideas off the drawing board*: 'Dummy equipment designed in my unit was soon being manufactured at locations all over the country. Our production line at Manchester was actually producing a pneumatic tank every three hours and if the enemy had taken to using rubber bullets we could have staged a first-class war.'

will benefit from all this in their new raincoats and waterproof beach bags. There will be wool nearly as light as silk, admirable for air travel because it does not crease.[4]

The CoID clearly had high hopes for this display and entrusted its design to Beverley Pick, the experienced exhibition and industrial designer. Pick had worked for the MOI in the Display and Exhibition Division during the war, and was involved in one of the earliest post-war propaganda exhibitions, designing the previously mentioned, hugely popular, *Victory over Japan* exhibition that opened London in August 1945. His clients pre- and post-war included BOAC (British Overseas Airways Corporation) and London Transport, but he is perhaps most remembered for his Regent Street Christmas light displays, in particular his stunning snowflake display for Christmas 1955.[5]

Pick was one of the designers singled out to be interviewed by the broadcaster and journalist Wynford Vaughan-Thomas for the BBC Home

Service at the opening of BCMI. He was asked to explain the purpose of the exhibition as a whole, and replied:

> Well, I would say the idea of it is to present to the British public the first of the post-war goods and to tell and show people that to produce goods to be functional, work well and be efficient isn't all that we have to do. We have to make them look good, we have to make them attractive and pretty to look at, and see in your home .... I'm sure that anyone who has a little money to spare and will want to buy new things in the near future will feel more in seeing the way they've been presented to them.[6]

Pick's design for 'From War to Peace' was highly theatrical, and he was praised in a contemporary journal for using a 'technique reminiscent of the stage or film set' [34]. *The Times* was also struck by the realism and drama that Pick had managed to convey: 'The exhibits will be picked out by pencils of light like the beams of a searchlight'.[7] Entering the display, visitors found themselves in a recreation of a blitzed London street during a blackout, dimly lit, and full of familiar wartime imagery. It was an emotive, graphic representation of the civilian war experience, one that would have struck a chord with many of the visitors. A scale model of a barrage balloon (produced especially for the exhibition by Dunlop Rubber Co. Ltd), steel shelter, gun turret, camouflage gun [35] and section of a real Spitfire were set against Morris Kestelman's powerful London skyline mural [37].[8] Here the reality of post-war Britain – nowhere near recovered – was profoundly illustrated, as explained in an article in *Display, Design & Presentation*: 'This introductory section is meant to be symbolic and rightly it aims at atmosphere all the time'.[9] The atmosphere so described was reinforced by text panels near exhibits: behind the fake inflatable anti-aircraft gun were the optimistic words, 'Out of destruction comes new life. The inventions of war have made possible the greater enjoyment of peace.'

The installation of the Spitfire plane in the display was described by James Gardner in his autobiography. He explained that it arrived one morning on the back of a trailer, straight from Biggin Hill, 'its wings folded as though still emerging from the chrysalis'. At first Gardner did not think he had the need for such an exhibit, then remembered he had asked for a tail of a crashed plane to be included as a key feature in the 'From War to Peace' section. He recalled: 'That Spitfire stood on its low-loader for days, but the Air Ministry, who had spent thousands to produce her, now treated her as disposable. In desperation we cut off her tail and shoved the rest on the district refuse tip'.[10]

It was a self-assured decision on behalf of the organisers to start the BCMI experience in such a sober way and, not surprisingly, the gloom

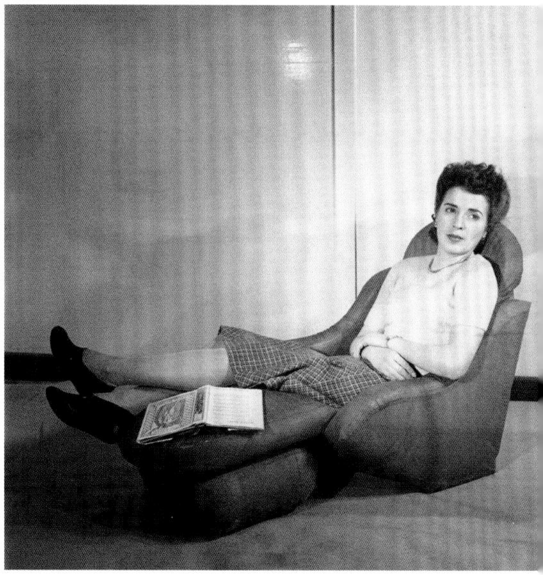

36. Developed from the same techniques used in making the inflatable guns, the chair shared their portable lightweight properties. It was to be made available for the domestic market from January 1947. The journal *Display, Design & Presentation* (September 1946, p. 14) commented that although some of the products displayed were clearly connected with their wartime counterparts – including the personal, portable Romac radio derived from equipment dropped by parachute to the Underground movement in France, Belgium and Holland – what was more typical were items that had no connection with its wartime uses, 'The company which made the wonderful dummy rubber guns which fooled the Luftwaffe is exhibiting a post-war luxury product in the form of an inflatable armchair to go in your rucsack! [sic]'.

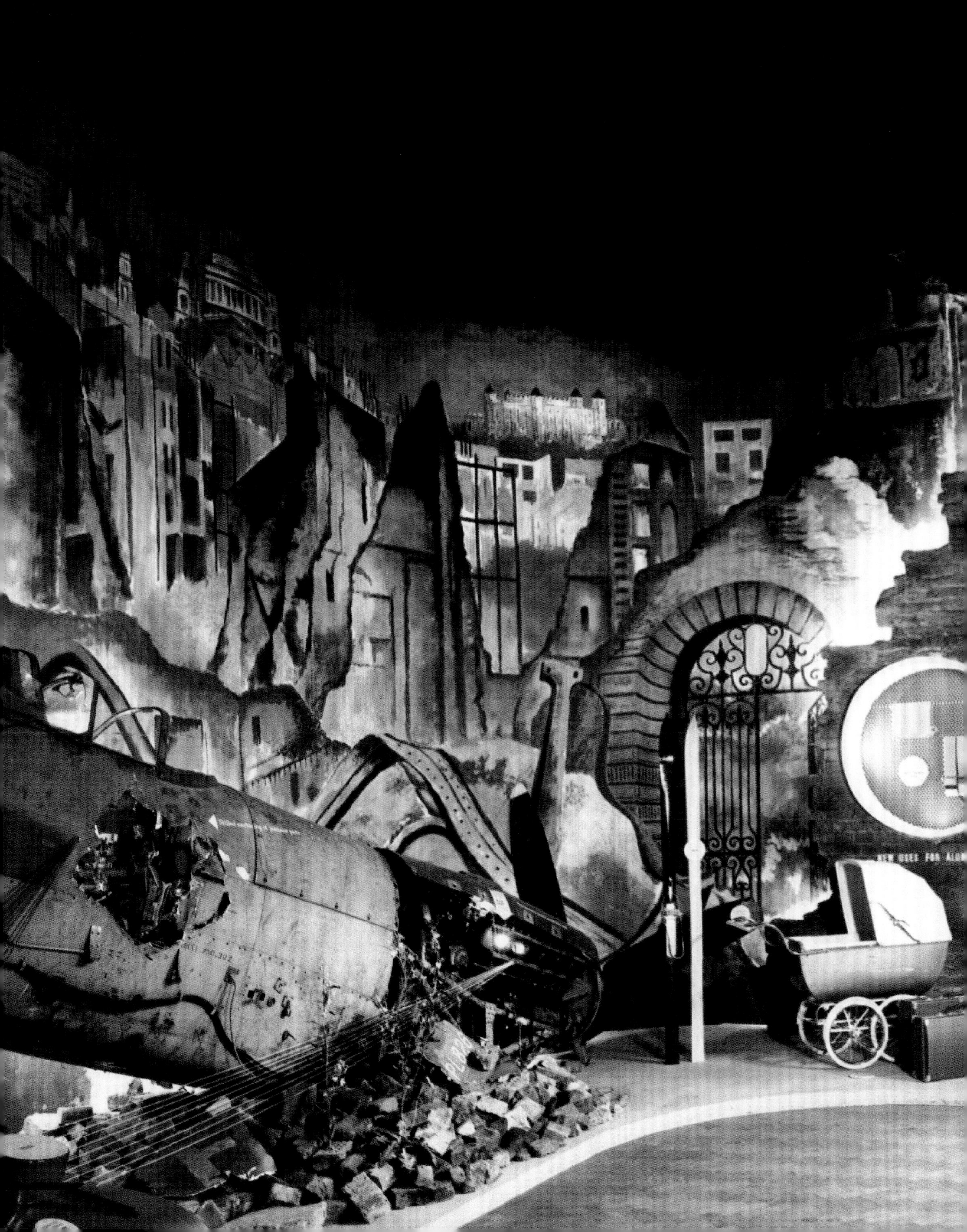
NEW USES FOR ALUM

of this introductory room provoked some criticism. Amongst feedback received by the CoID was a letter from Leslie Gamage, Managing Director of The General Electric Company. Concerned about the mental welfare of visitors entering 'From War to Peace', he wrote: 'The general arrangement is reasonably good although the approach, though possibly symbolically accurate, through comparatively dark entrance to lighter display is somewhat gloomy and may have a bad psychological effect.'

But the darkness was intentional – it was making a point – and in his reply S.C. Leslie (Director of the CoID) wrote: 'The brightness of the peacetime displays comes with extra effect after the dim opening with its war time settings'.[11] As noted in the *Supplement to the Board of Trade Journal*,

> Here, side by side with their war-like origins, will be shown some twenty domestic articles which have been evolved from the new processes, new materials and new techniques developed during the war. Beams of light will pierce the blackout to reveal, against a vivid background of bomb-shattered London, industry's newest contribution to domestic comfort and convenience.

A little prematurely, bearing in mind rationing would continue for a number of years, the display also promised a pay-off for wartime shortages to the British housewife: 'In 1940 the housewife gave up her pots and pans to provide the material for much-needed Spitfires. In the "War to Peace" display she will see beside the exhaust stub of a wrecked Spitfire the saucepan that will be returned to her' [39].[12] Briefing notes for the guides at the exhibition explained, 'It is important to note that in every case not one single article in this Section was in production or could even have been in production in the form in which it is shown prior to the War'.[13]

Products were displayed in their newly manufactured (peacetime) form, alongside their wartime predecessors [38]. An article in the *Daily Telegraph* picked out domestic products whose manufacture was made possible by research developments initiated, or accelerated, during the war [36],

> The inflated rubber dummy 6-pndr gun was the precursor of a comfortable garden seat; tank periscope prisms have become ornamental bowls and vases; the Hurricane ammunition box is now an all-aluminium suitcase. And the search for the perfect Spitfire exhaust has issued in better pots and pans, so completing the circle started by the housewife's gift of her aluminium ware in 1940.[14]

37. Morris Kestelman's blitzed London street scene mural, with St Paul's prominent in the centre, was a dramatic backdrop to the machinery of war that was also displayed in 'From War to Peace' (Section 1). In the foreground is part of a spitfire aircraft. The pram illustrates the domestic uses of aluminium after the war. Morris Kestelman, who trained as a stage designer, had exhibited a painting of a similar composition described as 'tumbled steel girders and broken walls' at an exhibition on the theme of the affect of war on Britain, organized by the Artists' International Association, that took place at Charing Cross Station between 16 September to 9 October 1941. The exhibition was covered by an article in *The Times* 'Artists and the War on Britain. Exhibition of Pictures' (17 September 1941), p. 6. Kestelman's BCMI commission led to him taking part in the Society of Mural Painters first exhibition held by the Arts Council in 1950. It appears Graham Sutherland might have originaly been considered to carry out the mural for the 'From War to Peace' section as his name appears on a list of passes required by mural artists. [V&A Archive: MA/28/75/6].

Lines Brothers manufactured the Hurricane ammunition box during the war, and were asked if they could provide the aluminium pram, an example of which is poignantly displayed amongst the blitz scene. Strong yet lightweight, aluminium suitcases produced by The Heston Aircraft Co. Ltd, were also displayed next to the pram. Other post-war uses of aluminium focused on the pressing issue of housing and 'beating bombers into bungalows was the modern version of turning swords into ploughshares'. According to *The Times* ('Aircraft aluminium for houses', 20 November 1945, p. 2), 'From thousands of tons of aluminium scrap recovered from obsolete aircraft 50,000 bungalows would be produced in the next year'.

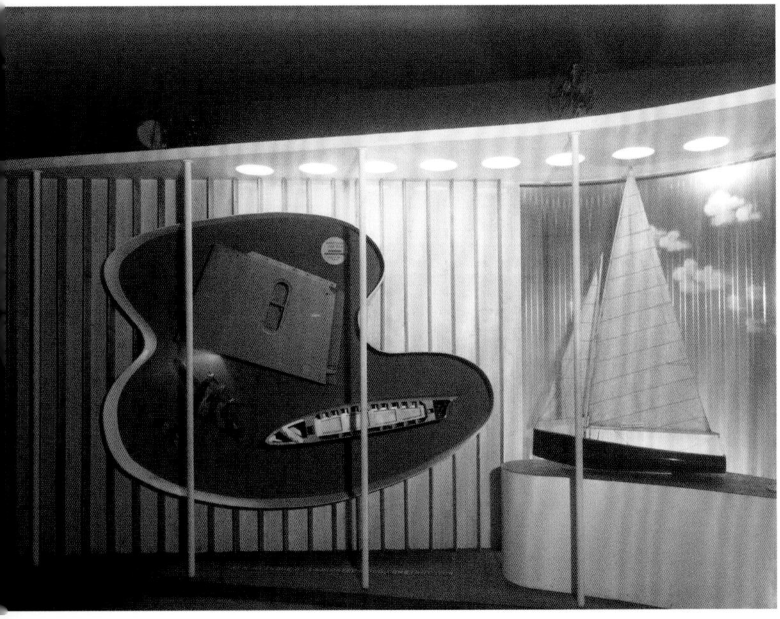

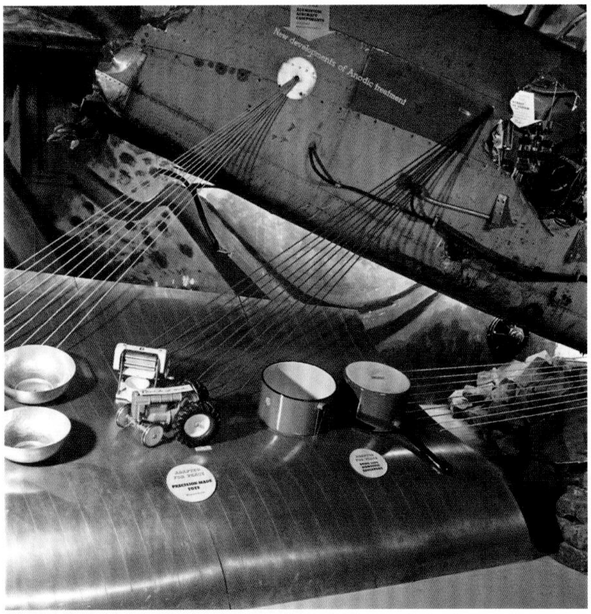

38. A section from an actual Mosquito aircraft wing together with a small model of a Mosquito were displayed next to a model of a yacht, to demonstrate methods of wood construction and the varied uses of plywood.

39. The war-time development of a process to treat the exhaust stub of a Spitfire to improve its strength and ability to withstand extreme temperatures for longer, known as Anodic treatment, was used in the domestic setting for coating pans – increasing their lifespan and preventing burning and cracking. Toys, including a tractor and mangle, made using precision engineering which had been developed during the war, were also displayed on part of an aircraft wing.

Great advances made in the development of Perspex and plastics during the war were a key feature in the display. Glamorous women's shoes made in rose pink PVC made by H. & M. Rayne Ltd (then Royal Shoemakers), were displayed alongside a pilot's hood for the cockpit of an aircraft, together with more everyday products, including a plastic tray, plastic handle to an electric iron and a cotton wool container.

The success of the display was examined in a report on the exhibition organization by S.D. Cooke. Alongside James Gardner and R. Dudley Ryder, Cooke had taken over the responsibility for both the 'What the Goods are made of' and 'From War to Peace' sections. He noted that, with one exception, this success was made possible by the 'remarkable co-operation of the industries approached to make and loan items for the display'.

Despite the apparent willingness of industry to lend their support, the scarcity of raw materials caused real problems. Although aluminium was used widely during the war in the manufacture of aircraft, good examples of peace-time applications for this relatively new material were limited. Only two examples could be shown – a pram specially made by Lines Brothers, the British toy manufacturer – and suitcases [37]. Unfortunately, suitcases originally used in the display were of such inferior quality that they provoked criticism, perhaps not surprisingly, from members of the leather trade, and were replaced by improved versions.

However, the greatest difficulty was in sourcing well-designed plastic products. Companies whose products were of reasonable quality were

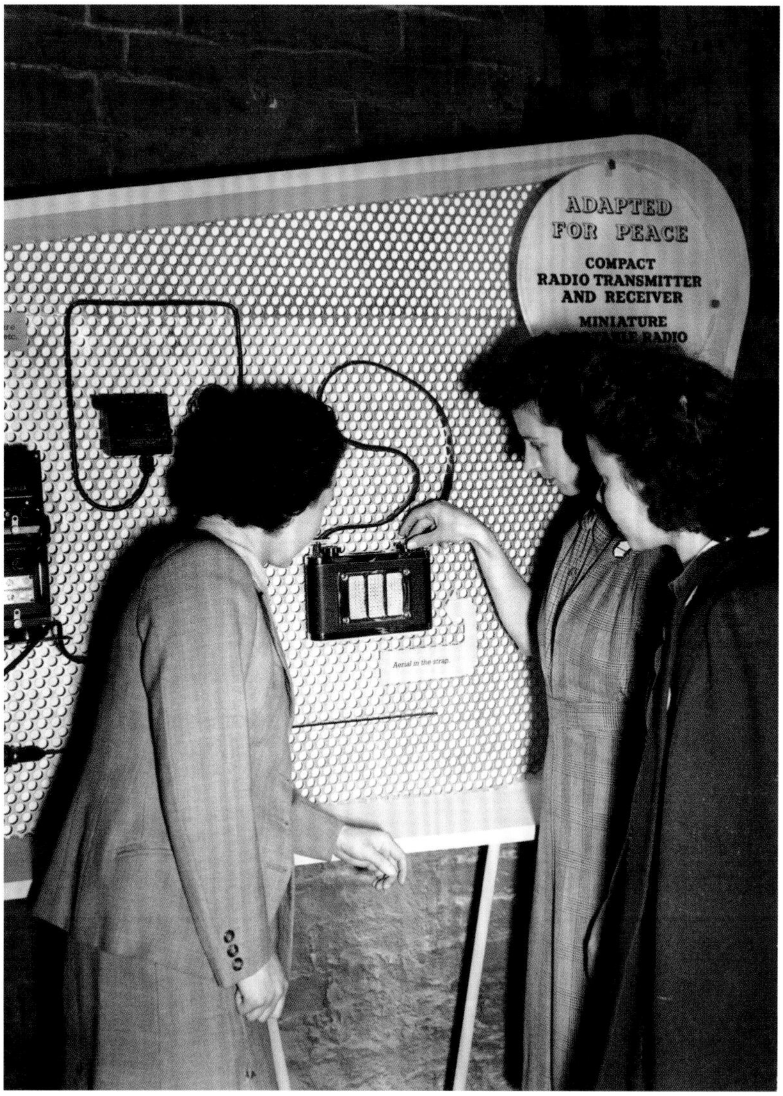

40. An example of the portable Romac 106 radio illustrated the peacetime use of miniaturized valves, described in *The Times* (21 August 1946, p. 8) as 'developed from the type of wireless communication instrument used by underground agents during the war'. *Vogue* also highlighted the radio in its 'Shophound' feature devoted to BCMI (October 1946, pp. 64 and 72). There it was described in great detail as a portable radio and fashion accessory: [extract] Shophound has music wherever she goes – her secret a Romac "personal" 106 receiver, the size and shape of a folding camera, 9 ½ ins. by 5 ¼ ins., the weight of a fairly full handbag – approximately 4 ½ lbs. The flat shoulder strap contains the aerial. The well-designed all-metal body, finished in ripple black and polished chrome is good and tough, as indeed it would have to be when carried by a shoulder strap. It is simple to tune, and fool-proof to fit with a new battery which, by the way, is an Ever Ready standard size obtainable everywhere. The radio covers the medium wave band. The Perspex dial, printed in two colours, gives a clear reading of the wave-length scale and the fourteen station names. These sets, which cost £18, including purchase tax, are being stocked by all the big stores and leading radio shops. Some lucky people may be able to walk out of the shop with one; others, however, may need to wait a little time before getting one'. [extract] The glamour and success of the radio was not lost in an article by Edward Chichester, The Marquis of Donegall, in *Newnes Practical Mechanics* (January 1947, p. 130), 'One of the first that we come to, photographed as often as a film-star, is the Romac first British Midget Portable Radio. I feel that most of my readers must know about it, as it has had so much publicity. I have one, and believe me, it is better than any American'. For further information on the Romac radio, see Jonathan Hill, *Radio! Radio!* (Bampton 1993), p. 173, fig. 659, p. 167; also Richard Chamberlain and Geoffrey Rayner, 'Domestic Equipment and Product Design', *Austerity to Affluence. British Art & Design 1945–1962* (London 1997), pp. 97–98, cat. PD18.

selected for inclusion in the plastics section of the 'War to Peace' display, but had to make changes to both the design and colour of their products to make them acceptable for display.[15] It was perhaps ambitious for the organisers to hope to show good-quality peacetime uses for materials pioneered only a few years before, and one outcome of BCMI was the recognition that further research and development in manufacturing processes was essential.

NOTES

1 V&A Archive, Exhibition File: AAD/19774/14, Britain can make it: Information for Guides. War to Peace Plan Key "C".

2 Design Council Archive at the University of Brighton Design Archives hereafter DCA: DAC PAX/P1/The Council of Industrial Design. Proposed Exhibition – Summer 1946. Draft Plan (2).

3 V&A Archive. Information for Guides (cited note 1).

4 Graham Sanford, 'The best news for seven years about things to buy summed up in – Four little words. Britain Can Make It', *Daily Mail* (27 August 1946), p. 2.

5 For the Victory over Japan exhibition see Misha Black (ed.), *Exhibition Design* (London 1951, p. 166, no. 2; '"Victory over Japan" exhibition', *The Times* (21 August 1945), p. 6; '7,000,000 See War Exhibitions', *Display* (December 1945), p. 26; Beverley Pick, *Display Presentation. Exhibitions, window and outdoor displays* (London 1957), pp. 16, 38–39 (he refers to 'documentary exhibitions' on page 38); Ruth Artmonsky, *Exhibiting Ourselves. Fifty Years of British Exhibition Design* (London 2014), pp. 147–50. For Hutton's mural, see Margaret Brentnall, *John Hutton. Artist and glass engraver* (Philadelphia & London 1986), pp. 58–59. Pick's Regent Street display included 1,200 alloy stars, silver-coloured and covered in glitter dust, reported by Pathé News 'Stars in Regent Street', 8/12/55, Film ID 549.34; Callum Murray, 'Recalling the lights fantastic', *The Times* (4 December 1990), p. 20. See also Beverley Pick, *Display Presentation. Exhibitions, window and outdoor displays* (London 1957); Beverley Pick, 'The New Conception of Exhibitions', *Display* (May 1946), pp. 28–30.

6 BBC Written Archives Centre, Wynford Vaughan-Thomas, 'Tour of Britain Can Make It', Home Service (24 September 1946), Film T566.

7 *Display, Design & Presentation,* 'War to Peace' (September 1946), p. 14; 'New Methods of Display. Showing Best from British Industry', *The Times* (12 June 1946), p. 3.

8 The military equipment was all lent by Air Ministry and caused considerable logistical difficulties.

9 'War to Peace', *Display, Design & Presentation* (September 1946), p. 14.

10 James Gardner, *The ARTful designer. Ideas off the drawing board by James Gardner* (London 1993), p. 134. The spitfire was taken into the museum through a window, which caused damage to the window cill. Report on the state of the building after the exhibition. V&A Archive, Exhibition File: AAD 4/12–1977, Britain can make it: Report dated 5 February 1947.

11 DCA/911/Letter from Leslie Gamage, General Electric Company to S.C. Leslie 2 October 1946; reply from S.C. Leslie to Leslie Gamage 2 October 1946. Reproduced in Patrick J. Maguire and Jonathan M. Woodham (eds.), *Design and Cultural Politics in Postwar Britain. The Britain Can Make It Exhibition of 1946* (London and Washington 1997), pp. 194–5 documents 32 and 33.

12 *Supplement to the Board of Trade Journal for 28.9.46, Britain can make it. The exhibition of British post-war products*, p. 4. For a discussion of many of the items in the War to Peace section see pp. 4–5 and fig 6.

13 V&A Archive. Information for Guides (cited note 1).

14 'Munitions inspired designs for the home. "Britain Can Make It" show', *The Daily Telegraph* (24 September 1946), p. 3.

15 V&A Archive, Exhibition File: AAD/1977/4/14A, Britain can make it: Confidential account of the preparations and organization of the two special sections in the 'Britain Can Make It' Exhibition – 'War to Peace' and 'What the Goods are Made of' 23 January 1947 produced by S.D. Cooke.

DIANE BILBEY

What the Goods are Made Of

> Britain CAN make it. First you will see the materials from which things are made. Later when you see the goods, decide – Is the material right? Is it well used? Look, and judge for yourself.
> Repeat: 'Britain CAN make it ...' etc ....[1]

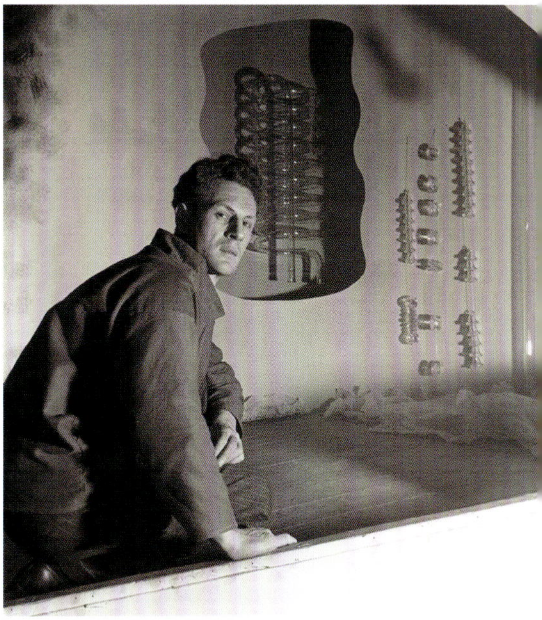

41. John Beresford-Evans, industrial designer, was responsible for the display of cameras, binoculars, thermometers and barometers at BCMI. He was also the designer of the glass display in the 'What the Goods are made of' section. Here he is shown in front of the partially installed glass display at BCMI. Beresford-Evans later became a member of the Design Research Unit, 1958–63.

'What the Goods are made of' was one of the introductory displays at *Britain Can Make It* (BCMI) [42]. Its aim was simple, to educate and inform the visitor about the products that they would see throughout the exhibition by explaining the raw materials used to make them. It also sought to emphasise that 'the basis of good design is the materials used'.[2] Designed by James Gardner and Basil Spence, the display attempted to make this somewhat dry subject exciting by demonstrating the beauty of the materials – their colour and their texture – as well as to explain their application in the consumer products the ordinary man or woman on the street might buy.

Following on from the display illustrating the transition from war to peace, 'What the Goods are made of' 'was divided into seven component areas covering metals, textiles, rubber, glass, timber, plastic and paper. A different designer was commissioned for each section, the brief being to demonstrate to the public the unique characteristics of each material, and it resulted in some of the most Surrealist displays in the entire exhibition.[3] As one caption to an official photograph of the glass display explains, 'Materials that are shown throughout the Exhibition in the form of finished domestic articles are demonstrated in abstract fantasies designed to show their extraordinary versatility' [41, 43]. An article in *Display* highlighted the creativity of the displays, concluding that 'Most imaginative of the introductory settings are seven devoted to raw materials'.[4]

Greeting visitors as they entered into this display area was 'Mr Designer', the aluminium mannequin figure used throughout the exhibition. Although mainly used as a wayfinding device, versions holding a megaphone would appear to 'speak' to visitors as they passed by.

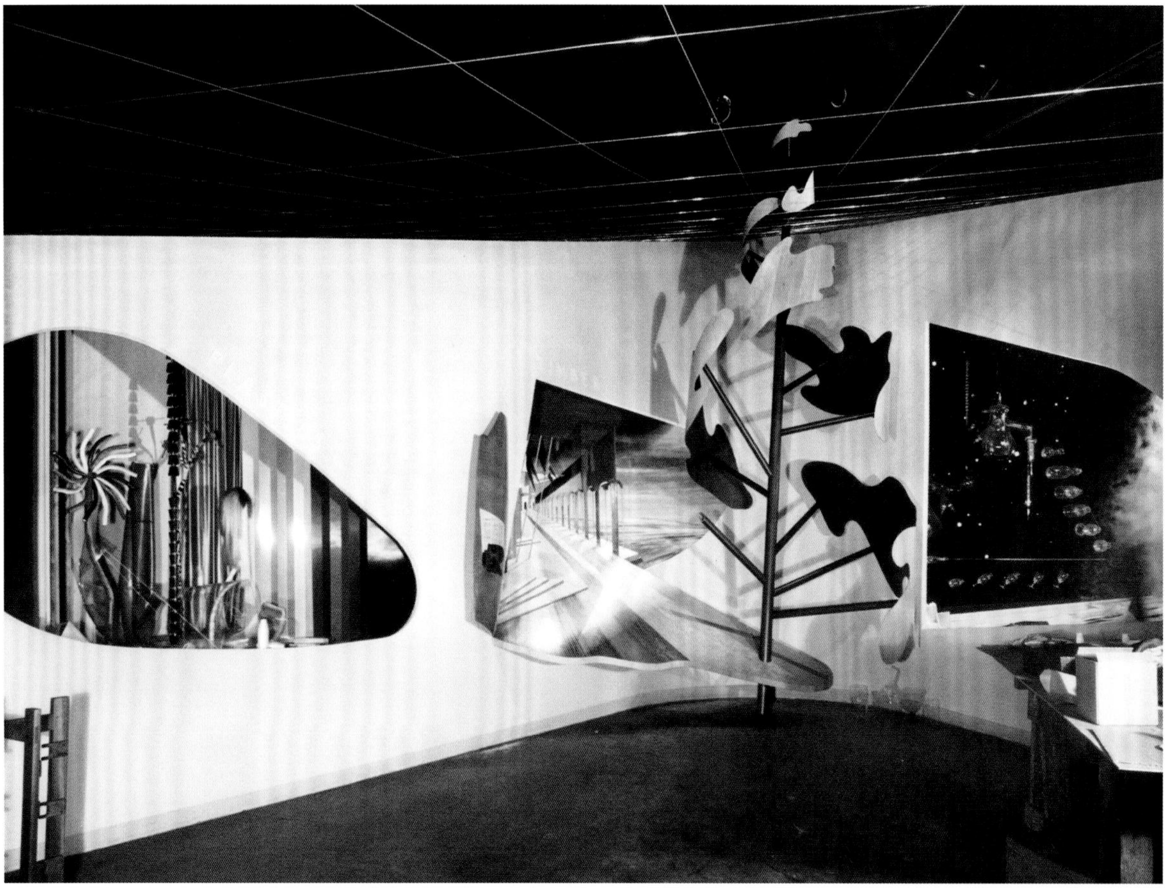

42. A view of the 'What the Goods are made of' section showing displays illustrating the use plastics, wood, and glass

The 'talking' version of 'Mr Designer' was positioned at the start of 'What the Goods are made of', with a salutary, brief summary of the primary aims of the display: '*First you will see the materials from which things are made. Later when you see the goods, decide – is the material right ...*' [see **85**].

A shortage of raw materials

Along with others from the Council of Industrial Design (CoID), S.D. Cooke was asked to write a report after the close of the exhibition on the problems, as well as the successes, he experienced whilst working on BCMI. Cooke, who had worked alongside James Gardner on 'What the Goods are made of', reflected on the huge workload that had been demanded of those working on the 'War to Peace' and 'What the Goods are made of' sections, describing 'the clicking of typewriters and the constant ringing of telephones adding to the general air of "Dante's Inferno"'.

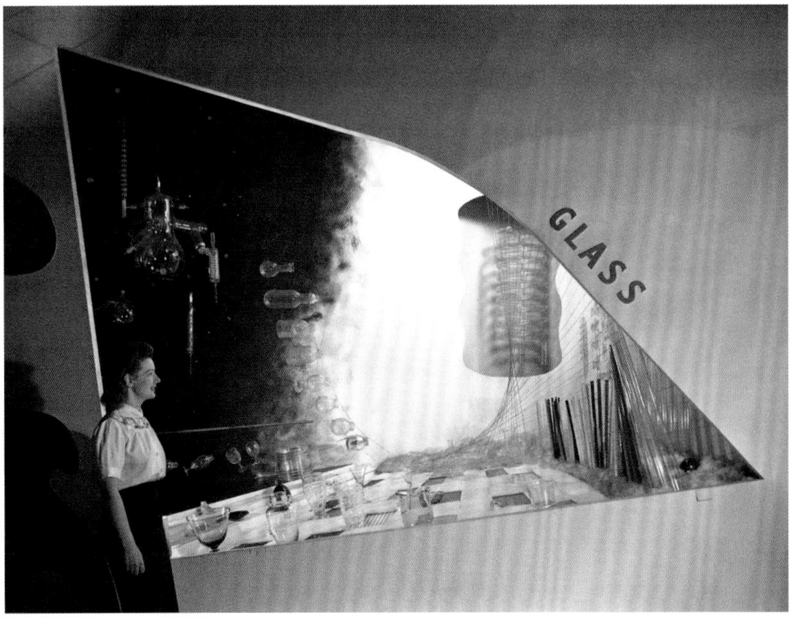

43. Information provided to BCMI Guides explained the content of the glass display designed by John Beresford-Evans: 'The Glass display consists of a very large number of different types of glass, including sheet glass in many decorative forms, glass tubes and rods used for lighting and neon tubing, together with examples of glass wool which is used for heat insulation and electrical insulation, and various blown glass shapes of all kinds. Lumps of glass in the raw state are shown together with one or two fine examples of glass domestic articles – one series showing how a wine glass is made, and the other showing how a thermos flask is produced. A set of brilliant cut-glass chessmen are also exhibited.'

44. The timber display was one of the most talked about in the whole exhibition, and James Gardner's masterly use of perspective, created an almost Surrealist work of art. Plywood – used to such great effect during the war in aircraft production – was the main component. A variety of different types of wood were used to create a marquetry style effect, giving the illusion of depth and texture to the abstract architectural composition. In this display, Gardner sought to illustrate the diverse uses of this material, 'as pliable as paper yet does not tear'. To demonstrate this versatility, flags in the display were made from plywood a mere 132nd of an inch thick and foliage for the tree was carved in two ways, using a development from wartime construction. Information provided to BCMI Guides explained: 'This is a composite display of some of the rarest and newest types of wood and it is remarkable after six years of war, during which the import of wood to this country practically ceased, that British Industry can produce such a wonderful exhibit. Almost every rare and decorative wood is shown, and the whole structure gives an effect of a street with houses and buildings composed of timber of every kind. A wooden "Tree" stands in front of the exhibit. The leaves are made of plywood bleached on one side white and stained green on the other side, and the mast or trunk of the tree is a very new development of plywood from which very strong tubes can be made.'

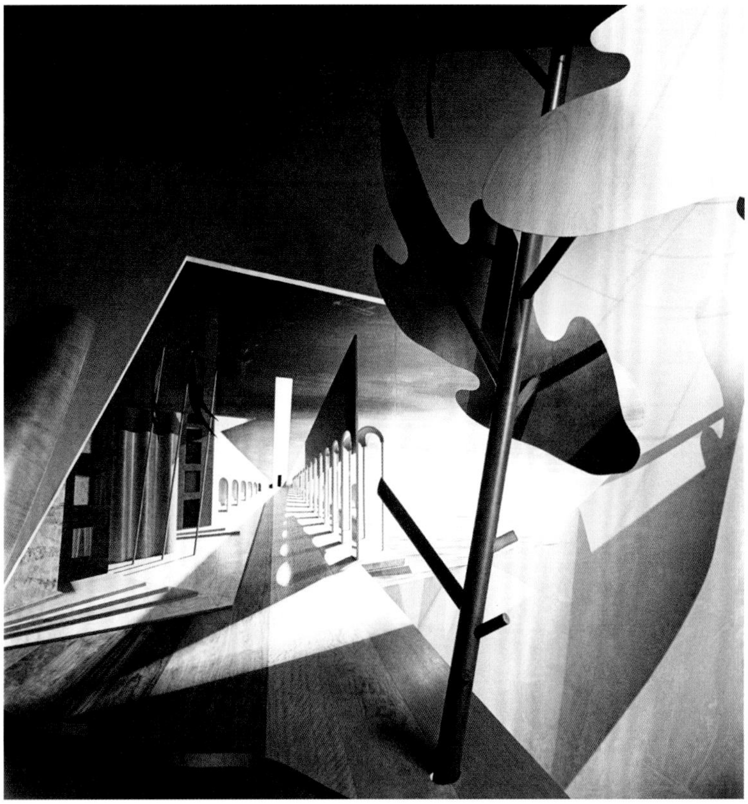

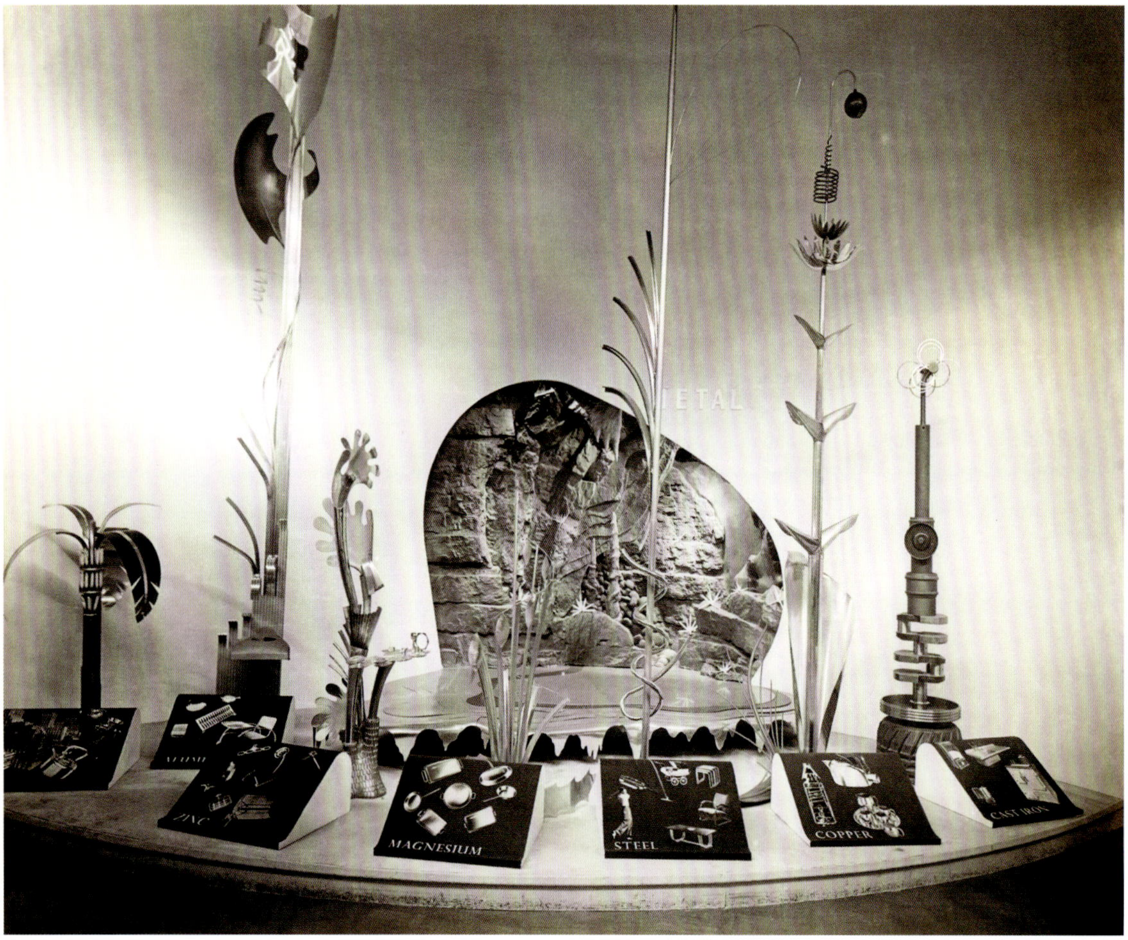

45. In the metals display, James Gardner aimed to demonstrate the malleability of metals by forming them into elaborate, fantastical 'trees'. Each tree, aimed to illustrate the qualities of the particular metal from which it was made. Information for BCMI Guides explained that in the background to the display, a stainless steel sheet metal lake 'flowed' out of a 'symbolic cavern... supposed to show metallic ores in their natural setting inside a cave... The various ores are not real, but the display sparkles with metallic colours and rather weird lighting, and is merely staged as a dramatic effect.' In the foreground, the 'pool' into which these metals flowed was made up of stainless steel, sheet iron, copper, and brass. Examples of products made from the materials were laid out in front of the various trees. Finished off by a waterfall, made of polished aluminium, this was an ambitious and highly imaginative display. By taking inspiration from the natural world to demonstrate metals, their properties and uses, Gardner added a Surrealist and attention-grabbing touch to this display. The journal *Display, Design & Presentation* (November 1946, p. 13), described the whimsical qualities of the display: '"Metal" shows magnesium, copper, aluminium and other metals as plants from which the potential products blossom forth; leaves are displayers for items such as piston rings'. The tallest 'Tree' in the display – the 'Aluminum Tree' – included long stretches of aluminum that appeared to sprout from it. In this design, Gardner emphasized the key qualities of aluminium – that it was lightweight, and could also be highly polished. The 'Magnesium Tree' illustrated the changing use of the material from military and wartime, to domestic and peacetime purposes. Information provided to BCMI Guides, explained, 'Most of us know it best, or worst, as an incendiary bomb, and it is remarkable to note to-day that frying pans and articles which are used directly over flame are made of this extremely light and strong metal. Guides are likely to be asked questions as to why it does not catch fire immediately and burn with great fierceness, this is entirely due to the fact that it is thick and cannot be ignited in this form so that the articles are 100% safe and are very well designed'. To illustrate the versatility of the material, the 'Steel Tree' was formed from steel tubes of varied lengths, including the finest of tubes used in hypodermic syringes. The 'Copper Tree' was described as a material that 'can be easily worked and looks very beautiful when polished, but its use to-day in the home is largely confined to brass articles and copper kitchen utensils'. The 'Cast Iron Tree' display designed by Derek Bridgwater was described as of an 'engineering nature' showing items including a crankshaft. It illustrated the more practical uses of cast iron, which were mainly to be found in the home in products such as cookers, cooking utensils and baths.

For its designers, 'What the Goods are made of' posed some of the most difficult challenges of the entire exhibition. In almost every case, there were acute problems in obtaining the actual raw materials. Despite the importance of the exhibition being explained, some of the various trade associations flatly refused to cooperate, although most showed a willingness to play their part. One of the most accommodating was the textile trade, and Cooke reported that 'In no case was the question of coupons raised by manufacturers, who were only too willing to co-operate if they had the material'. The Timber section (designed by James Gardner) proved to be the easiest of all to organize, and also 'aroused more comment than any other in the Section', which was not surprising given its Surrealist references [**44**; see **front cover**]. The company of William Mallinson & Sons Ltd initially, 'perhaps somewhat reluctantly', made and constructed the Timber display to Gardner's design. They later wrote to the CoID to express sincere thanks for their involvement and stated that 'they were convinced that it had done them a very great deal of good.'

There were seven 'trees' in the Metals section made from tin, aluminium, zinc, magnesium, steel, copper and cast iron, and examples of relevant products were displayed alongside the raw materials [**45–47**]. Manufactured by The Tin Research Institute, the 'Tin Tree', with its bunch of bananas made out of tooth-paste tubes, became the target for unwanted attention from visitors. Cooke reported that the public 'took great delight' in destroying the bunches of bananas, which had to be frequently replaced [**47**]. Behind the row of metal trees, the 'cavern' feature, towards the back of the display, caused further problems: 'The pool flowing out of the cavern was composed of metal sheets in various metals involved, and this item presented more difficulties, as the supply of sheets was almost non-existent'.[5]

Visitors left this section well-informed as they entered the Main Hall, a window-shopping paradise enticingly called 'Shopwindow Street'. But, before they left, exhibition guides were encouraged to reinforce to visitors the connection between the raw materials they had just seen and the selection of goods that they were about to see, and could, one day, possibly even buy.

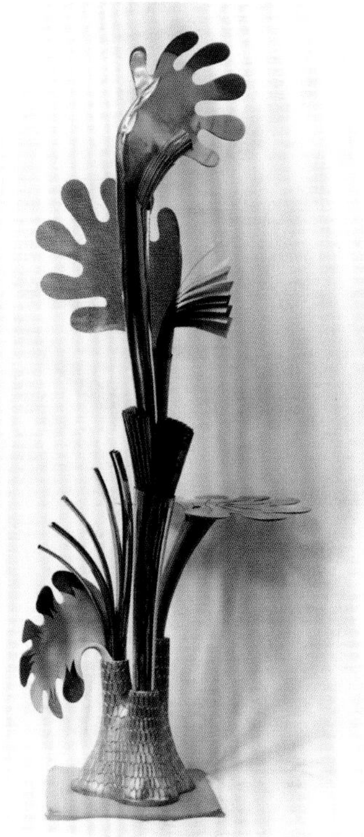

46. Although Gardner designed most of the trees, the 'Zinc Tree' was designed by R.Y. Goodden, who was also responsible for the display of travel and leisure goods later in the exhibition. The 'Zinc Tree' aimed to show how zinc was used to galvanize metals to stop them from rusting.

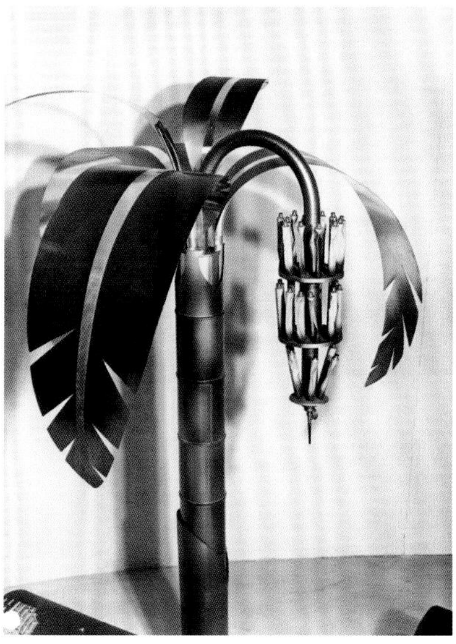

47. The 'Tin Tree', designed by the Product Design Unit, demonstrated the significant advances made in tin-plating made during the war, to create packaging that prevented contamination of food or drinks.

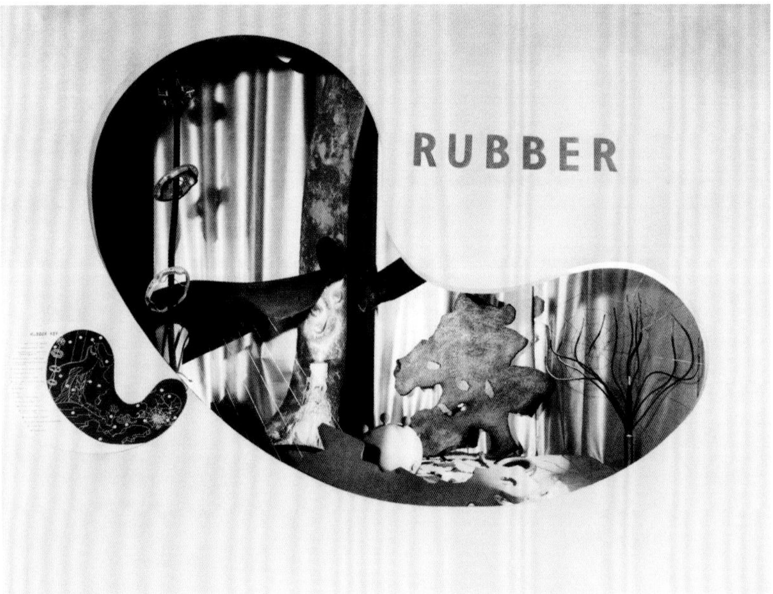

48. The rubber display, designed by the architect Stirrat Johnson-Marshall, was put together by Dunlop Rubber Co. Ltd. During the war, Johnson-Marshall was posted to Farnham Castle in Surrey, a centre for research and development in camouflage, where he had helped design inflatable tanks and guns. Information provided to BCMI Guides explained: 'Here we have a symbolic display of Rubber in the raw state emerging as a liquid from the little "tree" on the right and flowing on to the floor. This is known as Latex and it is then formed into materials which you will recognise as Crepe Rubber, Foamed Rubber, and other sheet forms. Part of the floor is painted with rubber paint which is used for treating steel work to prevent rust. The submarine cable "tree" is composed of wires insulated with rubber and the weird shape on the left is a piece of steel, covered with sheet rubber used in industry where strong acids contact metal, which would otherwise be dissolved. The end of the shape is equipped with a rubber brush which is wiping an industrial rubber boot. Other items are the spider's web and a household rubber glove which inflates and deflates automatically. The tree-like form on the left-hand side is a series of examples of rubber bonded, or stuck, to metal used in the universal joints of a motor-car and other devices for smoothing out shock, used in motor-car transmissions etc. A piece of table-tennis bat sheeting, a piece of batting glove rubber sheeting used on the back of cricket gloves, and a series of articles in rubber, including golf balls, tennis balls, squash balls and football bladders, which are easily recognised.'

49. Designed by Albert Smith, this display sought to demonstrate the various uses of paper and in so doing created another Surrealist display. It included printing paper, wallpaper, parchment, and compressed paper used in forming new styles of packaging for foodstuffs. Information provided to BCMI Guides explained: 'This display shows a series of compressed paper shapes, together with a woven piece of fabric which is actually the material out of which Lloyd Loom chairs are made. This is not generally recognised as being made of paper. The remainder of the display consists of a very large selection of small pieces of fine quality British papers of every type.'

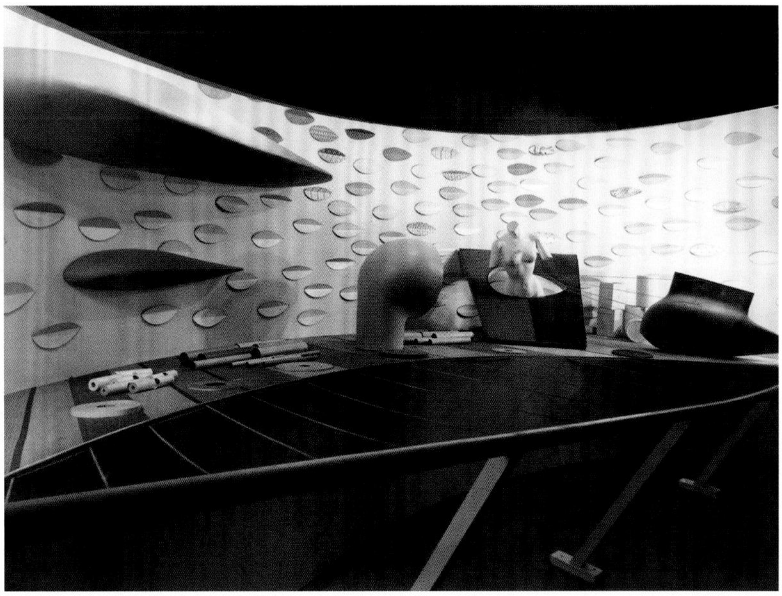

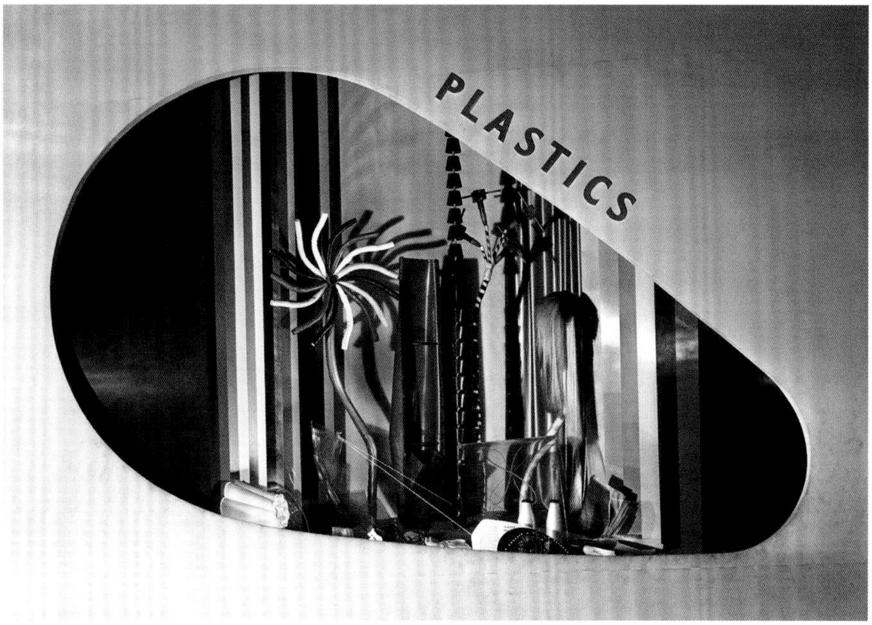

50. The role and experience of the designer in making best use of this immensely adaptable material was explained on a panel accompanying the plastics display: 'There are over forty kinds of plastics and many ways of moulding. The efficient designer knows which materials and processes will suit each different job'. John Hutton, who was responsible for the futuristic mural in the 'Hall of the Future', also designed this display. The possibilities that plastic offered in manufacturing were bought to the fore at BCMI. As well as in the dedicated plastics display in 'What the Goods are made of', plastic products were shown in many other sections of the exhibition, including 'From War to Peace', 'Shopwindow Street' and in the 'Hall of the Future'. The journal *British Plastics* wrote that plastics, particularly PVC and Perspex, were widely used in the exhibition, featuring them in articles in its September and October issues. Information provided to BCMI Guides explained: 'The plastic display consists of a series of building boards in colour made of plastic. These are known as Laminated boards and are made in layers. The outside is either in colour or is faced with very thick timber. The back of the draping is composed of poly-vinyl-chloride sheeting in a blue-grey shade which you will recognise as the material you don't like in the shops. The central display stands on a clear Perspex sheet on which a specially bent and curved sheet of Perspex, engraved with a design, stands. It will be noted that one of the characteristics of the display is again extreme lightness and strength. On the left-hand side of the base is a "tree" composed of extruded tubes in colour. Next to this stands a plastic air chute and some wing tips of a Lancaster aircraft, and in the centre of the display is a series of moulded plastic electrical insulators. On the right-hand side of the display is an extruded tubular stalk out of which falls a cascade of pure nylon filaments. Next to this "tree" is the Radar "tree" which consists of a series of extremely high quality technical mouldings used in radar sets during the war. Cables from these form the stalk. In front of the display is a series of extremely interesting plastic domestic articles, including a pure nylon stocking in the white – that is before it is dyed and sown up – and a pair of real British nylons of extremely fine texture. Alongside these are shown cast resin blocks from which buckles for belts are cut, various plastic bowls, a pair of plastic evening shoes which it should be noted are only for wear on special occasions indoors, some woven plastic material and other simple articles including a lamp shade, showing how it is formed from sheet, a pair of men's braces etc.'

NOTES

1 Commentary played at the entrance to the exhibition via a speaker in a 'Mr Designer' mannequin figure.

2 The National Archives (TNA) INF 6/595: Commentary Britain Can Make It issue no. 12. A copy of the film (Britain Can Make It no. 12) is held by the British Film Institute and can be accessed via BFI Player.

3 As in each section of the exhibition, details of the designer and manufacturer relating to this display were published in the accompanying catalogue. See *Britain Can Make It Exhibition Catalogue* (London 1946), pp. 147–52; *Britain Can Make It Exhibition Catalogue Supplement* (London 1946), p. 237. Information issued to exhibition guides to help with public enquiries is held in V&A Archive of Art & Design, Exhibition File: AAD/1977/4/14, Britain can make it: Information for Guides. What the Goods are Made of. Plan Key D.

4 Design Council Archive at the University of Brighton Design Archives (DCA): caption to one of the photographs showing the glass display. 'Britain Can make It', *Display, Design & Presentation* (November 1946), p. 11.

5 V&A Archive, Exhibition File: AAD/1977/4/14A, Britain can make it: Confidential account of the preparations and organization of the two special sections in the 'Britain Can Make It' Exhibition – 'War to Peace' and 'What the Goods are Made of' 23 January 1947 produced by S.D. Cooke.

DIANE BILBEY

Shopwindow Street

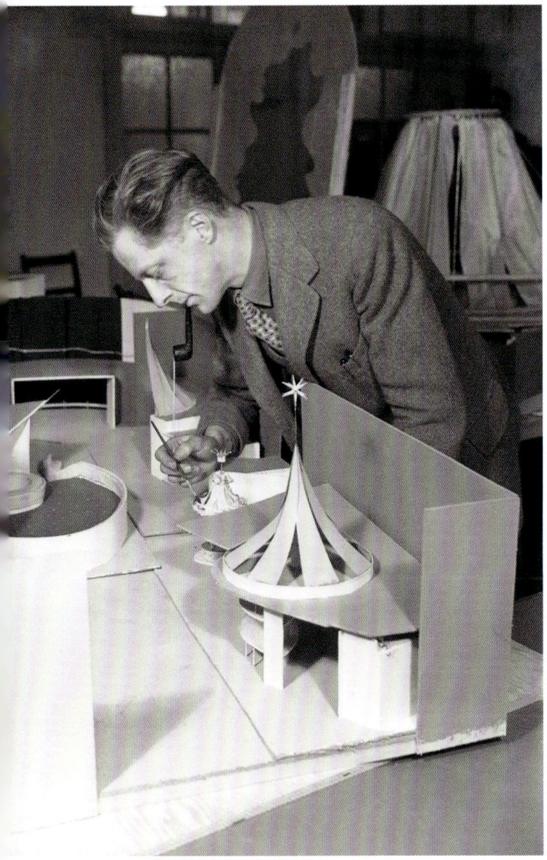

51. James Gardner working on the model for 'Shopwindow Street'

On our new shop window the world looks down.
Britain Has Made It.[1]

For a so-called 'nation of shopkeepers', it's not surprising that 'Shopwindow Street' – the Main Hall of the exhibition – was one of the most popular features. Designed by James Gardner, 'Shopwindow Street' housed a variety of desirable consumer products, in a setting created using modern techniques in construction, display and lighting [**51**]. Information provided to teachers bringing groups of students to the exhibition stressed that 'Shopwindow Street' was 'not only one of the principle [sic] sections, but is designed to emphasize the importance of display as such'.[2]

'Shopwindow Street' could only be accessed once visitors had been through areas of the exhibition covering the introductory sections, 'From War to Peace', and 'What the Goods are made of': such was the pre-defined route set out by Gardner [see **7, 187**]. These two sections were housed in corridor-style galleries, that opened out into the much larger exhibition space given over to the window-shopping paradise that was 'Shopwindow Street'. It was cheerful, fun and airy, a marked contrast to the atmosphere created by Beverley Pick in the 'From War to Peace' section, replicating a black-out and blitzed London street scene. 'Shopwindow Street' lifted the mood – the darkness of wartime imagery 'was replaced by brilliant lights, decorations and colours, with a shop window presentation of small commodities. Here in fresh colours and design will be newly-manufactured pottery and glass, radio and television, cutlery, toys, kitchen equipment, and many other goods.'[3] 'The Designer's intention here was to reproduce a night street scene and the "sky" was purposely darkened and clouded to give this effect, and to provide a contrast with the brilliantly illuminated shop windows by the light of which a catalogue could easily be consulted'.[4]

The use of the name 'Shopwindow Street' was an excellent ploy by the organisers to draw in a British public deprived of any real shopping opportunities during the war. An article in *The Architects' Journal* noted: 'Window Shopping is a traditional amusement, and at the V. and A. you will find it as delightful and stimulating as ever'.[5] The name 'Shopwindow Street' might also be a nod towards J.M. Richards's publication entitled *High Street,* published in 1938, and colourfully illustrated by Eric Ravilious with the different types of shop fronts found in an average High Street.[6]

Set against a dark blue sky and 'lit with a faint afterglow of sunset', silvery Perspex 'clouds' suspended from the existing structure of the V&A helped create an atmosphere of unreality, more like a film set than an exhibition hall. Gardner wrote about the unforeseen problems he faced with the installation of the clouds, recalling comments made by a workman on site: 'Wires for suspending lower "clouds" will have to pass through the upper "clouds", which are already in position. "It may look okay on the drawing, Mr Gardner, like it always does, but how to you reckon we're supposed to get at them up there?"'[53].[7]

In 1936, Gardner had visited the *International Surrealist Exhibition* that took place in London, and later reflected that 'Images of it had stayed in the mind', and there is no doubt that Surrealist influences can be seen in his designs throughout the exhibition. Gardner later suggested that his particular use of 'décor' and 'Barnum and Bailey' theatrics had given him the reputation of being 'a rather airy-fairy designer'.[8] Oversized display

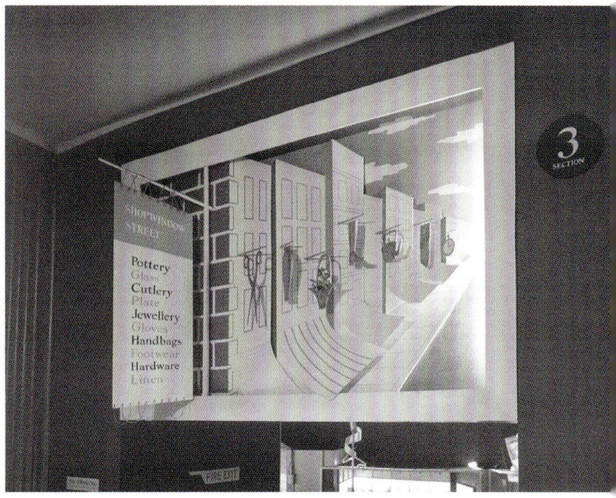

52. Entrance to 'Shopwindow Street' (Section 3)

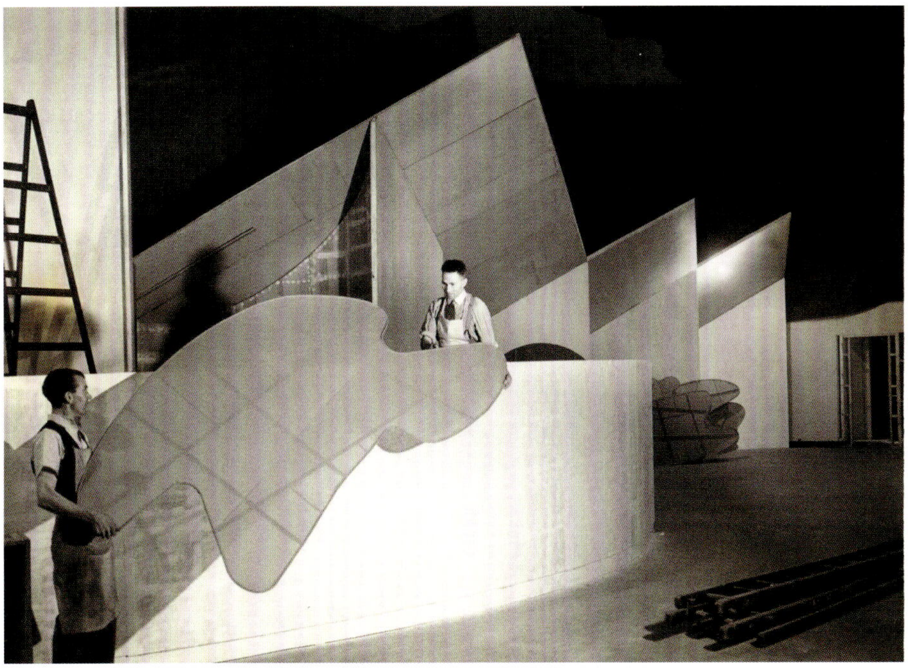

53. The Perspex clouds feature being installed in 'Shopwindow Street'

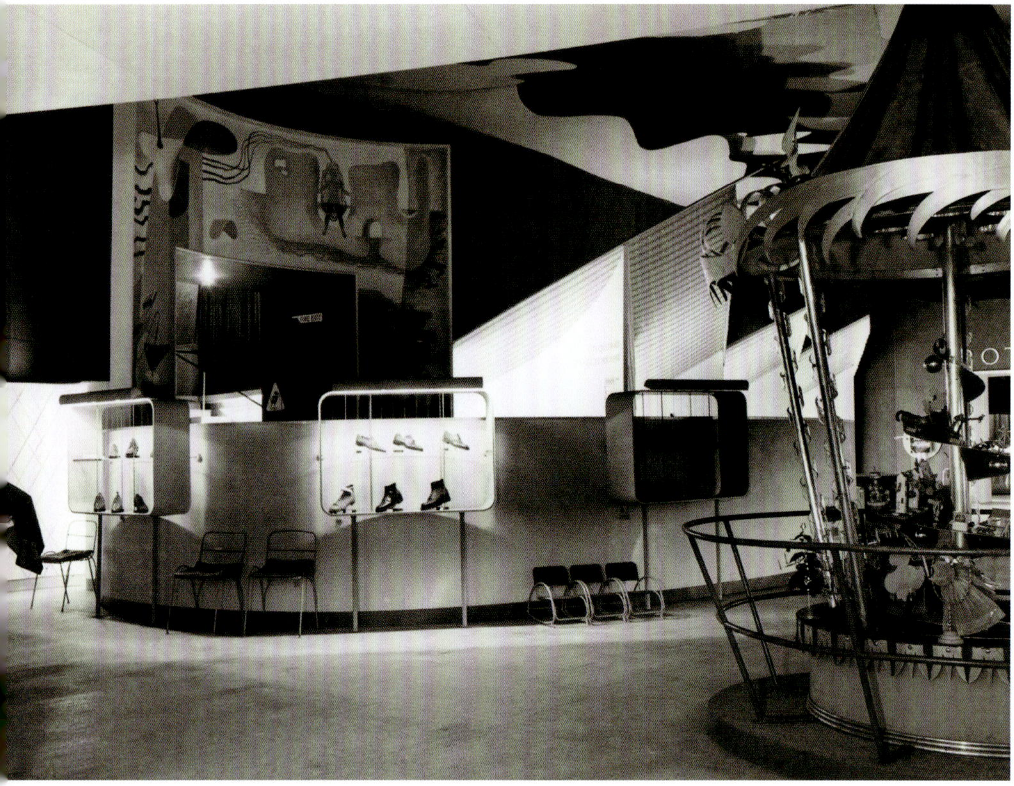

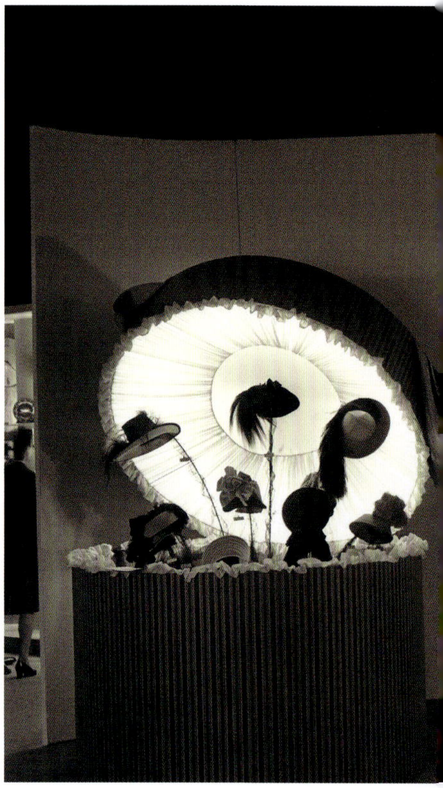

54–56. Views of 'Shopwindow Street', showing William Crosbie's mural depicting 'The Voices of the Underground' during the war (left); the hosiery and hat section with Jack Miller's mural depicting a park scene, providing the background to the truncated legs on display (middle); and the mural by Duncan Grant and Vanessa Bell above the doorway in the background (right)

features – including a giant hatbox to display a selection of women's hats and a huge pair of a woman's hands directing the visitor to enter one of the display alcoves – added to the sense of surrealism (and humour) that was evident in much of the display. A plaster reclining nude figure by Raoh Schorr, draped with black and white lace, must have caused some concern during the royal opening, as Gardner noted that 'someone had discreetly covered it up with a dustsheet'.[9] An abstract mural by the Scottish artist William Crosbie, depicting 'The Voices of the Underground', framed one of the doorways, providing a striking post-war tribute and a suitably modern backdrop to the King as he gave his fifteen-minute opening speech [**54**; see **72**, **130**]. Visitors were greeted with another brightly coloured mural, by Vanessa Bell and Duncan Grant, above the doorway as they left 'Shopwindow Street' [**56**; see **123**]. Large-scale figures symbolizing 'Art' and 'Industry', side by side, were an appropriate metaphor for the aims of the Council of Industrial Design, the V&A, and for BCMI. The Quiz Bank, where visitors would collect their 'Quiz Coins' in order to cast their vote at various stages around the exhibition, was also located at the entrance, near the Crosbie mural. 'Shopwindow Street' was the hub of the exhibition.

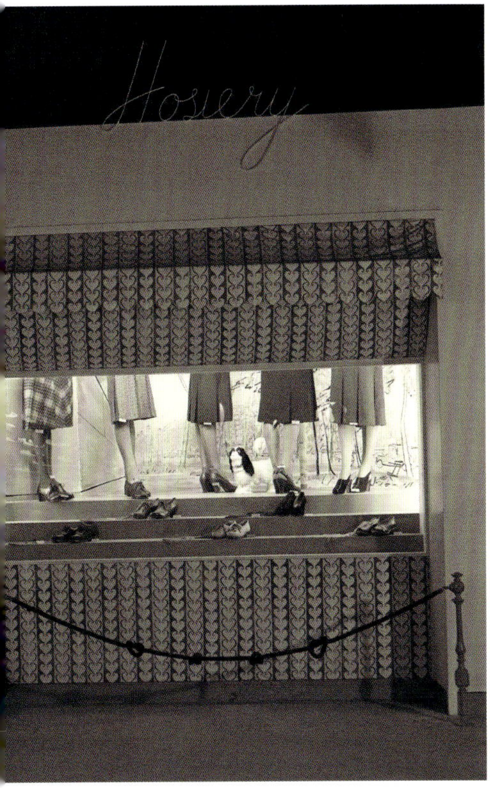

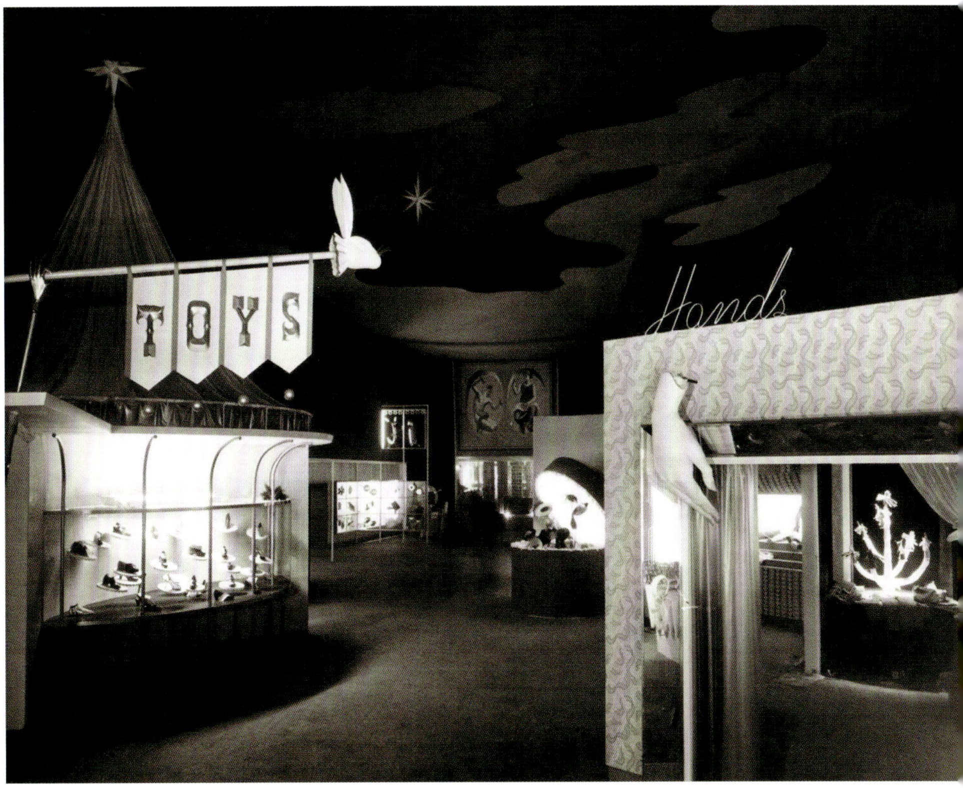

James Gardner devised 'Shopwindow Street' as a series of alcoves displaying a range of around twenty different commodities, including glass, ceramics, shoes, hardware, clocks, cameras, luggage and radios. These alcoves were brightly illuminated and 'tented with white and silver drapes, like floating parachutes'.[10] Music, supplied by the Decca Record Co. Ltd, was played throughout the exhibition, and Gardner noted that '[Shopwindow Street] all came alive in 3D, to the ominous sound of 'The Nutcracker Suite'.[11] In his autobiography, Gardner wrote about the problems he faced in trying to design a scheme that would look good even if there were no products to display – a distinct possibility. An additional practical problem arose in obtaining the materials necessary to build the shop window alcoves. These were dependent on large amounts of plywood being available, but it was a scarce and restricted product. According to an article in *The Architect & Building News*,

> The designers were politely told that this could not be, and they must find a substitute. After a lot of thought one was selected, and a licence to obtain it applied for. This was refused, taken to a higher authority and refused again, finally turned down by almost the highest authority of all. It was referred to

> one whose word in the Board of Trade is law .... The Great Man said (or so it is rumoured ...) 'I'm keen to help you fellows, but I can't release this type of board. It's Priority No. 1 for housing.' Pause. 'But I'll tell you what – couldn't you make do with quarter-inch ply?'[12]

Dr Wilfred F. Coxon (an expert on industrial processes and techniques) presented a series of documentary-style programmes for the BBC Pacific Service, relating to the concept of 'Britain Can Make It'. This was not merely concerned with the exhibition, but with the general theme of the post-war regeneration of Britain. Nevertheless one of Coxon's broadcasts featured the exhibition itself, and the diverse range of products on show in 'Shopwindow Street',

> Here we have delicately carved cut glass articles, pottery encased in a golden sheath, shoes and handbags made from leather and the newer plastic materials, cutlery even brighter and more stain free than we have ever known whilst there were a multitudinous number of kitchen gadgets to help the housewife in her daily tasks in the kitchen.[13]

The description of Britain as a nation of shopkeepers was also noted by R. Dudley Ryder (Chief Exhibitions Officer at the CoID), when he explained the value of good display in selling goods, 'We have been called "A Nation of Shopkeepers", but often we have been told that we do not know how to display our goods or make them appeal to the potential buyer'.[14] The journal *Display* regarded the presentation techniques of 'Shopwindow Street' as exemplary: 'True displayman's touch is James Gardner's breakaway from normal "stand" architecture, presenting exhibits from various trades in a "street" of shops'.[15]

Mass Observation (M-O) reported that 'although the whole section is extremely popular, far and away the most important individual item in it is the pottery'.[16] Children viewing BCMI seem to have echoed the interests of adults, among whom the pottery, glass and jewellery displays in 'Shopwindow Street' were the most popular exhibits: '... one teacher who brought a party of girls aged 16–18 years says that "tea and coffee sets were remembered and described in minute detail later". Nor did the décor go unnoticed; the roofing in Shop Window Street and the background of music were both approved.'[17]

Punch summed up the mass appeal of 'Shopwindow Street', that it

> seems to epitomize the spirit and execution of the whole exhibition. The setting is very fine. As the visitor comes into the section he sees the 'street' with its bright shop-windows filled with commodities of the near-future, stretching away under a dark-blue sky; if he is giving his mind to it he must

> realize, almost with a shock, just how much has gone into the creating not only of this single effect, but the show as a whole, and Industrial Design as a movement; the plan is the same throughout – an overall scheme calculated to throw the individual exhibits into prominence.[18]

The collective self-discipline of the 1.4 million people who went to BCMI resigned to the fact that they could not have the products on display must surely be commended. In an article in the *Daily Mail* entitled, 'Britain Can Make It, and – Europe thinks we're terrific', Alexander Clifford wrote (perhaps with an overly positive tone):

> Coming from abroad I expected the British people to be resentful when they saw the things they made but could not use. But they did not seem to be. I think I detected a faint scepticism – an instinctive suspicion that very few of these things would ever really be on sale. Mostly, there was just genuine admiration.[19]

One of the most popular sections of the exhibition, 'Shopwindow Street' was also one of the most ingeniously and ambitiously designed. But, with very few of the products displayed available for visitors to buy, nothing would alter the fact that window shopping was, for the time being, all they were able to do. As Lesley Whitworth remarks, 'Austerity added to the difficulty of locating good examples of product design at the same time as it increased the public's appetite to view them'.[20]

NOTES

1 Commentary from *British Pathé* short film entitled, 'Shop Window. Britain Had Made It', devoted to the 'Shopwindow Street' section at the exhibition. Filmed 30 September 1946.

2 V&A Archive, Exhibition File, AAD/1977/14/4, Britain Can Make It: NOTES ON "BRITAIN CAN MAKE IT" FOR TEACHERS.

3 V&A Archive, Exhibition File: AAD/1977/4/12, Britain can make it: Notes for the information of the Press (Summary), National Exhibition of Design "Britain Can Make It". Opening in September, dated February 1946.

4 Design Council Archive at the University of Brighton Design Archives, hereafter DCA: 911/Letter from N. E. Kearley to Charles Scougall, dated 1 January 1947.

5 Astragal, 'Note & Topics. Private View', *The Architects' Journal* (3 October 1946), p. 239.

6 I am grateful to Michael Twyman for suggesting this possible connection.

7 James Gardner, *The ARTful designer. Ideas off the drawing board by James Gardner* (London 1993), p. 137.

8 *Supplement to the Board of Trade Journal for 28/9/46. Britain can make it. The exhibition of British post-war products*, p. 6. The *International Surrealist Exhibition* took place between 11 June and 4 July 1936 at the New Burlington Galleries. Penny Sparke (ed.), *Did Britain Make It? British Design in Context 1946–86* (London 1986), p. 11.

9 Gardner (cited note 7), p. 135.

10 Sylvia Pollak, 'Round the exhibition', *Architectural Design & Construction* (October 1946), p. 269.

11 DCA/742/Special Features (constructional). Gardner (cited note 7), p. 133.

12 'Events and Comments. Another naughty story', *The Architect and Building News* (27 September 1946), p. 186.

13 BBC Written Archives Centre: Dr W.F. Coxon, 'Britain Can Make It (no. 4)', Pacific Service (10 October 1946). Dr Coxon's programmes began in August 1946.

14 R. Dudley Ryder, 'The "Britain Can Make It" exhibition', *Design '46* (London 1946), p. 12.

15 'Britain can make it', *Display, Design & Presentation* (November 1946), p. 14.

16 University of Brighton Design Archives: Design Council Archive: 903, Mass Observation report.

17 V&A Archive, Exhibition File: AAD/1977/4/14, Britain can make it: How a child saw the "Britain Can Make it" exhibition, report by Miss C.M. Dowler dated January 1947.

18 John Basil Boothroyd, 'Britain Can Make It. A Late Look Round', *Punch* (30 October 1946), vol. CCXI, no. 5523, p. 375.

19 Alexander Clifford, 'Britain Can Make It, and – Europe thinks we're terrific', *Daily Mail* (19 November 1946), p. 2.

20 Lesley Whitworth, 'Anticipating Affluence: Skill, Judgement and the Problems of Aesthetic Tutelage', in Lawrence Black and Hugh Pemberton (eds.), *An Affluent Society? Britain's Post-War 'Golden Age' Revisited* (Aldershot 2004).

ERIC TURNER

Silverware

> One of the most striking things to the layman who studies the British jewellery and silverware industry is the small attention which appears to be paid to design and the small honour in which designers appear to be held.[1]

In 1946, the same year as the *Britain Can Make It* exhibition, the Working Party on Jewellery and Silverware, appointed in March, presented its report to the government. While the exhibition was the public face of the government drive to prove that British manufacturing industry could revive and flourish in peacetime, the Working Party investigated a whole swathe of industries (including silverware), and reported on the traditional and present deficiencies of these industries and how and why they should be improved.

Industry working parties

These working parties were the initiative of Sir Stafford Cripps, the newly appointed President of the Board of Trade in the new Labour government. Cripps shared with other members of the Labour administration a frankly pessimistic view about the state of British industry. The working parties had been established in an effort to obtain some leverage on the development of industrial policy in the still vast private sector. Conservative opposition to these initiatives was strongly expressed, and echoed the manufacturers' fear that the evidence of the Working Parties would be used to interfere with management's right to manage, and boost trade union influence. Worse still, they feared that they would push industry nearer to state control, and take the first step towards nationalization. Winston Churchill, now in opposition and out of power, accused Cripps of putting 'Socialist nominees' on to his Working Parties. Cripps was faced with the delicate task of assuaging these fears, but was nonetheless determined to push ahead. He was forced to make it clear that there was no suggestion that the government wanted to impose solutions, but only wanted to stimulate industry to provide its own.

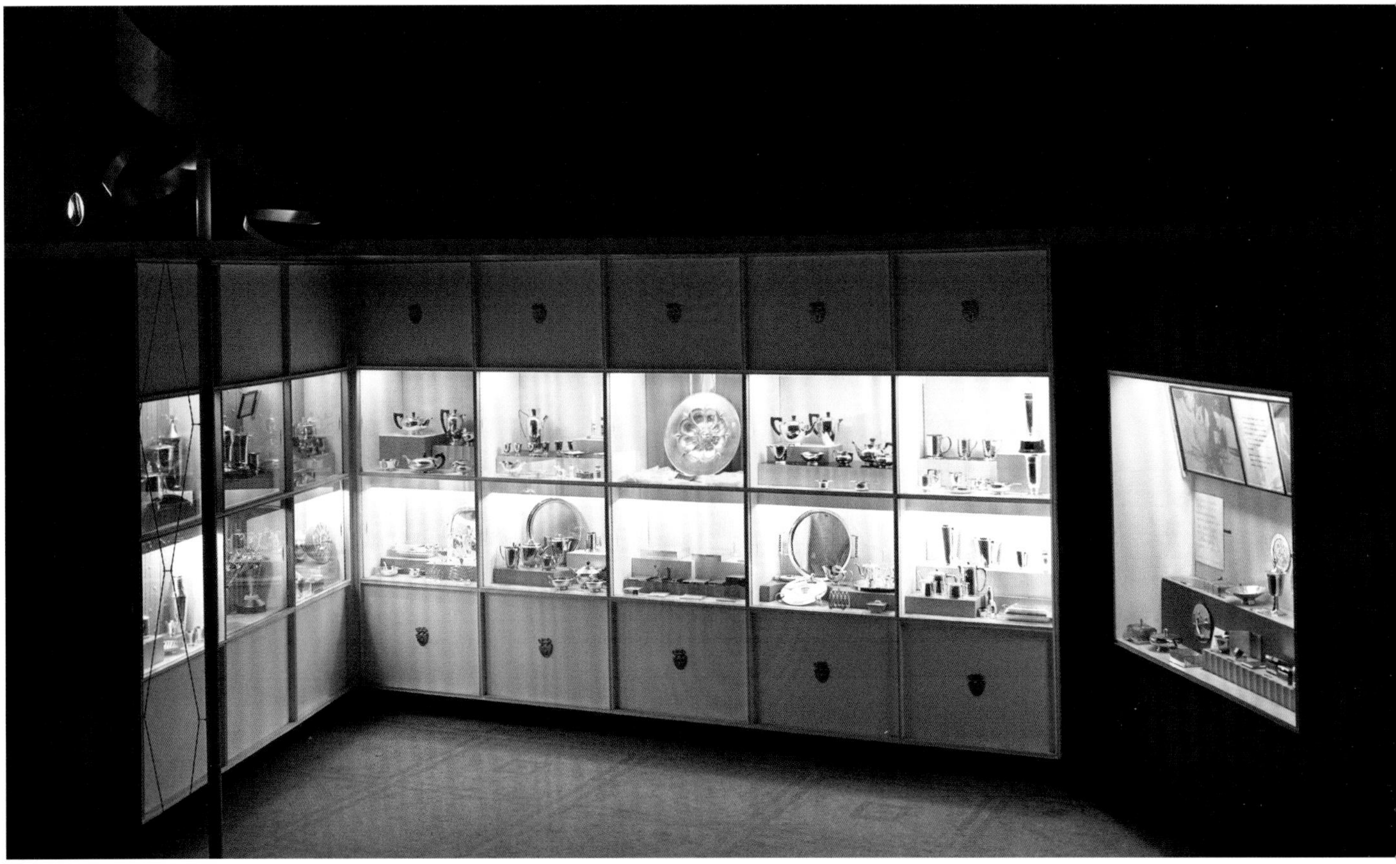

57. 'Shopwindow Street' display of silver and electroplate, designed by James Gardner, display of plate arranged by Ronald Dickens

The composition of the Jewellery and Silverware Working Party consisted of twelve members equally representing employers, trades unions and independent contributors. One of the independents was George Ravensworth Hughes, the Clerk of the Goldsmiths' Company (in other words, the Chief Executive), who throughout his career vigorously promoted the cause of good, contemporary design in silver and jewellery, and who was later to serve on the selection committee for this section in the BCMI exhibition, together with four other members from across the industry. Significantly, it did not include a representative from the Metalwork Department of the V&A. The then Keeper, Charles Oman, viewed the contemporary with Olympian disdain.

The report makes for sombre reading. Silver and jewellery workshops in all three manufacturing centres – London, Sheffield and Birmingham – with a very few notable exceptions, were described as ranging from 'bad' to 'very bad'. Ventilation was poor, light levels were inadequate, overcrowding was rife; conditions were too often insanitary and equipment inadequate and obsolete. Production flow layouts were generally inefficient. Consumer research was more or less non-existent, as was any cost or statistical analysis of profit margins, which prevented management from properly planning future investment. Workers' remuneration was inadequate, particularly at entry level, and it was noted that even the children of established craftsmen were neither tempted nor encouraged by their parents to join the trade. One chapter was devoted to the role of the professional designer. They were especially badly paid, if they were employed by management at all: they all too often regarded the designer as a 'frill and a frill they cannot afford'.[2] The function and status of the designer in British industry was to attract increasing attention over the coming years, particularly after Robin Darwin was appointed Rector of the Royal College of Art in 1948. He was determined to, and succeeded in 'bringing the designer out of the backroom'. But this lay in the future.

Of grave concern to the Working Party was the imposition of Purchase Tax. This was a tax introduced in 1940 with the obvious intention of raising revenue for the war effort but with the secondary purpose of diverting consumer spending away from luxury goods to concentrate on producing munitions for war. For jewellery and silver manufacturers the tax was levied at 100%. They had little choice. In April 1946, purchase tax was reduced for many goods to 33⅓% but not for silver and jewellery, which remained at 100%. Unlike VAT, which replaced it in 1973, purchase tax was a wholesale tax, applied at the point of manufacture and distribution, not at the point of sale. This prevented retailers from building up adequate stocks and discouraged overseas buyers. The report highlighted inconsistencies in its application. Luxury silk goods were now subject to the new rate of 33⅓% whereas silver and jewellery were not. Articles for the serving of food and drink made of ceramic, glass, stainless steel, copper or brass, for example, were now free of tax, but similar articles of silversmiths' ware were still subject to the highest tax rates.[3] Purchase tax at the rate it was applied to the silver manufacturing trade was crippling the industry, and the report called for its urgent revision and immediate reduction. This was the background to the silverware and jewellery trades in the run up to the BCMI exhibition.

The selection committee

As we have seen, the purpose of the exhibition was to promote good, modern design. The choice of potential exhibits was to be made by a small selection committee appointed by the Council of Industrial Design (CoID). Their remit was to include not only silver and jewellery but cutlery, watches and clocks as well. The committee was the sole adjudicator. Manufacturers could not, as with previous industrial fairs, purchase space on their own behalf. Nor could sales be made directly within

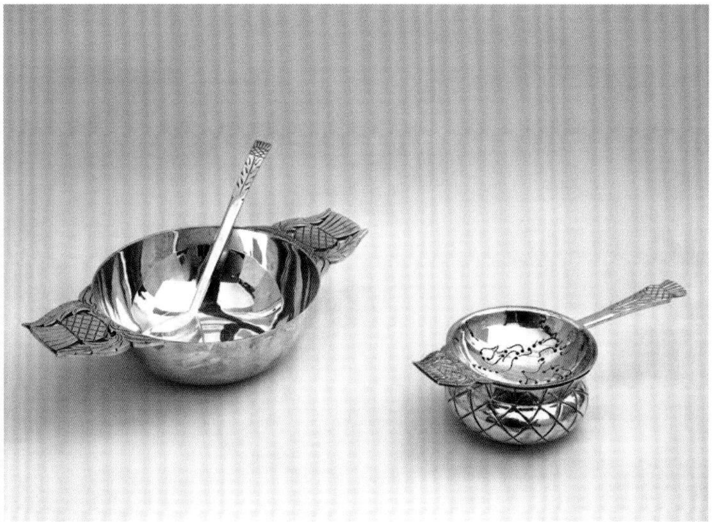

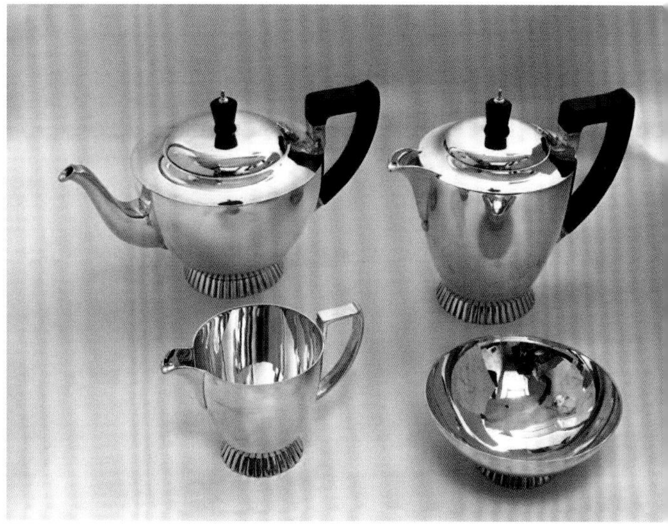

the exhibition. As with other industries to be represented, some silver manufacturers were slow to appreciate these unaccustomed restrictions and reacted either aggressively or with indifference. In a report on a visit to Birmingham, G.R. Hughes expressed surprise when he was assured by Southam of the Birmingham Jewellers and Silversmiths Association of approximately 1200 members that the response to the invitation to submit proposals had been 'quite good', only to discover that this amounted to ten or eleven letters.[4] Their visit to Jarrett and Rainsford met with open hostility. And a letter to Hughes from Walter Belk, of the Sheffield firm of Roberts and Belk, which had a deserved reputation for producing well designed, innovative silver, concluded by saying:

> I am taking little active interest in trade these days being tired of the strangleholds imposed upon production and where it may be sent. If we create for the home market in small but various designs we are out of it for export in mass, so the little man must either wait for better times or stand aside and see rubbish made.[5]

Despite these difficulties, a creditable 95 exhibits, including three items of pewter and a stainless steel tea service, were assembled. Understandably in the circumstances, the majority of the exhibits from the trade manufacturers were small items, such as a silver porringer and spoon from the silversmith Robert E. Stone [58], silver serviette rings, and cigarette cases from a variety of manufacturers. But the London firm of Wakeley and Wheeler presented an elegant tea service with fluted feet [59], and Mappin & Webb of Sheffield displayed a stylish entrée

58. Silver tea strainer and drip bowl, porringer and spoon, made by Robert E. Stone of London, 1946

59. Four-piece silver-plate tea set, comprising jug, bowl, teapot and water jug, designed and made by Wakeley & Wheeler, London, 1946

60. Silver entrée dish, designed by J.W. Creke, made by Mappin & Webb, 1946

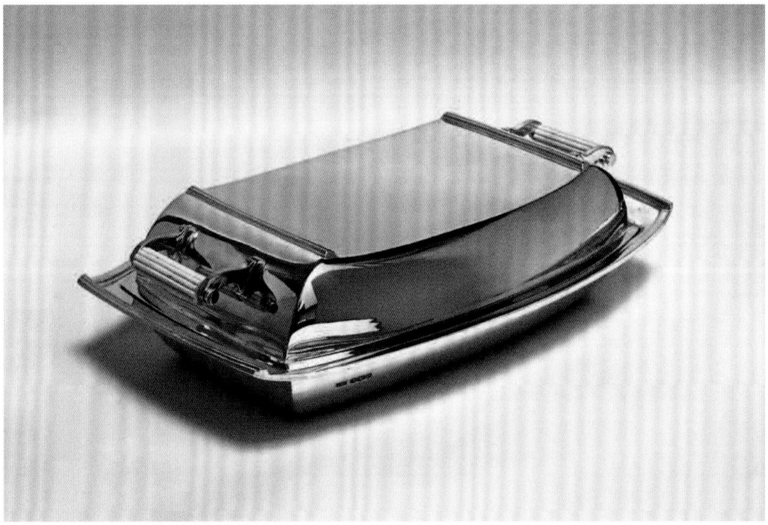

tray with ivory handles [60]. Purchase tax on all exhibits was temporarily suspended for three months and, although bullion was in desperately short supply, it could be obtained for exhibits under licence from the Board of Trade through the CoID. The cutlery selection was to include articles 'made in, or plated with gold, silver, platinum or rhodium. Ivory, or ivory with mounts will be included.'[6] No mention was made of stainless steel and none was included. Harry Brearley's invention of stainless steel in 1913 for the Sheffield steelmakers Firth Brown was largely ignored by the British tableware industry between the wars. Eventually Firth Brown established an outlet in Scandinavia, where such firms as Gense of Eskilstuna and Georg Jensen of Copenhagen started making stainless steel cutlery and quickly gained an international ascendancy in this market. The British once again had lost the initiative.

One of the conditions laid down by the CoID was that the 'name of the manufacturer and where possible the designer also, of each article be published' in the catalogue.[7] This was clearly an effort to identify and enhance the status of the designer, which had so concerned the authors of the Working Party report. In the case of the silverware selection this policy was inconsistently applied. Small studios such as Robert E. Stone in London and Joyce Himsworth in Sheffield both designed and made the items displayed and went on to make illustrious careers for themselves as artist-craftsmen. Large manufacturers such as Mappin & Webb and Gladwin Ltd. of Sheffield obliged. Others such as Adie Brothers and Elkingtons of Birmingham did not.

The organisers of the exhibition had a stated policy that the selection of exhibits was to focus on 'consumer goods in, or intended for, quantity production'. However it was recognized that a significant sector of the silver trade was formed by the individual artist craftsmen, who were at the forefront of technical and aesthetic innovation. There was a section, selected by G.R. Hughes, of silver cups from the Goldsmiths' Company's collection which displayed precisely these qualities. Needless to say, full credit here was given to the designer, maker and, where applicable, the distributor with full names and addresses.

The display

G.R. Hughes noted that 'the space available for the display of silver and jewellery was likely to be small', and indeed it was.[8] James Gardner, Chief Designer for the exhibition, faced a daunting task. In a very short time, he had to conceive a coherent and appealing scheme for the entire exhibition, which was to accommodate 5,000 objects, grouped within 32 divisions in an area of 90,000 square feet. An additional handicap was the uncertainty of what goods were to arrive, and his solution was ingenious. Many of the displays including the silver and jewellery were corralled in a series of alcoves, so that the visitor would not be immediately aware of the possible paucity of goods on display. For the layout of individual exhibits, Gardner had to delegate, and nearly 80 designers were employed. For the silverware and jewellery, this task was given to Ronald Dickens, who came from the newly established Design and Research Centre at Goldsmiths' Hall. The silver was displayed in 17 small vitrines in an elegant, if rather conventional manner [57]. Joan Evans described the selection as 'traditional yet modern' and that in contemporary silver there was 'less demand than there might be for exaggerated modernism in form and design'.[9] The commemorative silver from the Goldsmiths' Company gave the display added gravitas.

Some may have found the quiet, understated simplicity of the silverware uninspiring. Raymond Mortimer writing in the *New Statesman* certainly did: 'The weakest section is that devoted to silver, jewellery and plate. Here is craftsmanship but hardly a glimmer of invention or vitality.'[10] Perhaps such criticism explains why, years later, G.R. Hughes gave his involvement in the BCMI exhibition only a glancing reference in the catalogue of his valedictory exhibition at Goldsmiths' Hall.[11] He specifically mentions, and then only briefly, a dispute about the jewellery selection with colleagues from the CoID. But a detailed discussion of the jewellery component is the subject of the next chapter.

NOTES

1 Working Party Reports, *Jewellery and Silverware* (London 1946), p. 37.
2 Ibid., p. 39.
3 Ibid., pp. 75–78.
4 Goldsmiths' Hall. O.II.1.114: Report by G.R. Hughes on a visit to Birmingham dated 20 April 1946.
5 Ibid., Letter from W.P. Belk to G.R. Hughes dated 6 March 1946.
6 Ibid., CoID directive, *Jewellery and Allied Trades Participation*, 1946.
7 University of Brighton Design Archives. Design Council Archive (DCA)/No.12. Notice for the Guidance of Manufacturers Issued By The Council of Industrial Design in 'Britain Can Make It' 1946, National Exhibition of Industrial Design in Consumer Goods (London 1946).
8 Goldsmiths' Hall. O.II.1.114: Note of a meeting between James Gardner and Ronald Dickens dated 1 February 1946.
9 Joan Evans, 'English Silver', *Design '46* (London 1946), pp. 39–40.
10 Raymond Mortimer, 'Britain Can Make It', *The New Statesman and Nation* (28 September 1946), p. 220.
11 George Ravensworth Hughes, *The Worshipful Company of Goldsmiths as patrons of their craft 1919–1953* (exh. cat.), Goldsmiths' Company, London 1965, pp. 53, 55. An exhibition to honour the career and contribution of G.R. Hughes to the Goldsmiths' Company as Assistant Clerk, 1919–39 and as Clerk, 1939–53.

CLARE PHILLIPS

Jewellery

> Looking back over the past year, our first year of Peace, I am afraid many of us have been sadly disappointed, for Peace has not brought many of the blessings we anticipated. Shortage of stocks, materials and labour ... to say nothing of the controls, restrictions, form-filling, quotas etc., which many of us hoped would be a thing of the past by now.[1]

Britain Can Make It was greeted with limited enthusiasm by many in the jewellery industry, particularly those who with bombed factories and depleted workforce were struggling to fulfil export orders. Despite the gloom and the almost non-existence of a home market, foreign orders for British jewellery meant that the majority of commercial manufacturers in Birmingham in April 1946 had 'far more orders booked than they could conveniently cope with'.[2] War-time restrictions on precious metals had been lifted for these revenue-generating exports, although nine carat gold remained the highest standard allowed for the home market. By those with busy workshops the exhibition was largely viewed as an unnecessary distraction. As Claud Bunyard of Goldsmiths' Hall and the Industrial Liaison Officer for jewellery and plate was later to write, it was 'a sellers' market which ... obliged us to chase industry every inch of the way'.[3]

The Goldsmiths' Company (the ancient guild and assay office for London) had a central role in co-ordinating the jewellery and plate for the Council of Industrial Design (CoID). A combination of visits to manufacturers, discussions with the Birmingham Jewellers' Association and approaches to individual firms and craftsmen was set in motion. The files record increasing anxiety, however, at both the lack of awareness and the subdued response they were receiving from the trade: in April, just five months before the opening, of nine London jewellers contacted, four claimed to know nothing about the exhibition and only one was well informed.[4] A month later the secretary to the selection committee noted, 'Fine jewellery and decent fashion jewellery are still the bottlenecks; there should be more than enough silver'.[5]

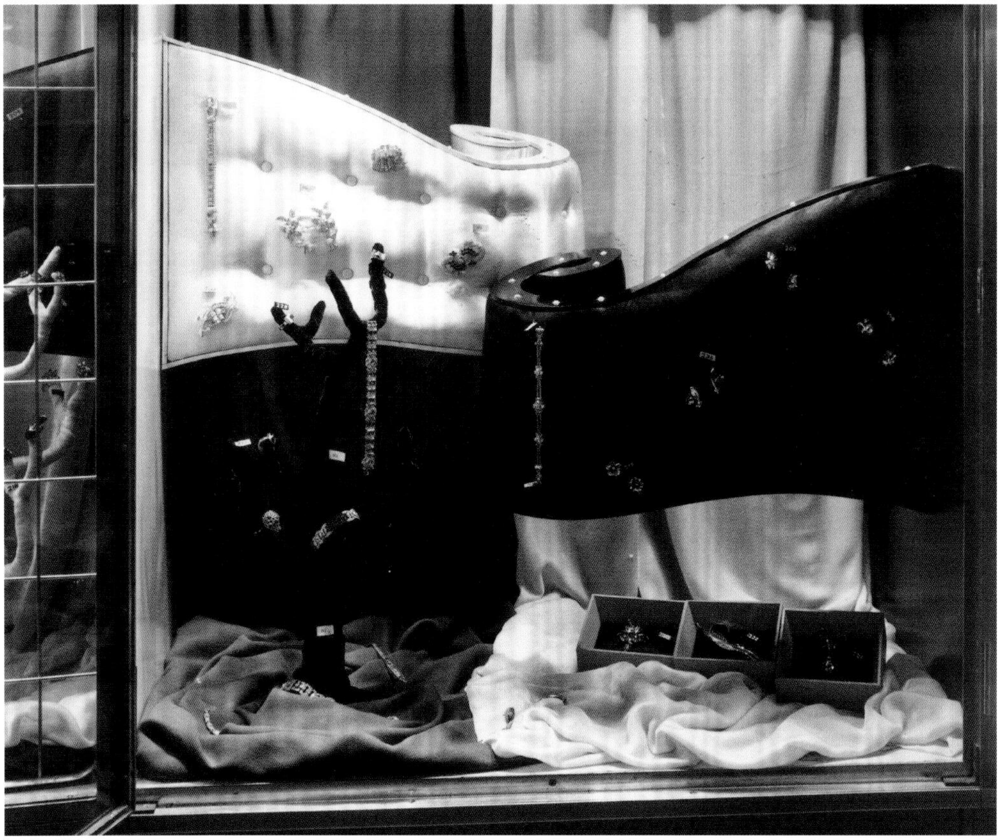

61. One of the displays of precious jewellery. Pinned to the white panel and hanging from the branches of the mount in front are pieces by M.P. Greengross, made with diamonds specially released by the Board of Trade's Gem Diamond Control; three brooches by Boucheron are in boxes lower right, a pair of Boucheron earrings above left; and, above centre, a pair of earrings by Schlesinger.

A measure which was to transform the precious jewellery section at BCMI was the agreement in May that up to £8,000 of diamonds would be made available for the creation of special exhibition pieces. These stones and also any platinum used to mount them were to be independent of jewellers' existing quotas and were in effect a loan from the Board of Trade's Gem Diamond Control. The maker would be allowed a small profit, Purchase Tax would not be levied, and their eventual export sale would be guaranteed. Both M.P. Greengross of Greville Street (who had previously declined involvement owing to 'the smallness of the diamond quota and the loss of their craftsmen's time') and Cecil Drayson of New Bond Street were able in this way to create lavish collections of diamond-encrusted jewellery.[6] These are listed in the supplement rather than the main catalogue, suggesting their late completion. Indeed the final selection of Drayson's pieces was not made until 11 September – for delivery on 20 September, just four days before the opening of the exhibition. It proved worth the wait, eliciting the comment 'every piece is ... better in design, and very much better in craftsmanship, than anything we have so far received'.[7]

62. Design submitted for approval prior to manufacture. The diamond and ruby bracelet converts to form two brooches and earclips, by H. Appleby for H.A. Byworth & Co.

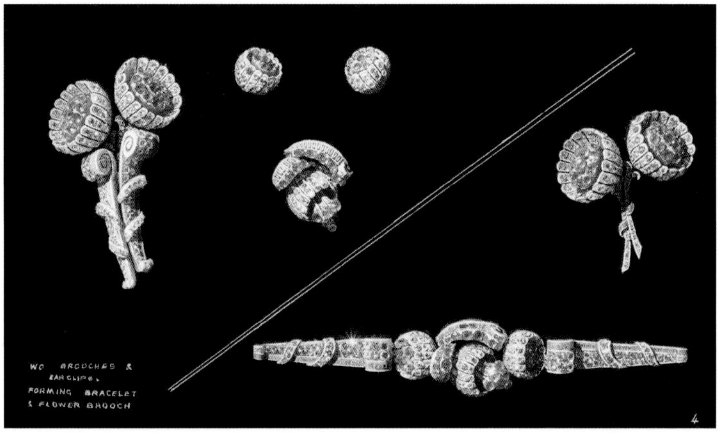

The jewellery, divided into three categories – precious, semi-precious, and inexpensive – was a glittering and highly diverse selection, and was valued at £40,000.[8] It was shown near the beginning of the exhibition, in 'Shopwindow Street'. The pieces were displayed 'beneath a ceiling of blue sky, flecked with sunset but lit with fluorescent daylight tubes' in 'small display cabinets ... on backgrounds of black or white satin or mounted on grotesque tree-like stands'.[9] A tiny label alongside each piece or group of pieces gave the number by which the pieces were listed in the catalogue [61].

The concentrations of precious stones within the export jewellery must have attracted considerable attention, although in design terms there was little new. The abstract lines of Art Deco remained a strong theme particularly in diamonds, tempered with the more swelling forms of the 1940s. Naturalism provided softer lines, although occasionally with a slightly exaggerated romanticism, as in Wolfe & Co.'s chestnut-leaf brooch with its diamond 'dew-drop'. Invisible settings featured in earrings by Drayson and by Asprey, while contrasting colours of gold were put to decorative use by Boucheron and Wolfe. Adaptable designs which could be taken apart to form several smaller jewels were commented on – notably H.A. Byworth's clip and bracelet [62]. At the end of each day these precious jewels were removed from the display and kept overnight in a safe.

Fox's of Bexley, in the semi-precious display, perhaps provided the closest approximation for British buyers to this as yet unattainable glamour. The three floral brooches they exhibited, with small clusters of peridots, tourmalines or aquamarines set within an outer frill of thin nine-carat gold petals, were an effective and economical means of achieving an impact. A more robust aesthetic was promoted by the chunky bracelets and 'snake' chains by Howitt's of Hatton Garden.

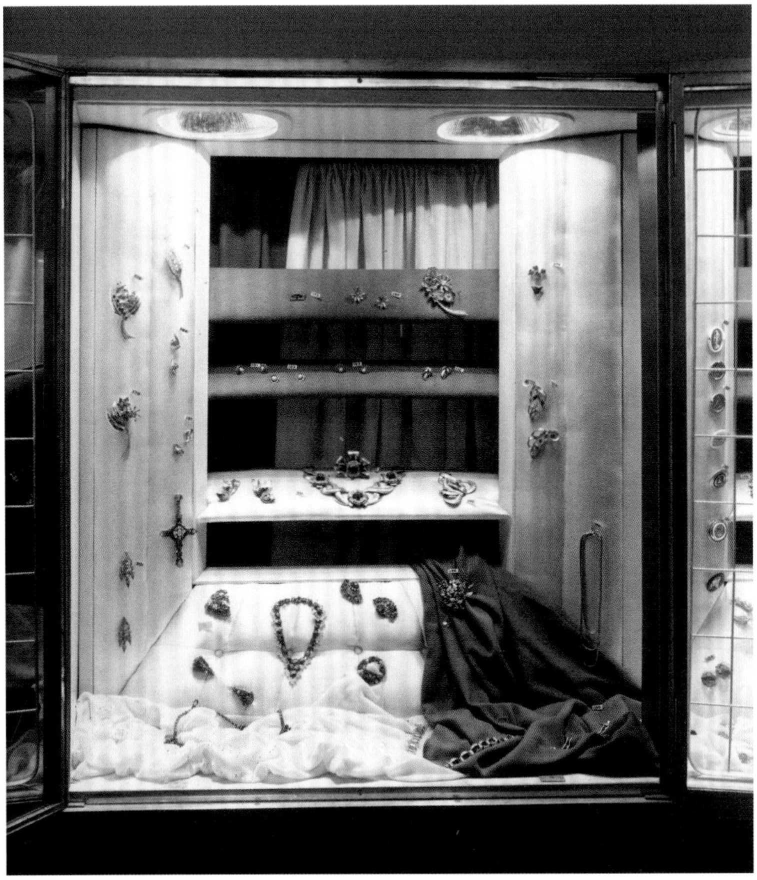

63. A selection of semi-precious jewellery in its showcase. The floral brooches upper left and centre are by Fox's of Bexley; lower right are pieces in nine carat gold by Howitt's; May Clark's necklace and brooch can be seen on the third shelf, her leaf brooches above and to the right; pieces by Dorrie Nossiter are on the base of the case with a pendant cross above to the left; the striated discs of a Catherine Cockerell necklace can be seen in the centre of the base, in front to the left the silk cord of her second necklace is just visible; circular earrings by Francis Harling are on the middle shelf. At the edge of the adjoining case can be seen a selection of glass jewellery by Bimini.

Both companies were making for the home market, and with immediate availability [63].

Jewellers with a more idiosyncratic style fitted awkwardly within the guidelines for the exhibition. The five pieces submitted by Sah Oved, including a hexagonal gold cuff with sapphire and diamonds which she valued at £500, were all rejected.[10] Perhaps her unconventional treatment of gold and precious stones offended the committee's technical assessor, or perhaps her designs were simply too unconventional. The unusual perhaps fitted most easily into the semi-precious category, which brought together a wide array of individuals and styles. May Clark's opulent swathes in hand-wrought copper-gilt were noted by *Vogue*, which illustrated her elegant palm-leaf brooches in its review of the exhibition (only one other jewel, by Boucheron, was illustrated).[11] Five pieces of Dorrie Nossiter's glittering encrustations of colourful gems, hand-set in silver (surprisingly listed as 'fashion jewellery'), can be identified from surviving photographs. Also included was work by Catherine Cockerell, daughter of the bookbinder Douglas Cockerell, whose jewellery has been described as 'of a slightly improvised nature using objets trouvés ... and materials of low value found by rummaging in suppliers' boxes in Clerkenwell Road';[12] and the neat, circular hand-beaten earrings made by the Hampstead jeweller Francis Harling, each set with a central blister pearl.

In his introductory essay to *Design '46*, Arnold Selwyn commented that 'the great advances made in plastics have given fresh scope to British jewellery design in the cheaper ranges, of which full advantage has been taken'.[13] Plastics formed a notable part of the inexpensive jewellery category, despite warnings of 'dangerous faults, as for instance, the colour running under tropical heat, warping, twisting and general softening'.[14] Bloxidge Bros. of Birmingham, one of Woolworth's chief suppliers before the war, exhibited four brooches. They had previously impressed the selection committee with their 'quite interesting designs of astonishingly cheap price range ... between 6d. and 1/-'.[15] Perspex, which had lately been used for aircraft windscreens, was transformed into chunky beads by Freitag, and a hand-engraved brooch by L.S. Mayer.

The largest supplier of 'inexpensive' jewellery was the glass manufacturer Bimini. Its founder Fritz Lampl had been a highly regarded glass artist in Vienna until he fled at the Anschluss in 1938. He rapidly established his business in London and by 1942 was exporting glass jewellery and buttons internationally.[16] 'Each button, clip or ornament is hand-made in glass, white or coloured, clear or opaque, and resembles a seal with the design picked out in gold silver or bronze.'[17] Lampl was not the only refugee to contribute jewellery to this celebration of British design. S.B. Schlesinger of Hatton Garden, who exhibited five pairs of earclips in the Precious Jewellery section, had fled Pforzheim in 1938.[18] The company registered its mark at the London Assay Office in 1942.[19]

The stipulation by the CoID that the designer or maker should be acknowledged was not universally observed. In this the exhibition's aim of highlighting design and the designer were at odds with long-established custom. The sparkling array submitted by Cecil Drayson is credited to G. Music & Sons of Hatton Garden; however, the Goldsmiths & Silversmiths Co., while agreeing to name the makers of their silver ware, did not do so for their jewellery and it was noted resignedly, 'this seemed too difficult a nut to crack'.[20] At the other end of the scale, John Lewis exhibited a pair of flowered ear-clips with no mention of designer or manufacturer.

The attractive and varied selection of jewellery presented in the exhibition was achieved within a tight timescale against a backdrop of conflicting incentives, genuine uncertainty within the trade, and a simmering antagonism between London and Birmingham. Birmingham's dissatisfaction with the exhibition had crystallised some months before the opening, when 'firms with a considerable reputation for design and fine craftsmanship were astonished to receive back their entire parcels of submitted articles because none were considered suitable'.[21] Presiding over a thriving export trade, they felt that their skill at balancing design with commercial viability was proven and should count for more than the views of any panel, however distinguished.[22] As bastions of popular taste they pointed out tersely, 'what sells like hot cakes in Bond Street does not necessarily cause much stir in Wigan; and Wigan, however one may deplore the fact, is the more representative of English taste as a whole'.[23] While this indicates a fundamental lack of understanding between the CoID and the trade, internal factors had also played a part in impeding and restricting the choice of work. Claud Bunyard was later to comment intriguingly on the dynamics of the selection committee: 'I am quite certain that the ultimate selection would have been different and larger in quantity, if each of the ladies in question had made their selections independently of the others'.[24] Perhaps it would have been impossible to please all sides at this time of change and uncertainty, particularly in an area of taste that is so personal. Bunyard optimistically commented, 'next time industry will be chasing us', looking to a future of greater impartiality and absolutes in design terms.[25]

NOTES

1 J.M. Buff, Chairman, speaking at the AGM of the British Watch and Clock Makers Guild. *Jeweller and Metalworker* (1 June 1946) p. 351
2 Goldsmiths' Hall, O.II.1.114: Report on visit to Birmingham by C.F. Bunyard, E.R. Fawcett and R. Dickens 20.4.46.
3 V&A Archive, Exhibition File: AAD/1977/4/14, Britain can make it: Letter from Claud Bunyard to Mr Newman, 13.1.1946, p. 1.
4 Goldsmiths' Hall, O.II.1.114: London 1946 V&A (BCMI) General 1946 May – June. On phoning nine manufacturing jewellers in London. 25.4.46.
5 Ibid., London 1946 V&A (BCMI) General 1946 May – June. Report on a visit to Birmingham 24.5.46.
6 Ibid., London 1946 V&A (BCMI). General 1946 May – June. Report of a meeting with Mr Matthias of the Diamond Control, 6 May 1946
7 Ibid., London 1946 V&A (BCMI). Letter Ronald Dickens to Clive Bunyard 11.9.46.
8 '"Britain Can Make It" Trade's products make fine show in the exhibition at Victoria and Albert Museum', *British Jeweller* (October 1946) vol. 14, no. 2, p. 36.
9 Ibid.
10 Goldsmiths Hall, O.II.I.114: Advice notes for silver and jewellery. The bracelet is possibly that included in Sotheby's London sale, 23 March 2005, lot 288.
11 'Britain Can Make It', *Vogue* (October 1946) pp. 60–67.
12 Alan Powers, Obituary of Catherine Cobb, *The Independent* (12 October 1995), p. 22.
13 Arnold Selwyn, 'Jewellery', *Design '46* (London 1946), p. 44.
14 Goldsmiths' Hall, O.II.1.114: London 1946 V&A (BCMI). Note for Mr Dickens 'Discussion between L.G. Chaffey, Plastics Federation, C.I. Bunyard, Council of Industrial Design and T.C. Eckenstein, General Celluloid Co. Ltd.' July 1946/26.7.46.
15 Goldsmiths' Hall, O.II.1.114: London 1946 V&A (BCMI). Report on visit to Birmingham by C.F. Bunyard, E.R. Fawcett and R. Dickens 20.4.46.
16 Bimini/Orplid scrapbook held in V&A Ceramics & Glass departmental library, CER LM.1453.
17 Mary Delane, 'New Costume Jewellery', *The Sunday Times* (22 September 1946), p. 7.
18 Jane Stancliffe, *Fantasy, fashion or imitation? An international survey of costume jewellery in the twentieth century its origin and subsequent development* (Royal College of Art 1987, unpublished MA thesis), pages 86 and 167.
19 The mark was registered on 20 March 1942, bk 24, p. 138. Thanks to David Beasley for this information.
20 Goldsmiths' Hall O.II.1.114: London 1946 V&A (BCMI) General 1946 August. Interviews with Jewellers etc. 12.8.1946.
21 Editorial by W.H. Leese, *British Jeweller* (November 1946) vol. 14 no. 3, p. 35
22 The jewellery and plate selection panel comprised Anne Scott-James, editor of *Harper's Bazaar*; Madge Garland, fashion buyer for the department store Bourne & Hollingsworth Ltd; Audrey Withers editor of *Vogue*; Arthur Robert Emerson, jeweller and silversmith, and instructor in plastics and industrial design at Central School of Arts and Crafts; and George Ravensworth Hughes, Clerk to the Worshipful Company of Goldsmiths.
23 Ibid.
24 V&A Archive, Exhibition File: AAD/1977/4/14, Britain can make it. Letter from Claud Bunyard to Mr Newman, 13.1.1946, p. 4.
25 Ibid., p. 1.

ALUN GRAVES

Pottery and Glass

Yes, we shall buy it, when we can get it.[1]

Pottery and glass proved extremely popular at *Britain Can Make It* (BCMI). Even in the context of the exhibition's runaway success, the Pottery and Glass displays in 'Shopwindow Street' were distinct hits. Arousing interest out of all proportion to its size, the Pottery section was reported to have been continually blocked by crowds, who queued patiently to see it.[2] Indeed, there appeared to be a particular appetite among the public for ceramics. While glass presented something desirable but largely out of reach, pottery was routinely stated to be among the items most needed.[3] This desire to see and in time acquire new pottery designs perhaps reflected a more central position for ceramics in the homes and lives of mid-century Britain than we might be accustomed to today. But it was also a response to the deprivations of wartime, which had starved the public of decorated pottery through strict restrictions on its sale, put in place in 1941 and not finally lifted until 1952. Decorated wares produced by the factories of Stoke-on-Trent were instead destined only for export, primarily to Canada and the USA, in a bid to bring foreign money into the economy. The British consumer, or 'British housewife' as she was routinely presumed to be, had in the meantime to satisfy herself with the plain ceramic wares produced under the Utility scheme.[4]

BCMI was to present something different. 'Decorated pottery will be seen again in England this September', proclaimed the Council of Industrial Design (CoID) at a press conference in July 1946, where examples of the selected pottery were unveiled. 'Developments in decorating and printing on china, bolder and less austere shapes, and many new delicate shades in self-colour ware' would feature, and the exhibition promised 'once again [to] show British pottery leading the world in quality and attractiveness'.[5] The selection for both pottery and glass was made by a panel comprising Elvie Macdermott, Director of

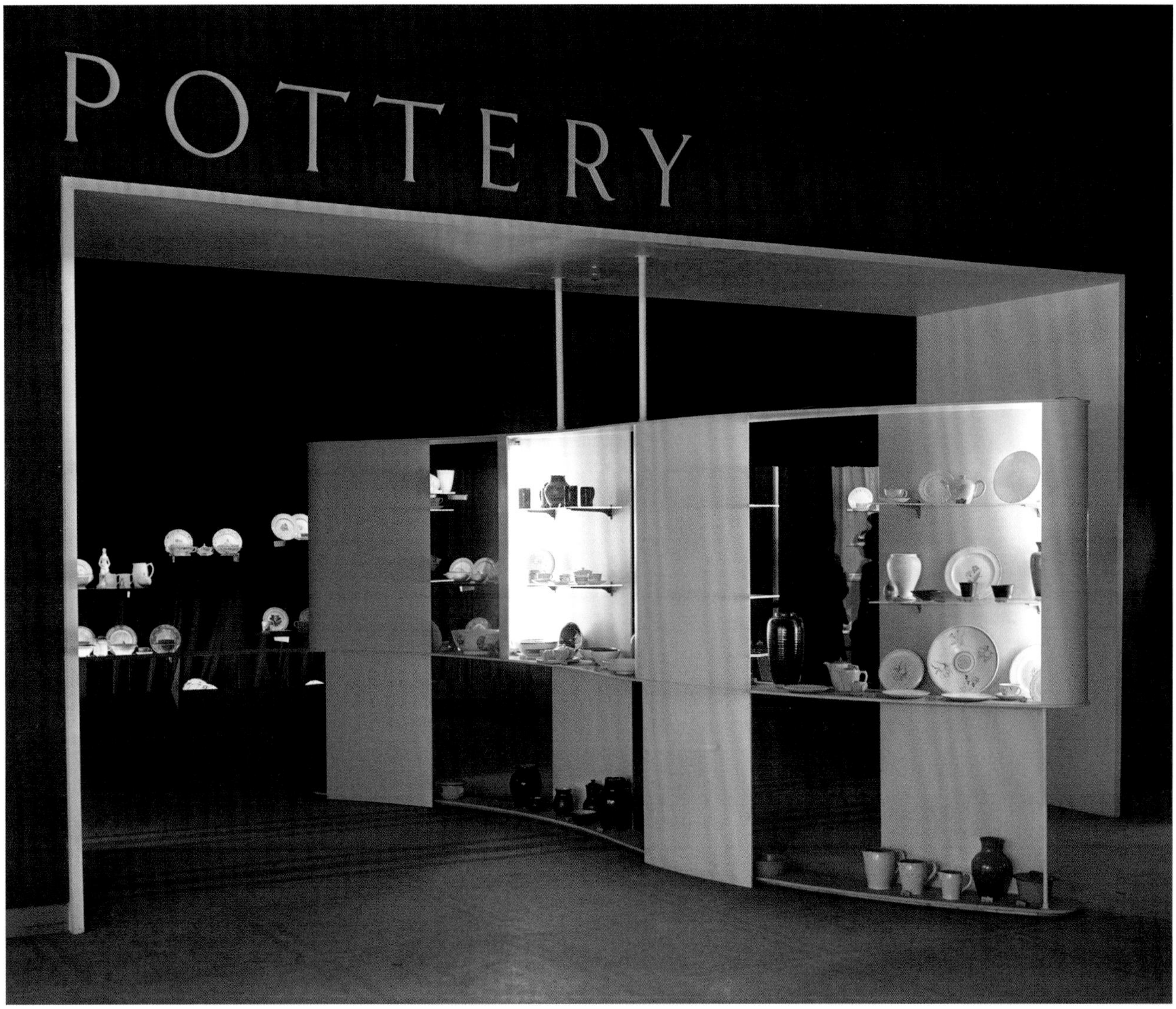

64. The Pottery display in 'Shopwindow Street', designed by James Gardner, the pottery arranged by Harry Trethowan

Buying in pottery and glass for the John Lewis Partnership, W.B. Honey, Keeper of Ceramics at the Victoria and Albert Museum, Harry Trethowan, Managing Director of Heal's, and Howard Robertson, former Principal of the Architectural Association.[6] As such it represented the views of the design establishment, views which did not always accord with those of the pottery industry. Writing in the journal *Art & Industry*, Trethowan noted that 'exhibitions always cause disappointment to those in an Industry whose works are not included', but maintained that 'a standard of quality in material and design is possible, and ... this standard can and must be adhered to'.[7] Trethowan was no doubt alert to the dissent in the industry over what constituted 'good modern design'. Such dissent

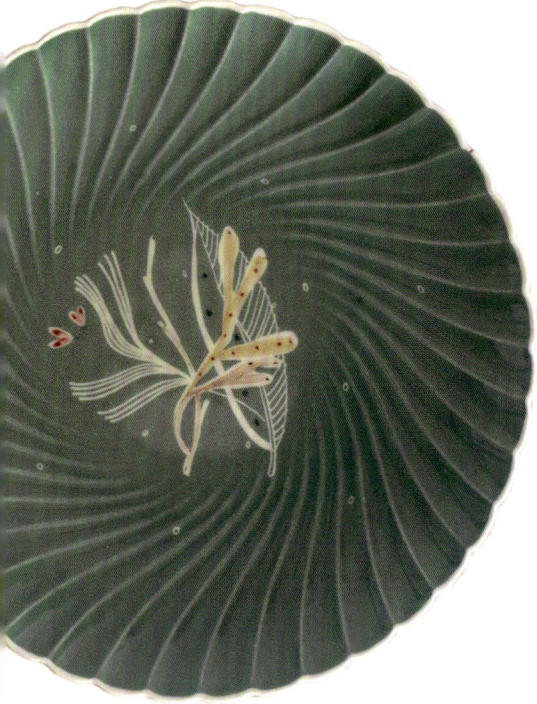

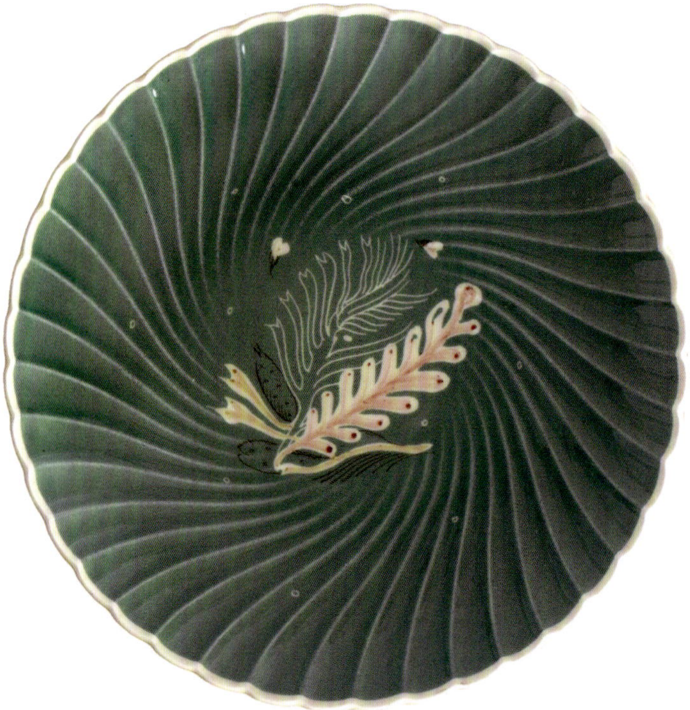

65. Spiral-shape dessert plates with seaweed patterns, designed by Susie Cooper for Susie Cooper Pottery, Staffordshire, designed about 1942, this example made 1949. Earthenware with incised and hand-painted decoration. V&A: Circ.146 & 147–1949

had been expressed at a meeting of the British Pottery Manufacturers' Federation, where concerns had been raised that the selection committee – as yet unappointed – might represent the tastes of 'a small too advanced minority'.[8] And, despite the exhibition's overwhelmingly positive reception, there were evidently some in the industry who felt it to be insufficiently representative [**64**].[9]

BCMI was primarily intended to be a display of post-war designs.[10] In reality, however, little new design work had been undertaken in the pottery industry during the war years.[11] As a consequence, most of the exhibits were pre-war designs that continued to find favour. Qualities of tradition and restraint remained widely valued in British ceramics, and, for the design establishment, this often specifically meant an association with late eighteenth-century styles. This was particularly evident in 'modern' Wedgwood tablewares, which proponents of Modernism like Herbert Read valued for their efficiency and geometric harmony.[12] For W.B. Honey, such continuity in the English tradition might 'serve also as a corrective against forced originality and freakish extravagance'.[13] Writing in *Design '46*, he argued that the search for 'a true machine aesthetic ... must govern all future developments in the industry', before going on to identify some recent tendencies in pottery design and decoration. The replacement of hand-throwing with casting and turning, he maintained, demanded that the designer work in the manner of 'an architect who designs a building'. Referring to the work of Keith Murray, he noted that this change should bring about a new type of shape design that stressed harmony and proportion. On decoration, he advised that 'the best modern freehand painting is limited to a few simple touches, or lines incised through layer of pigment .... Some sort of stylisation or limitation of palette is almost obligatory.' Lithographic transfers imitating detailed miniature painting, or printed outlines filled with a brush, should, he suggested 'be abandoned, however much [they] may be to the taste of an uneducated public'. But transfer-printing, when 'frankly employed' and based on 'honest undisguised engraved work', was 'entirely in keeping with the aesthetics of the machine', while retaining 'the touch of the artist's own hand'.[14] Among the pottery chosen to illustrate Honey's text were some of the relatively few examples of actual 1940s design exhibited: Eric Ravilious's 'Barlaston' commemorative mug for Wedgwood, decorated with lithographic transfers; and Susie Cooper's spiral-shape dessert plates, with decoration of seaweed or flowers incised through a coloured ground [65].[15]

While Honey's discussion of the requirements of the pottery industry foregrounded the role of the designer, he noted that there remained a place for the craft potter using the hand processes of throwing and freehand painting, whose work by necessity would be sold at high prices. Such work was important both for its own sake and for the part it could play in promoting aesthetic appreciation. The work of studio potters was, however, excluded from BCMI. Fritz Müller, the CoID's organizing secretary for the Pottery section, wrote apologetically to Elvie Macdermott to advise her of the decision, stating that it was considered their inclusion 'raised certain implications which might be much misunderstood by the manufacturing potters'.[16] In consequence he had written to Bernard Leach and Phyllis Keyes withdrawing earlier invitations to exhibit. Leach appears to have received the news with equanimity, replying that there was 'nothing to excuse', though he remarked, 'recrimination or jealousy seems to me very unnecessary as between factory and studio. Surely it is time we got a step on towards understanding and cooperation.'[17] The policy was nevertheless sanctioned at the highest level. A note from Stafford Cripps, President of the Board of Trade, to the CoID's Director, S.C. Leslie, insisted the exhibition include 'No "precious" stuff. All manufactured goods – not hand made.'[18] The exclusion of Leach was perhaps unwarranted. Concerned that studio pottery lacked the 'breath of reality', he had lately with his son David developed a range of robust domestic stoneware, termed Standard Ware. Not subject to the usual wartime restrictions, this had been retailed since around 1940, and the first mail order catalogue issued in 1946.[19] Handmade pottery was not, however, entirely absent from BCMI. Included were several pieces from

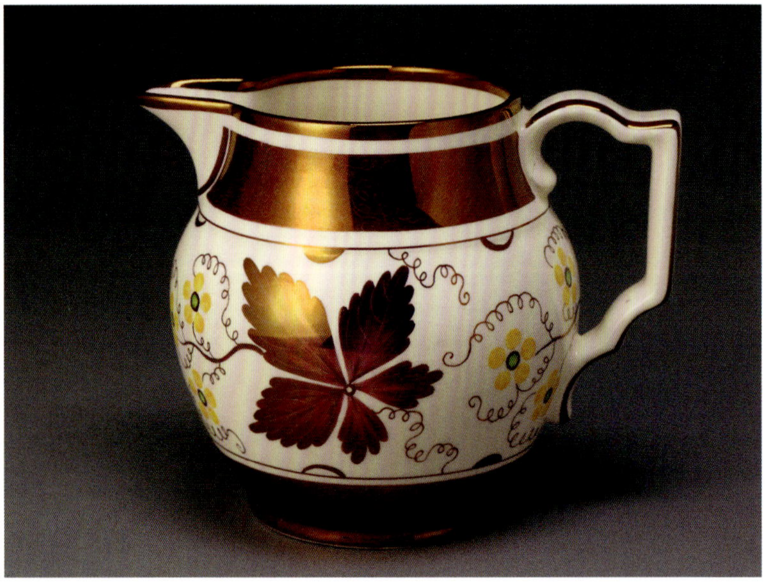

66. Earthenware jug painted in lustre, made by A.E. Gray & Co., Staffordshire, this example made 1948. V&A: Circ.489–1948

the Odney Pottery, recently established by former Leach student John Bew with support from the John Lewis Partnership, whom the pottery supplied.[20] Also shown were vases, bowls and animal figures made for Heal's in Agnete Hoy's studio at Bullers Ltd.

Such wares were some of the very few that were immediately available for sale on the home market. Others included those produced under the Utility scheme, such as Moorcroft's undecorated ivory tea and coffee wares. These found favour with the design establishment, one survey noting that the requirement for economy had here resulted in shapes 'which spring naturally from the material and the craft of the potter ... their stark white glaze [emphasizing] the sheer beauty of their forms'.[21] Yet they were among the least popular pottery items with BCMI's visitors. As a Mass Observation (M-O) survey noted, 'the dominant trend [in pottery] is away from Utility. People are searching for something delicate and colourful, which will not remind them of wartime products.'[22]

Though not a Utility product, one of the most widely admired pottery exhibits was of the practical kind. Everywhere it appeared, T.G. Green & Co.'s 'Cornishware' won approval.[23] To the designers of the furnished rooms – or at least those who selected the pottery placed within them – it was classless, equally at home in Maxwell Fry and Jane Drew's 'kitchen in a large, well-appointed house' as it was in Edna Mosely's 'kitchen of a cottage in a modern mining village' [see **111**].[24] Ceramics were shown widely in 'Furnished Rooms', but perhaps unsurprisingly they appear in their greatest numbers in kitchens and dining rooms. Kitchenware by the

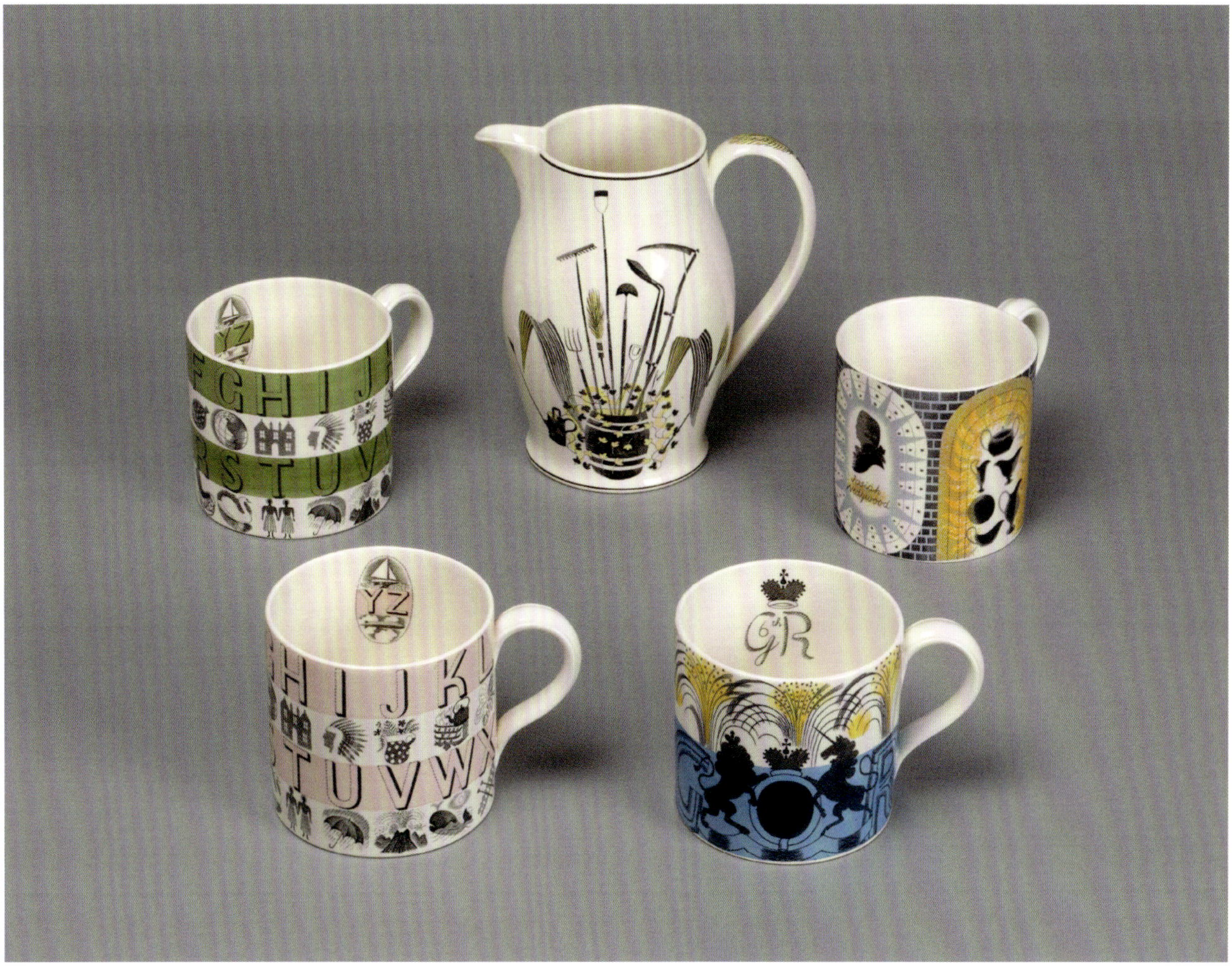

Worcester Royal Porcelain Co. took its place alongside the 'Cornishware' in Fry and Drew's kitchen, while brown stoneware jugs by Joseph Bourne & Son sit upon the worktop in Darcy Braddell's 'kitchen recess, in a small house'. Pountney & Co. tableware graces the dining table in Frederick MacManus's 'dining recess in a small modern house' [see **119**], while T.G. Green & Co.'s 'Polo' range appears cheerfully informal in a display of Utility dining room furniture. Elsewhere, lustre-decorated jugs by A.E. Gray & Co. add a more traditional touch to several rooms, including Elizabeth Denby's 'living room in a small house on a new estate' [**66**; see **115**], while the products of Josiah Wedgwood & Sons make numerous appearances [**67–69**], the most remarkable being in David Booth's 'dining room in a small suburban villa'. Like the other furnished rooms of BCMI, this came complete with imagined

67. Earthenware by Josiah Wedgwood & Sons with patterns by Eric Ravilious (V&A group Circ.470–1948; C.77–1971; Circ. 380–1939 and Circ. 465–1948; Circ.310–1961)

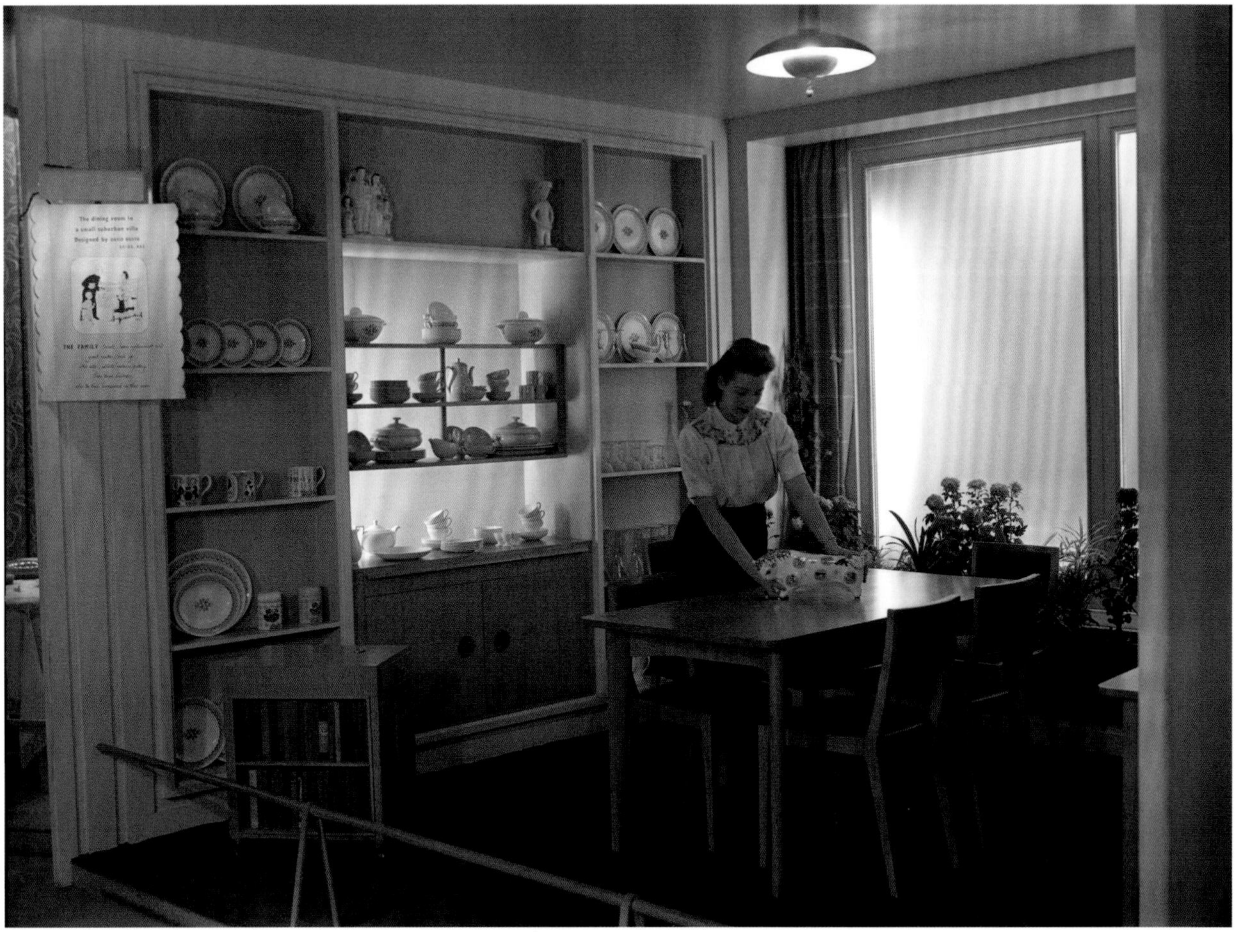

68. Dining room in a small suburban villa, designed by David Booth for an imagined curate and his family. A range of pottery by Josiah Wedgwood & Sons – including *Taurus* [69] – ostensibly collected by the curate's wife, is on display.

occupants – a curate, a keen naturalist and great reader; his wife, who collects modern pottery; and their three children [68]. For the curate's wife, modern pottery evidently meant Wedgwood, for her dining-room sideboard and shelves were well stocked with Wedgwood tablewares, including those decorated in Ravilious's 'Persephone' pattern, and self-coloured wares based on eighteenth-century shapes. Other designs by Ravilious are also present [67], including a large *Boat Race* bowl. Three figures by Arnold Machin meanwhile sit on a high shelf. Two of these, his *Bridal Group*, and his *Saggar Man* or *Placer*, were contemporary interpretations of Staffordshire flatbacks. The third was *Taurus*, a creamware bull decorated with lithographed zodiac symbols that Machin had himself designed. Lively and delightful, *Taurus* became an early icon of post-war British pottery, anticipating the ceramics of the 1950s [69].

The widespread display of pottery in the 'Furnished Rooms' perhaps helped to reinforce its practical and popular appeal. The situation with

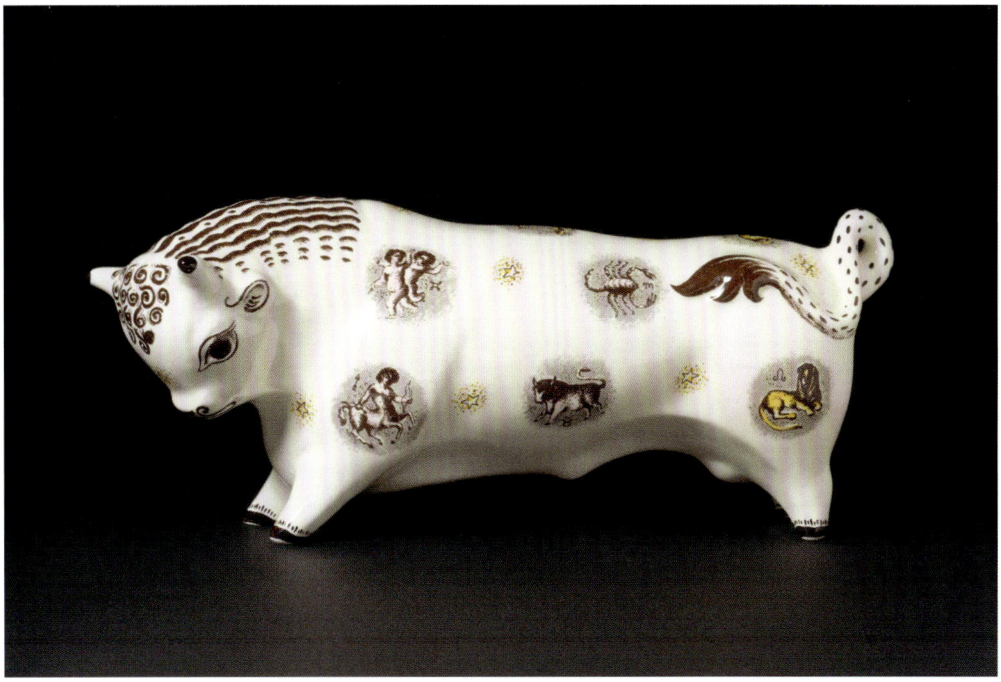

69. *Taurus*, modelled by Arnold Machin, about 1946, and decorated with a pattern of Zodiac symbols to his design. Made by Josiah Wedgwood & Sons, Staffordshire, this example made about 1950. Earthenware with lithographic transfers. V&A: Circ.276–1951

glass was, however, rather different. Although much admired, the display in 'Shopwindow Street' appeared glamorous and the products unattainable.[25] Glassware was meanwhile less in evidence elsewhere in the exhibition. Yet from the outset the CoID had been anxious to ensure that both expensive and cheaper glass was represented.[26] The preview of the glass selection had shown 'articles in both decorative and essentially practical styles' including 'examples of cut and engraved glass, and sets of jugs, decanters and wine-glasses made by a two-hundred-year-old technique', but also 'fire-proof glass for cooking dishes and casseroles ... and a variety of undecorated glass restricted to the home market for serving food and drink'.[27] And, indeed, the glass exhibited at BCMI included, for example, Pyrex ovenware by James Jobling & Co., and pressed glass by United Glass Bottle Manufacturers Ltd [71]. Yet the overriding impression was one of luxury.

This tendency reflected Britain's glass industry. While pressed glass and ovenproof glass were made in Britain, most cheaper blown glass was imported, primarily from Belgium and Bohemia. Unable to compete at the lower end of the market, British glass manufacture was largely confined to high-grade crystal. In this market, the issues of good design were contested. The problem, as identified by Nikolaus Pevsner, was that, unlike pottery, where 'the prevailing taste... [was] 1770, so to speak', in glass it was '1870'.[28] Extravagant cutting was thus the predominant

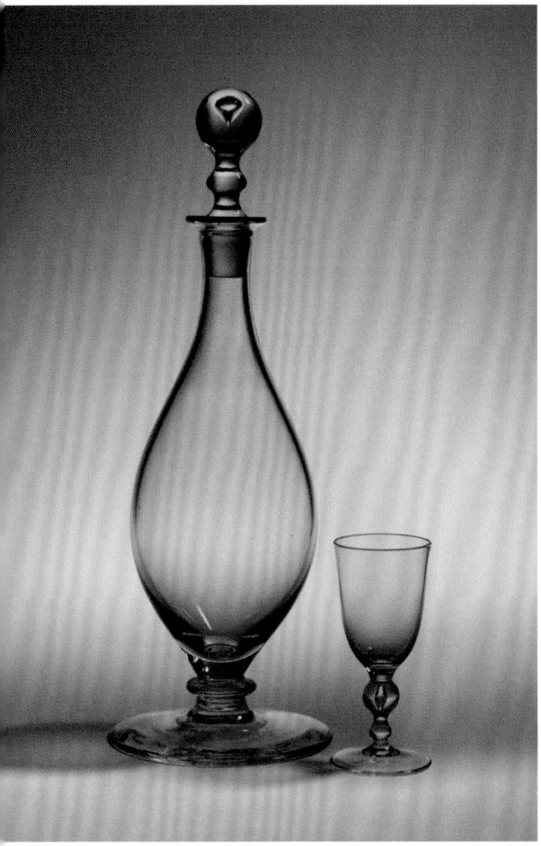

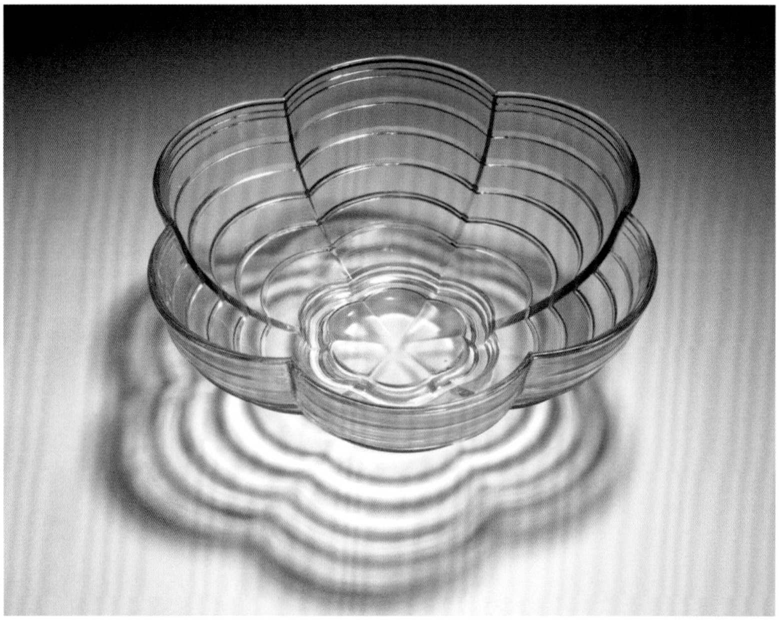

70. Decanter and sherry glass, service 9280, designed by James Hogan for James Powell & Sons (Whitefriars), London, designed 1940, this example made 1949. V&A: Circ.76&A, 79–1949

71. Pressed glass bowl, designed by A.H. Williamson for United Glass Bottle Manufacturers Ltd, London, this example made 1949. V&A. Circ.167–1949

feature of 'traditional' British glass. Only a handful of manufacturers, led by James Powell & Sons and Stevens and Williams, had begun in the 1930s to follow the modern Swedish model of leaving glass plain, or adopting shallow wider cuts, and fewer of them, so that the elegance of the forms and the transparency of the glass was more in evidence. By the mid 1930s a market for glass of the modern Swedish type had begun to develop in Britain, yet it still lagged some way behind the market for traditional crystal.

Mass Observation's survey of responses to the Glass section demonstrates this polarity in taste, perhaps unwittingly revealing their own prejudices in the process:

> Some took the objection to plainness further and insisted on having a more ornate type of article which they described as 'gorgeous', 'exquisite', etc.; they rejected anything that was at all plain. Others – perhaps rather fewer – still disliking anything at all clumsy, preferred more simple designs.[29]

So while the displays represented the more ornate, traditional styles in British glass – as well as more whimsical items such as lampworked animals by Pirelli Glass Ltd – they also included glass of more progressive modern design. Among this latter category were Keith Murray's designs for cut and engraved glass for Stevens and Williams, and James Hogan's elegant, unadorned table glass for James Powell

& Sons, including his wine service of 1940, which had been designed at the request of the Government for use in British embassies around the world [70].[30]

With the benefit of hindsight, the Pottery and Glass displays of BCMI can be seen to represent a moment of transition. In design terms, the industries were still finding their feet after the hiatus of wartime, and they continued to endure stringent regulations in terms of what could be sold where. There remained perhaps an over-reliance on tradition, and on the maintenance of a discernible national style. Yet within the displays were glimmers of a more confidently contemporary and outward-looking approach to design, which would in time respond to the desires and needs of the post-war consumer and to Britain's growing spirit of optimism.

NOTES

1 Design Council Archive at the University of Brighton Design Archives, hereafter DCA: 'A Report on "Britain Can Make It" Exhibition', unpublished survey prepared by Mass Observation (December 1946), p. 12.

2 Mass Observation (cited note 1), p. 8.

3 Ibid., pp. 2, 8.

4 Mass Observation Archive, University of Sussex: Britain can make it, SX MOA1/2/26, Box 2D. 'Notes on Pottery Exhibition', Council of Industrial Design press release (2 July 1946), p. 1.

5 Ibid.

6 Ibid. See also *Britain Can Make It Catalogue* (London 1946), p. 230.

7 Harry Trethowan, 'Pottery', *Art & Industry* (November 1946), vol. 41, no. 245, pp. 130–31.

8 DCA/516/'Minutes of a Meeting Held. at Federation House, Stoke-on-Trent' (20 December 1945), [p. 2]. The meeting was of members of the British Pottery Manufacturers Federation, which acted as one of the Industrial Advisory Committees reporting to the Council of Industrial Design on *Britain Can Make It*. See *Britain Can Make It Catalogue* (cited note 6), pp. 233–35.

9 DCA/516/This point was for example made by the Director of Paragon China in a letter to the Council of Industrial Design, declining their offer to exhibit the single example of the firm's products that had been selected (29 June 1946).

10 'The Design Exhibition "Britain Can Make It"', *Pottery Gazette and Glass Trade Review* (March 1946), vol. LXXI, no. 825, p. 159

11 See for example Julie McKeown, *Burleigh: The Story of a Pottery* (Shepton Beauchamp 2003), p. 167. Although creative design at Burgess & Leigh largely ceased during the war, new sample patterns were produced for and exhibited at *Britain Can Make It*. On this, see p. 173.

12 Herbert Read, *Art and Industry: The Principles of Industrial Design*, 2nd edition (London 1944), p. 69.

13 W.B. Honey, 'Pottery and Glass', *Design '46* (London 1946), pp. 23–24.

14 Historically, most transfer-printed ceramic designs were printed from engraved metal plates, but by the 1920s and 30s, colour lithographs predominated. However, these were often perceived to be of lesser aesthetic merit, in part because their design was more often carried out by the printing firm rather than the pottery. On this, see Nikolaus Pevsner, *An Enquiry into Industrial Art in England* (Cambridge 1937), pp. 79–80.

15 Ibid., 25–26.

16 DCA/516/Letter from F.J. Müller to E.H. Macdermott (4 July 1946).

17 DCA/516/Letter from Bernard Leach to F.J. Müller (11 July 1946).

18 DCA/312/Letter from Stafford Cripps to S.C. Leslie (10 August 1945).

19 Leach used the phrase 'breath of reality' in his pamphlet *A Potter's Outlook* (London 1928) in connection with studio pottery's then failure to serve 'the normal man or woman'. Early examples of the domestic stonewares that would coalesce into the Leach Pottery Standard Ware range were purchased by the V&A from Heal's in 1940.

20 Jonathan Blatchford, 'The Foundation of Odney Pottery Limited', http://www.johnlewismemorystore.org.uk/page/the_formation_of_odney_pottery_limited [consulted 26 January 2015].

21 'Britain Can Make It: Present and Future in Industrial Design', *Future Books : Industry, Government, Science, Arts, vol. 4: Transformation* (London 1947), p. 28. This illustrated survey accompanied an interview with the industrial designer Norbert Dutton.

22 Mass Observation (cited note 1), p. 9.

23 Ibid.

24 The 'Furnished Rooms' section of the exhibition catalogue (cited note 6), pp. 115–46, does not list the pottery or glass items exhibited within the rooms. Identification of these items has been made through photographs from the Design Council Archive, University of Brighton Design Archives.

25 Mass Observation (cited note 1), p. 2.

26 'The Design Exhibition: Explained to the Glass Industry', *Pottery Gazette and Glass Trade Review* (March 1946), vol. LXXI, no. 825, p. 167.

27 'Exhibition Glass: Growing Demand for Bottles', *The Times* (24 Aug 1946), p. 2.

28 Pevsner (cited note 14), pp. 84–86.

29 Mass Observation (cited note 1), pp. 3–4.

30 Lesley Jackson (ed.), *Whitefriars Glass* (Shepton Beauchamp 1996), pp. 79–80.

DIANE BILBEY

Design Quiz

> Does it do its job properly? Is it well made? Does it look well?

Encouraging the public to appreciate good design was a principal aim of the Council of Industrial Design (CoID), and educational tools like quizzes were useful to get that message across in a seemingly subtle way. The use of a quiz to encourage public participation and understanding of good design was part of a much bigger plan for, among other things, Design Centres, publications and exhibitions. It is perhaps not surprising that a 'Design Quiz' was included in *Britain Can Make It* (BCMI), combining stimulation of thought and discussion about products with entertainment and an element of competitiveness. There was a prize.

Although many of the products displayed in the exhibition might not have been available for purchase immediately, educating the British public on what constituted good design was preparation for a time, in the not too distant future, when hopefully they could be bought. The design of a product, even something as everyday as a saucepan, should be carefully considered by the potential consumer before making a well-informed choice by using the CoID's somewhat paternalistic guidance: *Does it do its job well? Is it well made? Does it look well?*

There was a precedent for the CoID using quizzes to engage exhibition-goers. Its stand at the ambitiously timed *Daily Herald Post-War Homes* exhibition, which opened in June 1945, took the form of a Design Quiz Competition, and gave members of the public the opportunity to register their likes on a wide range of domestic products. Entrants could win modest prizes when their selection matched most closely those of the judges.[1] With 37,000 entries, it was a huge success. In their First Annual Report, the Council recorded that public interest had surpassed all expectations and that 'an unusually high proportion of those who attended the exhibition' participated.[2]

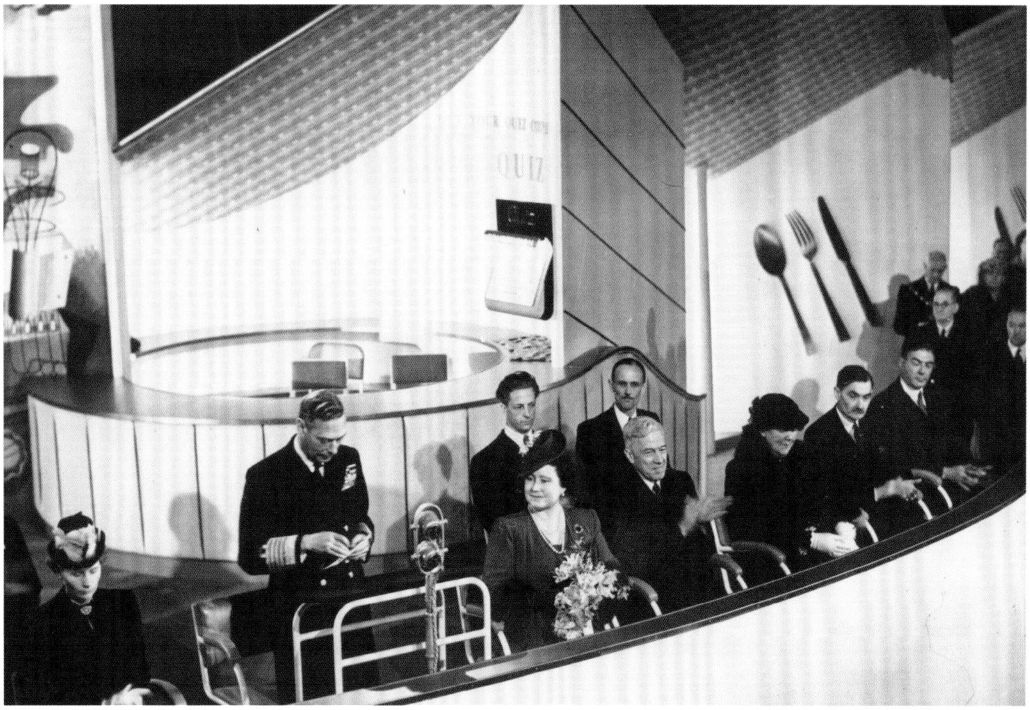

72. King George VI opens the exhibition on 24 September 1946. Visitors could collect their Design Quiz coins and envelopes from the Quiz Bank information desk behind.

Like the *Daily Herald Post-War Homes* exhibition, the *Daily Herald Modern Homes* exhibition that took place in the spring of 1946 also focused on the pressing needs of post-war housing. Here the CoID's Design Quiz was even more popular than on its first outing [73]. The quiz displayed photographs of two designs of the same everyday item,

> from powder compacts to milk bottles and door handles – and the competition is to decide which exhibit in each pair is the best design. It is not always the same kind of design that is at stake, of course, sometimes it is a matter of decoration sometimes of function. This is not simply a popularity poll and people will have to think hard and use their skill if they are to produce answers which will stand up to expert judgment.[3]

A review of the exhibition in a contemporary journal *Display* reported on its popularity: 'A check here showed that out of a hundred who arrived at this display thirty-five stopped long enough to study every design and read every word of copy. Eighteen of these took the trouble to fill in the ballot form to express their preferences for selected designs.'[4] As well as the CoID's quiz display and competition, a 'Quiz Kitchen' was also set up nearby in the exhibition. This demonstration kitchen, designed with the co-operation of Mrs Darcy Braddell, 'the

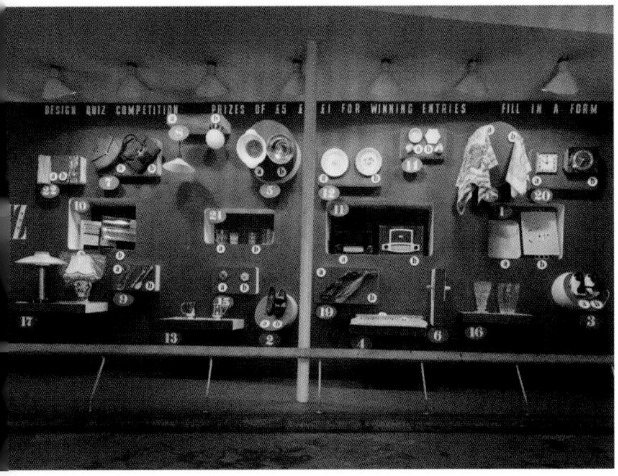

73. The Design Quiz exhibit organized by the CoID at the *Daily Herald Modern Homes* exhibition between March and April 1946

well-known expert on kitchen planning and design' – who for BCMI designed part of a living room with kitchen recess for an imagined storeroom clerk and his family in the 'Furnished Rooms' section – aimed to show 'an intelligent demonstration of the pitfalls of planning' a kitchen. Here people were urged to become a 'domestic detective' and to use common sense as well as taking into account important issues such as 'convenience, efficiency, safety and labour saving' to discover the features that would improve the kitchen design. The handbook to the exhibition helpfully included diagrams of the poorly planned kitchen as well as that of the improved layout. Clearly the public liked design quizzes, and these recent successes must have encouraged the CoID to take the format to the next level when it organized its own exhibition – BCMI – later that year.

The Quiz Box format

The inclusion of a 'Design Quiz' at BCMI was first formally discussed in a committee progress meeting in September 1945, when the exhibition had no name and was simply called 'Proposed exhibition Summer 1946'. There it was suggested that a Design 'Quiz' or 'Puzzle' might be included in the section explaining 'The Fundamentals of Design'. The proposal was that there would be two rooms, the first containing items or features which are

> incorrect from a Design point of view, visitors being asked to fill in a numbered form stating what they think are the incorrect features or articles .... The second room should approach the subject from the opposite point of view, namely, so many features or articles should be correct, the visitors again being asked to fill in a similar form.[5]

But it seems this early suggestion was discounted and replaced early in 1946 with the idea for Quiz Boxes. Ten Quiz Boxes (sometimes referred to in exhibition correspondence as banks, kiosks, stalls or machines) were installed around the exhibition. Designed by James Gardner, these took the form of simple but futuristic looking curved metal boxes mounted on to the walls of the various sections [74, 75]. The curved fascias were coloured matt white, the back fascias in bright scarlet.[6] Each Quiz Box displayed three black and white photographs of the same category of product, but all quite differently designed.[7] One was to show a traditional style object, or one similar to the type of product available pre-war. Another sought to illustrate how a particular material might be used in a functional way. The final image would show 'applied design' or would 'express a particular character or whimsy intentionally to be designed with a decorative as well as functional and "clean lines" aim'.[8] Beneath each of the three photographs were corresponding slots, so a visitor registered their choice by inserting a 'coin' or token into the relevant slot. Visitors were invited to collect an envelope containing ten quiz coins at the Quiz Bank, a reception area located at the entrance to 'Shopwindow Street' section of the exhibition, and were given instructions on how to use them by the Quiz Feature Officer [72].

At the same time as dropping a coin in the relevant slot, visitors could also record their choices on the reverse of the envelope in which their quiz coins were provided. The idea that a competitive element should be added to the Quiz section was suggested to James Gardner by the exhibition's typographer Edward Barr, who thought that the quiz scheme as it stood 'does not provide sufficient reason for the visitors' participation'.[9] Barr thought the solution to this would be to integrate a simple form on the back of the envelope containing the Quiz coins, so that the visitor would note down his or her choice at each Quiz Box. By then adding their name and address, visitors could put themselves forward for the opportunity to win a small number of cash prizes [76].

The Quiz Boxes and envelope did add a certain democratic element to the exhibition, with visitors 'voting' for their choices of what constituted good design by marking 'X' against the items they had chosen. In an article in *Architectural Design & Construction* one of its writers, Sylvia Pollak, suggested that the Quiz Banks were like 'ballot boxes': 'Into these the public are asked to drop plastic coins provided for them, selecting the design they prefer. This idea should stimulate critical appreciation though I fear that gaping crowds will be the rule.'[10] The Quiz Boxes were emptied daily by the Quiz Feature Officer and a record of the number of votes for each particular product was recorded by the CoID for later analysis of public taste.[11] The quiz was popular, sometimes too much so. Over-zealous participants, presumably cramming coins (or other unknown materials) into the slots of the quiz banks, often meant the quiz had to be 'suspended' [77].

If you were lucky enough you could be awarded one of the three prizes of £10, £5 and £3 (a worthwhile win of about £355, £177 and £106 in today's money). Rather tellingly the envelope explained that a prize would be awarded to the 'visitors whose choice most nearly agrees with that of the Selection Committee'. Clearly, no-one was expected to get all the answers 'right'. This assumption proved to be correct: for the duration of the Design Quiz not one person (out of the 120,000 participants) came up with the 'correct' answers that corresponded exactly with the Judges' selections.[12] Instead the Council selected the winning entrants by choosing those who had managed to get 80 per cent 'correct' answers, and in the order in which their envelopes were drawn.

Optimistically, and with a nod to the future, it was originally proposed that the Quiz coins be made of brightly coloured embossed plastic (a material pioneered during the war) and bear a motivational

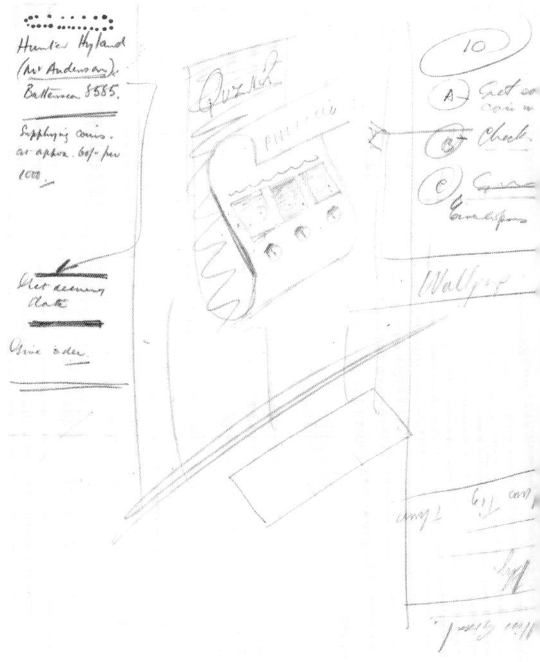

74. James Gardner's preliminary design for the Quiz Box feature at BCMI

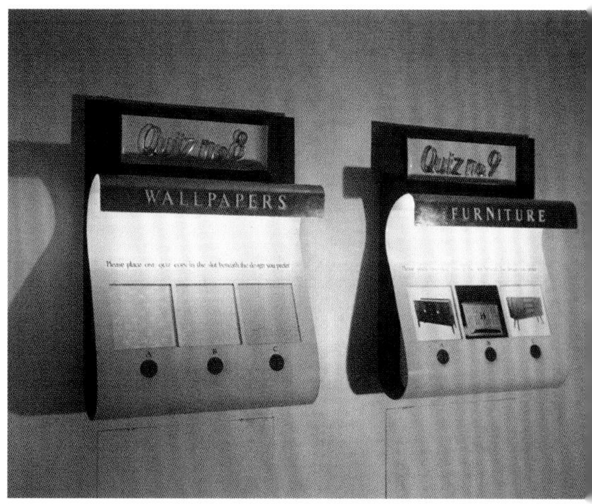

75. Two of the Quiz Boxes used at BCMI

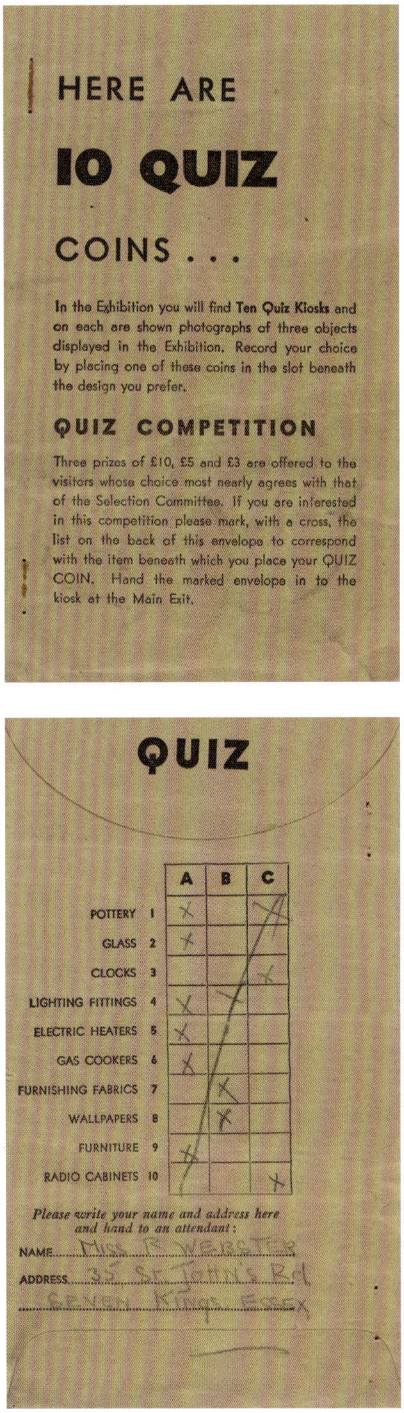
HERE ARE

10 QUIZ

COINS . . .

In the Exhibition you will find **Ten Quiz Kiosks** and on each are shown photographs of three objects displayed in the Exhibition. Record your choice by placing one of these coins in the slot beneath the design you prefer.

QUIZ COMPETITION

Three prizes of £10, £5 and £3 are offered to the visitors whose choice most nearly agrees with that of the Selection Committee. If you are interested in this competition please mark, with a cross, the list on the back of this envelope to correspond with the item beneath which you place your QUIZ COIN. Hand the marked envelope in to the kiosk at the Main Exit.

QUIZ

	A	B	C
POTTERY 1	X		X
GLASS 2	X		
CLOCKS 3			X
LIGHTING FITTINGS 4	X	X	
ELECTRIC HEATERS 5	X		
GAS COOKERS 6	X		
FURNISHING FABRICS 7		X	
WALLPAPERS 8		X	
FURNITURE 9	X		
RADIO CABINETS 10			X

Please write your name and address here and hand to an attendant:

NAME Miss R Webster

ADDRESS 35 St John's Rd

Seven Kings Essex

76. One of the Quiz coin envelopes, completed by Miss R. Webster from Seven Kings, Illford

message along the lines of 'Good Design is Good Business'.[13] Practical considerations, namely the shortage of materials and limited manufacturing capabilities, meant that the ambitious plastic coin idea had to be adapted. Aluminium alloy or Duralumin coins were considered as another option, but again this idea was shelved because of difficulties in obtaining materials. Finally, and pragmatically, 200,000 one-and-a-half-inch steel discs were ordered from the firm of Hunter & Hyland. Gardner wryly noted in a memo, 'The '1½" steel disc' is the washer with a hole in it, is it not?' These could be produced using a standard tool to save time and money. No superfluous decoration, no message, no embossing, although 250 polished coins were supplied for dignitaries attending the Royal Opening.[14]

Before these coins could be produced, officials at the CoID thought it necessary to check with the Royal Mint that such tokens would not infringe regulations. The response was that it would be preferred that the disc was adapted, to avoid any confusion with real coinage. One suggestion was that a hole should be stamped out in the middle, so that it would be clear it was not legal tender, and would prevent 'unprincipled people' who might try to 'use them for slot machines or pass them off in the dark'.[15]

Hunter & Hyland, the firm who manufactured the 'washers', were also contracted to collect the coins and re-fill envelopes on a daily basis for the duration of the Quiz. The Design Quiz ran from 23 September to 16 October 1946 and of the 200,000 coins produced only 16,500 remained at the end of the three and a half week quiz period. People must have found it too much of a temptation to take home the envelope and coins as a souvenir, or perhaps in austerity Britain the washers came in handy.

The colour penny guide to the exhibition included an explanation of the basis and motivation of the Quiz: 'The quiz device will give you an opportunity to exercise your judgement and record your preference between the style of design and another. By the time the Exhibition closes, the Council will be left with a useful record of public feeling about design.'[16] Revealing conclusions could be made by analysing the public's choice, and these were detailed and considered by the CoID in a report shortly after the close of the exhibition.[17]

Design Quiz at the BBC

The relationship between the CoID and the BBC was a strong one, with the BBC keen to broadcast both radio and later television programmes that supported the Council's activities. On 8 July 1946, before the opening of BCMI, the BBC broadcast a twenty-minute television programme on the subject of 'What is Good Design'.

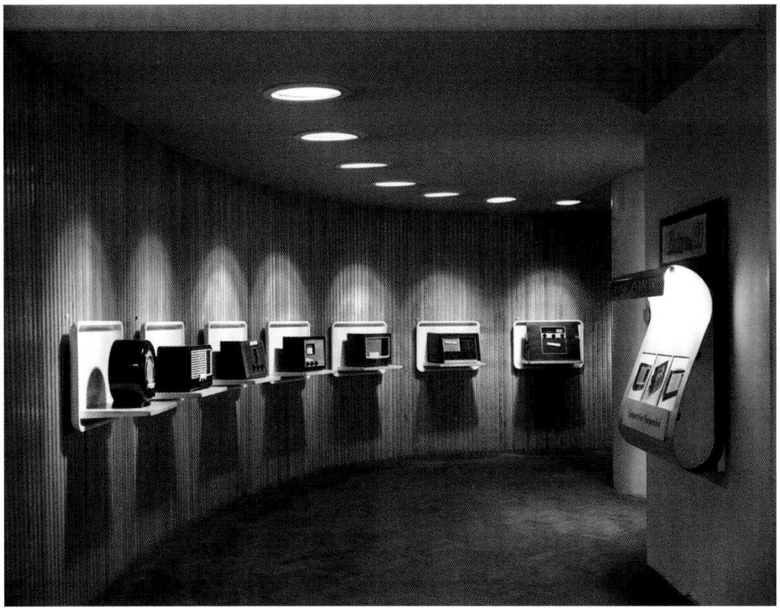

77. To the right, a 'suspended' Quiz Box in the radio and television section of the exhibition, designed by F.R.S. Yorke

It included examples of a variety of products including teapots, electric clocks and kettles, and also radios. Decca sent two radios for comparison. One was 'very modern and one, which is very old and of which they are themselves ashamed'. A further programme called 'Design Quiz' was shown a week or so later, in which viewers were asked 'to look in and test their taste with the experts'.[18]

Design Quiz publication

To supplement the Design Quiz feature of the exhibition, the CoID produced an accompanying booklet. One of the Council's earliest publications, *Design Quiz* reinforced the importance it placed on the use of quizzes as a means of interaction with the general public [**78**]. If you enjoyed participating in the Design Quiz at BCMI you could buy a copy of the booklet to take home, remind yourself of what you had learnt from the exhibition, and maybe try it out on your family and friends.[19]

Like the images of products attached to each of the Quiz Boxes, the *Design Quiz* booklet (designed for the CoID by the Design Research Unit) asked the reader to look at three photographs of each product, weigh up the pros and cons, and decide on which design they liked best. Echoing signage throughout the exhibition, it suggested readers should ask themselves: *Does it do its job well? Is it well made? Does it look well?* and could then check their answers at the back of the booklet against those made by a panel of three 'experts' – the artist Barnett Freedman, the architect Hugh Casson (who would later be appointed as

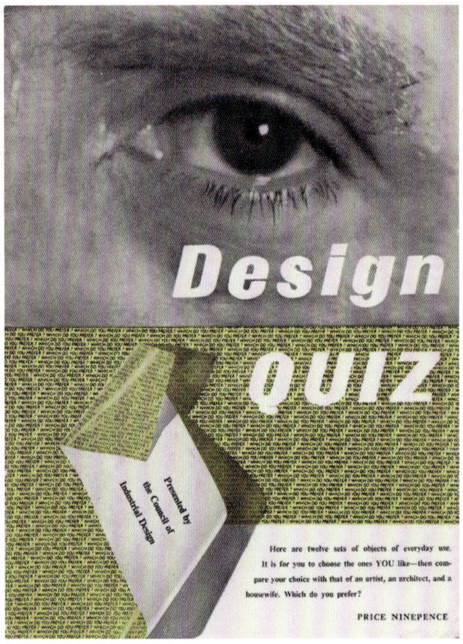

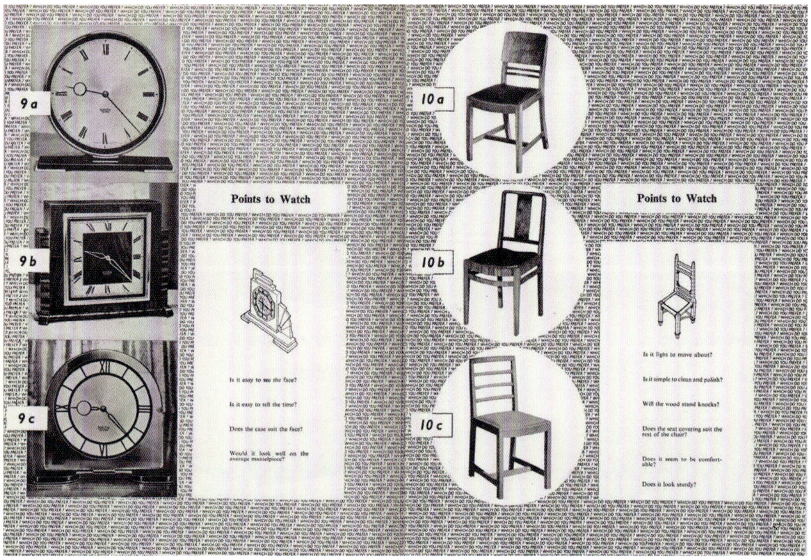

78. The front cover of the CoID's Design Quiz publication designed by the Design Research Unit (DRU), and inside pages highlighting the design of clocks and chairs

Director of Architecture at the Festival of Britain) and a housewife, Mrs Mary Harrison.[20] The instructions on how to use the booklet noted that

> The sets of three photographs are not meant to show objects which are – either in the opinion of the Editor or the experts – very bad, medium, and very good designs. In these designs there is no absolutely clear 'right' and 'wrong'. Even the experts did not always agree entirely, and, like the Brains Trust, they frequently said, 'It depends what you mean by good design ...'.[21]

The Design Quiz observed

Mass Observation (M-O) – who were commissioned to gauge public opinion by interviews and overheard comments – reported that an impressive five hundred copies of *Design Quiz* (at 9d per copy) were sold from the bookstalls in the exhibition each day. As part of their report on BCMI, M-O investigators informally talked through the booklet with a number of visitors, asking them to identify the designs they liked the best, and recorded the results. These showed that experts and the public agreed on certain categories – the armchair, lampshade, gas fire, and electric fire – but there was disagreement on other product categories, covering glass, a cup and saucer, saucepan, teapot, wireless, clock, chair and wall lamp. Despite these differences of opinion, most were only slight, and M-O concluded that there was not the extreme disparity between public and experts that perhaps might have been expected. There was only one category out of the twelve in which the public and experts disagreed profoundly, that of the upright wooden chair.

In this category only 6% of the public agreed with the experts' choice (c) – to most people it just looked uncomfortable. A woman described by M-O as an 'art student' commented, 'It looks as though the bars at the back will be uncomfortable'. Another participant, a man of 50, remarked, 'It's too rigid and not shaped to the body. I think that a chair must be shaped to the body'.

M-O concluded that the public seemed to understand two out of the three of the CoID's principles that an object should 'look like it should', and 'function as it should'. The public did not seem to consider the third point, that it should also look good, as being of importance. 'But the public generally accepts either the utility or the aesthetic point, and very rarely judges an item by both.'[22]

News of the recently published *Design Quiz* was reported in an article in the *Sunderland Daily Echo*, where it was described as a new 'game'. The writer focused on an important aspect, that good design need not be available to an elite few, and should be affordable for all. There was no need for good design to equate to highly priced goods:

> This is all part of the Council's educational campaign. A few years ago when home production was still a dream, such champions of design as Lord Woolton and Dr Hugh Dalton were telling us that good design must be bought within the level of the poor man's purse. The shopkeepers who only buy what they can sell have since shown us that taste has nothing to do with money. You only have to go from a South London suburb to Knightsbridge to have this truth illustrated. In the suburbs you can pay 47s [shillings] for a poor imitation parchment shade gaudily painted with flowers. In Knightsbridge you can find a plain shade which would pass as real parchment for 31s [shillings].[23]

The CoID also produced its own report on the conclusions that could be drawn from the public's engagement with the Design Quiz, entitled 'Indications of public taste as shown by the votes cast by visitors'. Although the Quiz only ran for the first three and a half weeks of the exhibition, a total of 120,034 votes were cast. Miss C.M. Dowler of the CoID prepared a thorough analysis of the results from each of the ten Quiz sections. In her report she noted that the Quiz pictures shown to the public were roughly divided into two categories – one in which fashion or decorative interest was the predominant issue, including wallpaper, furnishing fabrics, pottery and glass; the other in which function was more important, such as light fittings, electric heaters, radios and furniture. In the first category, Miss Dowler noted that

> There is a noticeable trend away from any plainness which is reminiscent of austerity, the least votes of all were given to the plain glass jug and tumbler …

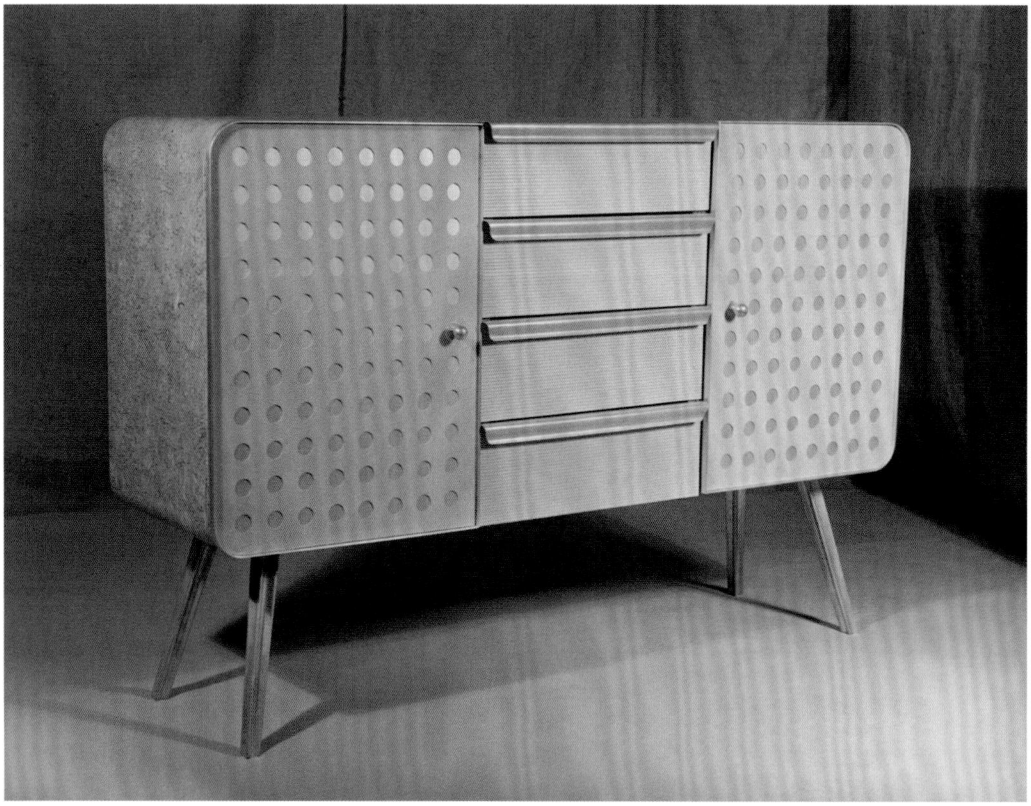

79. Clive Latimer's sideboard in plywood birch veneer and aluminium, manufactured by Heal's & Son Ltd – one of the objects included in the Design Quiz

> In the second class ... strong considerations of function seem to govern the preferences ... in the case of furniture and radios the desire is for something practical and pleasant but unobtrusive to the eye.

Quiz no. 9, the furniture category, was represented by a sideboard. Here the least popular choice was the modern sideboard with Duralumin detail (aluminium sheet alloy) produced using Redux bonding and plywood birch veneer, designed by Clive Latimer for Heal's & Sons Ltd [79].[24] It was described by Miss Dowler as 'somewhat modernistic .... It is of light coloured wood, has a cupboard on each side of the drawers, ornamented with cut-out circles and rather tall legs.' It attracted only around 3,000 votes, whereas a more conservative, traditional dark wood, plainly styled sideboard, with cupboards on either side of the central drawers, received 7,000 votes. The British public was perhaps not yet ready for such contemporary design.

In conclusion, Miss Dowler suggested that had the images of products been in colour (not black and white as they had been) then the public may have found it easier to choose, but that 'The clearest inference we can draw seems to be that the public wants in its home nothing

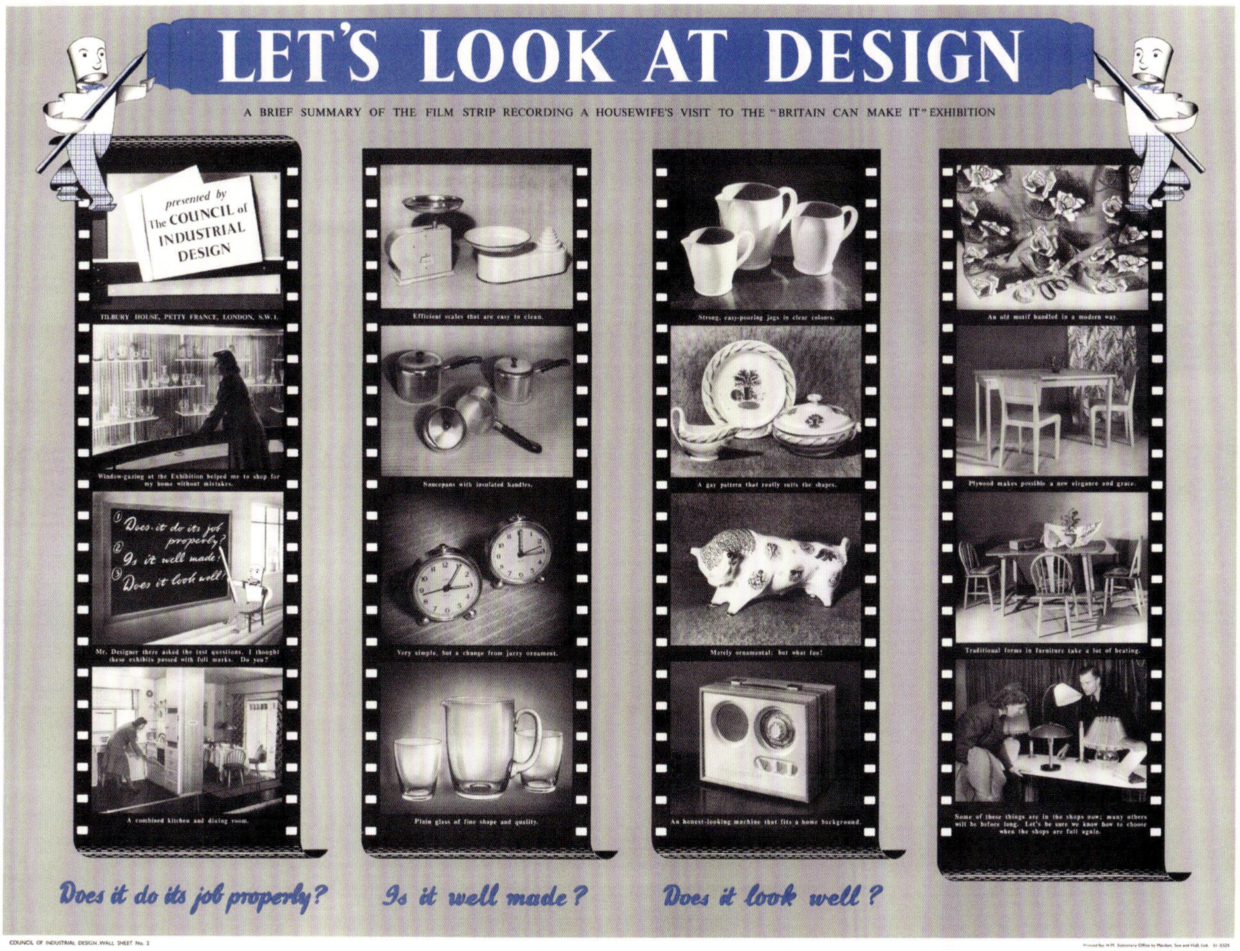

80. The CoID's 'Let's Look at Design' wall sheet recording a housewife's visit to the exhibition, 'where she learns how to choose well-designed furniture and furnishings for her home. Finally she watches a desk lamp being designed and made'. These posters could be obtained from the Information Division of the CoID at a cost of 2s 3d. An accompanying film strip could also be purchased.

revolutionary in design but will insist on an escape from Utility plainness and any labour saving device or means of giving additional comfort will be popular'.[25]

Let's Look at Design

The message that design was as important a consideration as price and availability when buying for the home was reinforced by further initiatives from the CoID, including *Let's Look At Design*, a short propaganda film strip recording a housewife's visit to BCMI, and an accompanying wall-chart in which a cartoon-style 'Mr Designer' character guides the reader through an array of objects, suggesting ways in which the housewife could decide on what she should buy for her family [80]. Other publications from the CoID included a series of books, *Buying for Your Home*. The one devoted to furnishing fabrics included three 'quiz' pictures of chairs, with sample fabrics, and suggested that people should consider 'whether the fabric suits the shape and style of the chair, imagine how the chair would go in your own rooms or other rooms you know; imagine the curtains you would choose to go with them; ask yourself if you would like to go on living with them'.[26] The CoID's *New Home* booklet also included a Shopper's Quiz, where shoppers had to make a choice from three items in groups that included kettles, radios, tables and clocks. 'There may not be a "best" because tastes differ, but on page 32 you will find notes that will help you to check your judgment.'[27]

Mixed reviews

The Design Quiz idea was not universally acclaimed by critics. When Herbert Read, a founding member of the Design Research Unit, wrote about his visit to BCMI in an article for the BBC magazine, *The Listener*, his general comments on the exhibition were acerbic, criticizing the way people were forced to move around exhibitions (via one predetermined route) as a journey through an 'intestinal tract'. He also downplayed the value of the Design Quiz:

> Since the education of the public is a further purpose of the exhibition, this may not be a bad policy. Visitors are encouraged to exercise their discrimination, and 'Design Quizzes' are placed at intervals along the tract where they can record their preferences. But it is to be feared that something analogous to Gresham's Law in economics prevails in aesthetics – in a free market the bad designs drive out the good. One has only to listen to the comments of passers-by. Public taste is deeply corrupted, and will need some action more drastic than an exhibition to retrieve it. Good design can be created by individuals, but public taste is the anonymous growth of a tradition.[28]

The pre-eminent role of the designer in leading public taste was voiced in a review of the exhibition in *Display*, from a writer taking a 'Foreign observer's viewpoint': 'An exhibition innovation which really caught my eye was the "Design Quiz". An interesting experiment this, but whilst one must know the truth about public taste, the designer must lead the public, not wait for their approval.'[29]

The Council's methods to increase public engagement with good design might be seen, with modern sensibilities and a twenty-first century perspective, as somewhat patronizing. But the reality seems to be that it worked, as tens of thousands of visitors voluntarily participated in the Design Quiz and presumably enjoyed doing so. And, although the analysis of it was not exactly scientific, the results indicate that the public's taste was not so far away from that of the 'experts': it was tired of basic utility, but not yet ready for radical change.

NOTES

1 The *Daily Herald Post-War Homes* exhibition took place between June and August 1945 at Dorland Hall, Regent Street, London.
2 *The Council of Industrial Design, First Annual Report 1945–1946* (London 1946), p. 20.
3 The *Daily Herald Modern Homes* exhibition took place between 26 March 1946 and 25 May 1946 at Dorland Hall, Regent Street, London. *Daily Herald Modern Homes exhibition catalogue* (London 1946), pp. 51–52; 'Modern Homes', *Display, Design & Presentation* (April 1946), pp. 31, 35.
4 'Exhibitions. "Modern Homes". "Daily Herald" show is one of the best since the war. But there are loopholes for the critic in its brilliant façade': *Display, Design & Presentation* (May 1946), p. 16.
5 Design Council Archive at the University of Brighton Design Archives (hereafter DCA) 312/. The idea for a quiz was considered at the onset when the proposed CoID summer exhibition was mooted in a document of 10 September 1945, 'Proposed Exhibition – Summer 1946'. For further information on the design quiz, see Jonathan Woodham, 'Design and everyday life at the Britain Can Make It Exhibition, 1946: "stripes, spots, white wood and homespun versus chintzy armchairs and iron bedsteads with brass knobs"', *The Journal of Architecture*, vol. 9 (Winter 2004), pp. 471–72.
6 DCA/742/Letter from James Gardener to F.R.S. Yorke dated 5 July 1946: '"Britain Can Make It" Exhibition, Victoria and Albert Museum. Quiz Kiosks.' The Quiz boxes were made by Beck & Pollitzer, one of the main contractors for the exhibition. It is likely they sub-contracted this out to J. Starkie Gardner Co. Ltd, who provided metalworking for BCMI. James Gardner almost certainly designed the Quiz boxes, though the Design Research Unit (DRU) is often cited as the designers. This confusion has possibly arisen because the DRU designed the accompanying *Design Quiz* publication for the CoID.
7 The ten Quiz Banks categories were Quiz No. 1 Pottery – a cup, saucer and plate; Quiz No. 2 Glass jug and tumbler; Quiz No. 3 Clocks; Quiz No. 4 Lighting fittings; Quiz No. 5 Electric heaters; Quiz No. 6 Gas cookers; Quiz No. 7 Furnishing fabrics; Quiz No. 8 Wallpapers; Quiz No. 9 Furniture (sideboard); Quiz No.10 Radios. The photographs were taken by Michael Wickham, who was the exhibition's official photographer and one of the Exhibition Officers. The only section to show colour photos was Quiz No. 7 for Furnishing fabrics.
8 DCA/742/Council of Industrial Design. Outline for Quiz Scheme, dated 3 April 1946.
9 DCA/58/. The suggestion was put to the Council's 21st Progress Committee meeting on 3 July 1946. Barr also worked as assistant to James Gardner on the 'Books and Printing' Section of the exhibition.
10 Sylvia Pollak, 'Round the exhibition', *Architectural Design & Construction* (October 1946), p. 269. Plastic coins were not used in the end, as discussed further in this essay and in footnote 13 below. Coins made of plastic for the Design Quiz are also, confusingly, cited on the penny colour guide to the exhibition.
11 The Design Quiz did not run for the whole period of the exhibition (it ran for just three and a half weeks from 23 September until 16 October 1946).
12 Mr Alan Jarvis and Mrs C.G. Tomrley were judges. Alan Jarvis was Public Relations Officer for the exhibition and Mrs Tomrley gave design advice and was responsible for producing the list of designers involved in the exhibition. Jarvis also wrote a series of books – The Things We See – published by Penguin.
13 Mr Mesling contacted Ernest & Henry Ltd for the supply of plastic quiz coins (and they produced a die) but the idea was rejected because the cost of producing plastic coins was too high. Mesling suggested a few plastic coins might be made for VIPs and the Royal Party. DCA/Memo from C.A. Mesling to R. Dudley Ryder 30 July 1946, 'Quiz coins'. The slogan 'Good design Good business' was the title of a later publication by John Gloag, first published by the Council of Industrial Design in 1947.
14 DCA/742/Memo from James Gardner to C.A. Mesling dated 23 August 1946, 'Quiz Coins'; Memo from C.A. Mesling to R.D. Ryder dated 22 August 1946, 'Quiz coins'; Letter to Mr Anderson of Hunter & Hyland Ltd from R. Dudley Ryder dated 11 September 1946; V&A Archive, Exhibition File: AAD/1977/4/14: Britain can make it: Hunter & Hyland statement of account sent to the Council of Industrial Design.
15 DCA/742/Letter from T.W. Kenner, Royal Mint to C.A. Mesling, CoID, dated 31 July 1946.
16 Britain Can Make It Exhibition Guide. 1d colour guide.
17 V&A Archive, Exhibition File: AAD/1977/4/14A, Britain can make it: 'Design Quiz. Indications of public taste as shown by the votes cast by visitors to the "Britain Can Make It" Exhibition between 23 September and 16 October 1946'.
18 BBC Written Archives Centre TVART1 Council of Industrial Design 1946–1954, File 1. The television programme devoted to a Design Quiz was broadcast on 19 July 1946. See also Michelle Jones, 'Design and the Domestic Persuader: Television and the British Broadcasting Corporation's Promotion of Post-War "Good Design"', *Journal of Design History* (2003), vol. 16, no. 4, pp. 307–18.
19 Although similar to the Design Quiz in the exhibition, the booklet illustrated twelve sets of everyday domestic objects rather than the ten in the exhibition. The categories of objects were also not exactly the same as those displayed in the exhibition itself, though they were closely similar.
20 Mrs Harrison might not have been the typical housewife depicted. One Mrs Harrison is noted on a list of governmental ministry departmental assessors to the CoID as being from the Ministry of Health: V&A Archive, Exhibition File: AAD/1977/4/12.
21 The Design Research Unit designed the *Design Quiz* publication (London 1946), unpaginated. The Brains Trust was a popular BBC radio (later television) programme with a panel of experts who would attempt to answer questions sent in by the public.
22 DCA/903/Report by Mass-Observation on Britain Can Make It Section C and B, typescript dated December 1946, p. 18.
23 'Design for Living', *Sunderland Daily Echo and Shipping Gazette* (20 November 1946), p. 2 .
24 The sideboard is included in Gordon Russell, 'Furniture', Council of Industrial Design. *Design '46. Survey of British Industrial Design as displayed at the "Britain Can Make It" Exhibition* (London 1946), p. 97 illus. Richard Chamberlain and Annamarie Stapleton, *Austerity to Affluence: British art & design 1945–62* (London 1997), pp. 13, 19, illus. p. 12. An example of the Latimer sideboard is held in the collections of the Carnegie Museum of Art, accession no. 2002.21.1.
25 V&A Archive, Exhibition File: AAD/1977/4/14A, (cited note 17).
26 Mary Shaw, Council of Industrial Design, *Buying for Your Home. Furnishing Fabrics, no. 1*, (London 1946), p. 25.
27 Council of Industrial Design, *New Home* (London 1946), pp. 30–32.
28 Herbert Read, 'Britain Can Make It. Herbert Read on the exhibition at the Victoria and Albert Museum', *The Listener* (3 October 1946), p. 430. Gresham's Law can be simplified into the idea that bad money drives out good. In this instance, Read is suggesting that poor public taste undermines good design.
29 Machiel Wilmink, 'Britain Can Make It. 5th viewpoint. The Foreign Observer's', *Display, Design & Presentation* (November 1946), p. 23.

PAUL RENNIE

Dressing the Goods: Packaging & Post-War Reconstruction

A pack(age) should be handy to store and carry, Attractive to look at, and easy to recognise.
—*Announcement at the entrance to the Packaging section at* BCMI

Introduction

'Modern' packaging was effectively invented during World War II. It is not the case that there was no packaging before 1939; there was, but it was much less widespread, and only the largest manufacturers were able to justify the added costs to their products. For practical purposes and as far as the majority of people were concerned, most goods were sold loose. Shops used bags and boxes, supplied by independent manufacturers, for the convenience of their customers. In his account of the growth of packaging Alec Davis describes this as

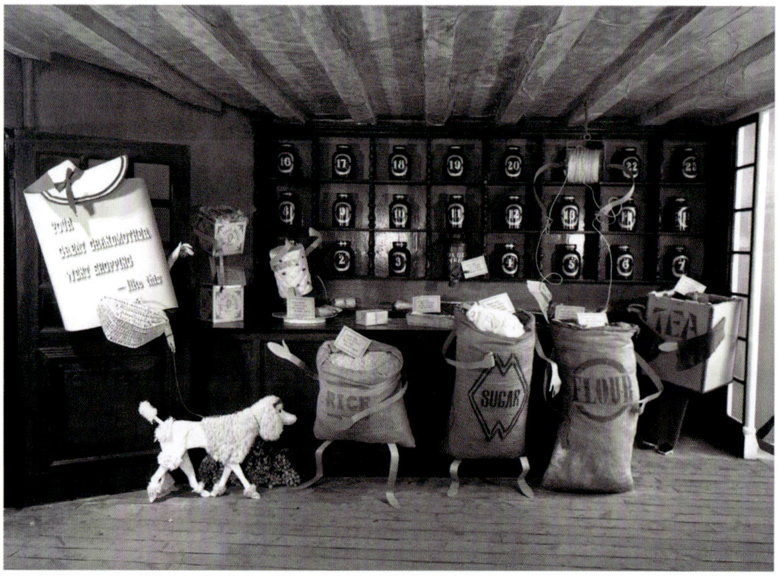

81. 'How great grandmother went shopping', historical display from the 'Dressing the Goods' section of BCMI

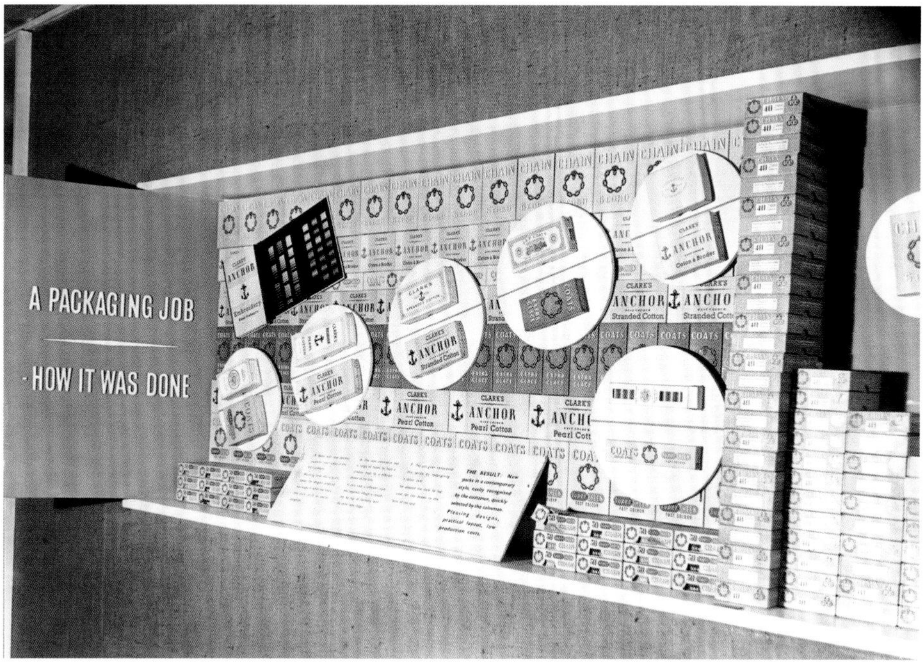

82. A Packaging Job, vocational display from the 'Dressing the Goods' section of BCMI

the story of packaging before packaging was invented, seeing its historical development in terms of materials – first glass bottles, followed by metal boxes, foil tubes and folded paper cartons.[1]

Writing in *Design '46,* John M. Ryan noted that the number of designers in Britain specializing in package design 'able to earn a reasonable living' could be counted on the fingers of one hand.[2] Ryan probably knew each of these designers personally, as during the 1930s he had been editor of *Shelf-Appeal,* a magazine for emerging professionals working in packaging and promotional material.

The *Britain Can Make It* (BCMI) exhibition proposed a new beginning for packaging and design. Packaging design would, it was suggested, become more like graphic design and would take its place amongst a group of co-ordinated design activities that bridged the gap between manufacturing and consumption.

The items for display in the 'Dressing the Goods' section of the exhibition were selected by a committee comprising the historian and design advocate John Gloag; the sculptor C.D.O. Pilkington Jackson; Dr G.L. Riddell, Director of Printing and Allied Trades Association; and Ryan. The displays for the section were designed by F.C. Ashford (who was also the designer of the air-conditioned bed in the 'Hall of the Future' [see **174**]), and the technical adviser on the section was Albert Smith, also the designer of the display illustrating the qualities and uses of paper in the 'What the Goods are made of' section of the exhibition.

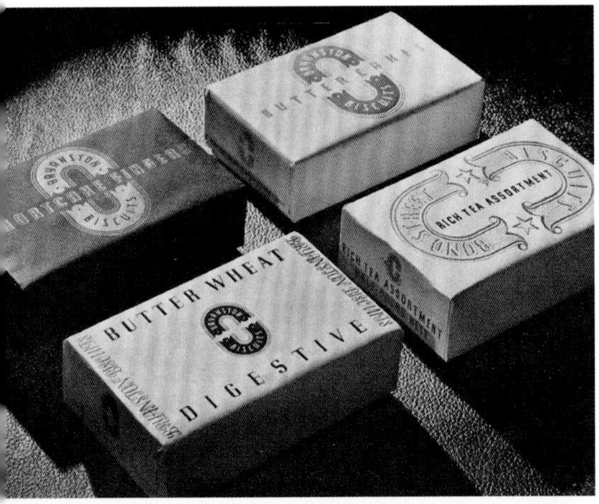

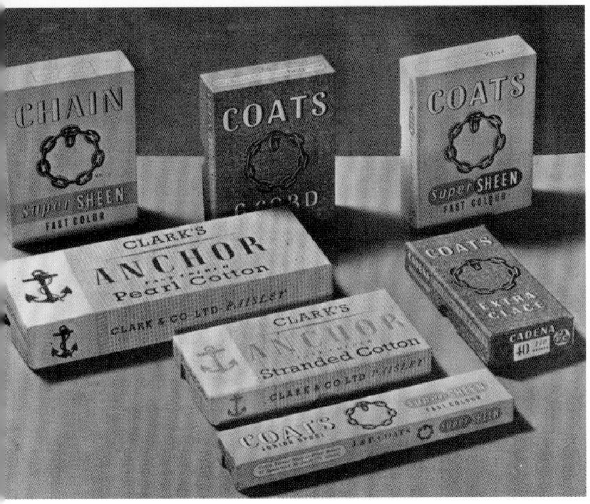

83. Bryanston Biscuit packets by Milner Gray (reproduced from *Designers in Britain*, 1947)

84. Selected cotton reel boxes, for J. & P. Coates, designed by Barnett Freedman (reproduced from *Design '46*)

The displays began with examples of pre-war packaging and showed how materials and design could be combined, so that packaging could become a practical and pleasurable element in the consumer experience. There were also more specialized sections, 'Planning the Pack' and 'A Packaging Job', that described design as a decision-making and problem-solving process [82].

The exhibition also provided a context for a number of related events. These included a conference on packaging, organized under the auspices of the Council of Industrial Design (CoID) in November 1946, and the publication of various surveys that promoted British design internationally. The conference speakers spoke on a variety of subjects including 'Technical Packing and Packaging', 'Scientific Research and Packaging' and 'Packaging from the Factory Angle'. The designer Milner Gray spoke on 'The Designer in Action'. The main purpose of the conference was to convince delegates of the 'scientific' or evidence-based benefits of good packaging.[3]

In March 1946, the editors of the Swiss-based international review of graphic and applied arts *Graphis* published a survey of British design. The review also featured the Society of Industrial Artists (SIA), illustrating examples of up-to-date packaging design.[4] Once the war was over, the SIA was immediately active in promoting the professional status of design activities, and in supporting packaging design as an effective source of value and profit – both culturally and commercially.[5]

An unexpected consequence of the war had been the promotion of a more collegiate relationship between military specialists. This was especially the case in relation to the planning and execution of large-scale combined operations. Implicit within the conception of these integrated projects was a form of resource allocation based on scientific management and evidence gathering. This revealed itself through measurable improvements in efficiency and effectiveness.

A similarly new approach to design was exemplified during the immediate post-war period by the Design Research Unit (DRU). Marcus Brumwell and Herbert Read had conceived the DRU as a creative agency of specialists, able to provide a complete service to whatever scale of organization. The founding partners of the agency were the designers Misha Black and Milner Gray, along with the architect Kenneth Bayes. It was testament to Milner Gray's skill as a designer, and to his professional significance, that his packaging designs were well represented within the BCMI exhibition [83]. The 'Mr Designer', as illustrated throughout BCMI, was, accordingly, transformed from commercial artist and tradesman to the professional colleague of architects and planners [85].

DRU's first offices were in the Bedford Square home of the printer Lund Humphries, the publisher of *The Penrose Annual*. The Annual was first published in 1895 and continued until 1982. The war caused a short break in publication, but when it resumed the journal provided a wider forum for the presentation of new developments in printing and design.

The economic situation in Britain remained difficult after the war. Nevertheless, the context of austerity provided the circumstances in which British designers could take risks and experiment. British printers and manufacturers become more adept at using foils, transparencies and plastic. The Packaging Committee member at BCMI, and speaker at the accompanying conference, Dr G.L. Riddell, was the senior representative of the Printing and Allied Trades Association.

A number of the names associated with packaging design in the 1940s are familiar from their poster design and illustration before the war: these include Barnett Freedman and Edward Bawden, but also Reynolds Stone and Eric Fraser [**84**].

BCMI proposed a development in consumer behaviour whereby the pleasure and enjoyment associated with the purchase of pleasingly designed and attractively packaged goods became more widely acknowledged. Within the context of post-war austerity, and against a cultural legacy of Victorian values, this was an important shift in the move towards the consumer society and leisure economy of the 1960s.

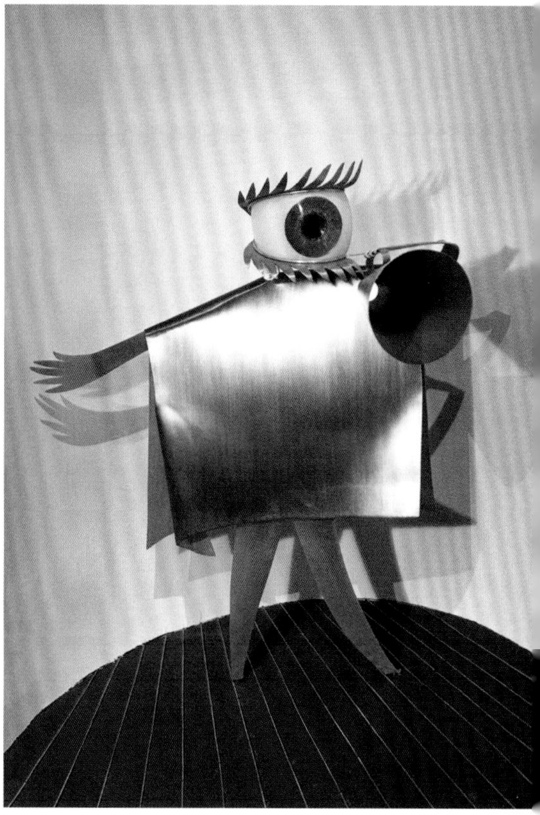

85. The figure of 'Mr Designer' appeared throughout BCMI

Logistique de Guerre

The British context of the war, including re-armament, storage, supply and distribution, highlighted the importance of appropriate packaging. During the war, the natural wastage associated with damage and loss of loose goods became unacceptable. This was true in relation to both the economic value of the war effort and to the evident harm in distributing damaged supplies to fighting troops.

Most military packaging was entirely utilitarian in nature, comprising metal boxes and wooden crates, with stencil lettering to identify contents and destination.[6]

The conference papers prepared in association with the BCMI exhibition made extensive use of the military experience. Implicit in the appeal to efficiency was a military concern for hygiene and discipline. The post-war significance of these themes in relation to civic reconstruction was reflected in the display 'What is well packed, is well protected'.

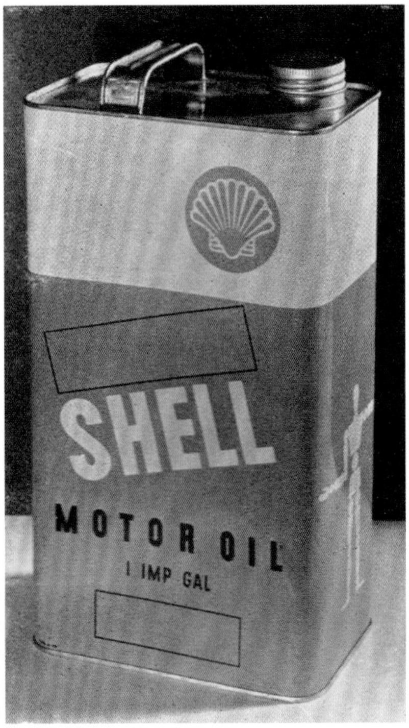

86. Metal can for Shell Motor Oil, by Metal Box Ltd (reproduced from *Design '46*)

Economic and efficient, practical and convenient

Against the backdrop of an export-led recovery, design was at the forefront of experimentation with new types of packaging material and printing.

The transfer of economic activity from military to civilian life marked a process of 'turning swords into ploughshares'. In packaging terms, this was exemplified by new materials that combined lightness and strength. Moulded plastics, clear plastics and aluminum supplemented traditional carton packaging.

The industrial conglomerate ICI was at the forefront of developing new materials and finding new markets for their products. Interestingly, they had commissioned the artist Edward Wadsworth (famous for his surrealistic pictorial arrangements of the tools and materials of modern life) to record and respond to their new materials. In 1948, the journal *Alphabet and Image* featured a number of these paintings of plastics.[7] At the same time, the Aluminium Development Association was active in promoting the potential of this lightweight metal through the distribution of samples [86].

The considerations of how to pack items together for shipping also led to products that could be stacked and displayed more easily in shops. This provided a significant impetus to the development of self-service stores in Britain during the 1950s and 1960s.

The standardization of packaging also ensured economic consistency through a more rigorous consistency of portion control. The BCMI exhibition helped convince British manufacturing and retailers that packaging was both efficient and economical.

At the same time, the consumer began to experience the benefits of modern packaging in terms of convenience and practicality. In addition to providing an assurance as to the integrity of the product, modern packaging provided a robust platform for labels and branding. Accordingly, modern packaging became associated with mass-market consumerism and its characteristics of choice, economy and convenience, expressed through branding and graphic identity [87]. The first food retailer to embrace the idea of co-ordinated identity, expressed through shop environment, display, products and advertising, was the Scottish Mac Fisheries group. Between 1952 and 1959, their visual identity was designed by Hans Schleger (known as 'Zero'). This included colour-coded packaging and the production of a consistent and coherent range of point-of-sale messages.

Conclusion

The inclusion of packaging within the displays of BCMI marks an important development in the conceptualization of the British consumer market as an expression of the affluent society.

Developments in printing technology, particularly in relation to photo-mechanical offset lithography, allowed graphic designers to specify their requirements with increasing precision, and to achieve more consistent results. Colour printing and the combined use of photographic, typographic and calligraphic elements helped re-position packaging design as part of graphic design. Accordingly, the packaging environment began to change more rapidly and to impact positively on the shopping experience in post-war Britain.

The systemic alignment of manufacturing, design and consumption expressed through the exhibition could not, for various reasons, be realised in Britain until much later. Nevertheless the professionalization of the design process took root and contributed significantly to the emergence, during the 1960s, of a vibrant retail environment.

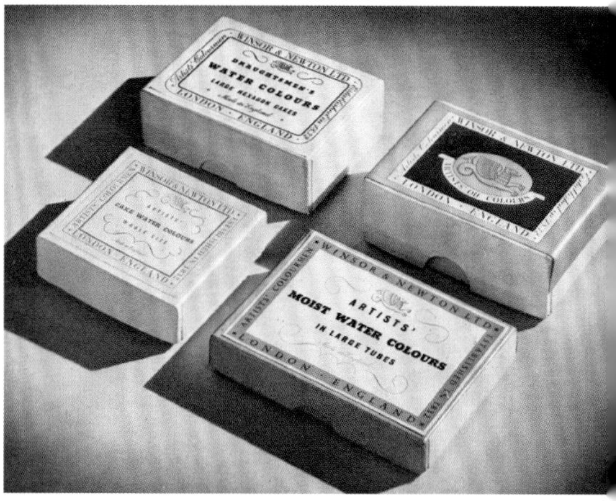

87. Winsor & Newton artists' supplies packaging, designed by Jesse Collins (reproduced from *Designers in Britain*, 1947)

NOTES

1 See Alec Davis, *Package and print: the development of container and label design* (London 1967). For Davis, containers, their materials and the labelling of products, were separate entities whose integration and co-ordination remained eccentric until modern times. For further reading, see Noel Carrington (ed.), *British Achievement in Design* (London 1946); Herbert Read, *The Practice of Design* (London 1946); Joe Brumwell, *Bold Ties Bold Ideas* (Truro 2010).

2 John M. Ryan, 'Packaging', *Design '46* (London 1946), p. 127.

3. The conference comprised a programme of speakers including Albert Smith, Dr G.L. Riddell, E. Richardson, Milner Gray and John M. Ryan. The accompanying publication, *Illustrated Notes on Technical Packing and Packaging and on the Design of Sales Packaging*, reprinted from the full report of the Conference on Packaging held in London in November 1946 (The Council of Industrial Design 1946), included many photographs of the damage associated with loosely shipped goods. A copy is held in the University of Brighton Design Archives, Design Council Archive. The conference took place at 23 Knightsbridge on 29 November 1946. A copy of the conference agenda is also held in the Design Council Archive.

4 Peter Austen, 'The British Exhibition', *Graphis* (1946), no. 14, Special Number GB "England", pp. 212–19; Peter Austen, '"Britain Can Make It". The Council of Industrial Design', *Graphis* (1946), no. 14, Special Number GB "England", p. 258; F.L. Billows, 'S.I.A. Society of Industrial Artists', *Graphis* (1946), no. 14, Special Number GB, "England", pp. 250–57.

5. See Peter Ray (ed.), Society of Industrial Artists, *Designers in Britain*, no. 1 (London 1947), pp. 44–51; *Designers in Britain*, no. 2 (London 1949), pp. 115–23; *Designers in Britain*, no. 3 (London 1951), pp. 47–59.

6 See Martin van Creveld, *Supplying War. Logistics from Wallenstein to Patton* (Cambridge 1977).

7 Douglas Newton, 'Paintings of Plastics by Edward Wadsworth', *Alphabet and Image* (December 1948), no. 8, pp. 63–69.

EDWINA EHRMAN

Women's Fashions

> … the fashion designer needs to be up and about earlier than the designer in other fields … he must make a point of knowing what is going on, not only in his own country but in all countries of fashion vitality … the whole of life is the dress and textile designer's raw material and inspiration.[1]

Fashion was an important element in the *Britain Can Make It* (BCMI) exhibition. Of the 90,000 square foot exhibition area, a quarter was allocated to dress fabrics, men's and women's fashions and accessories, and clothing for children, from infants to teenagers. The organizers believed that fashion's inherent glamour would attract female visitors and provide a focal point within the exhibition. Textiles were also an important British export market. Even in 1946, in spite of the difficulties the industry faced, they were still the largest category, accounting for 20% of British exports.[2] A cautiously optimistic advertisement published by Debenham & Co. in September 1946 detailed the challenge.

> We are told that it takes at least 9 months to convert a factory from war to peace production. In addition to this, men and women have had to await release from the Forces and need training or refresher courses before they are ready for full productive capacity. There is so much leeway to make up but there are signs that during the coming Autumn and Winter we shall see definite improvement in the supply position. So, soon perhaps, we may have news for you … and remember British Woollens, Silks and Rayons are WORTH WAITING FOR.[3]

This essay focuses on the women's fashions. There were about 100 outfits, chosen to represent clothes appropriate for town and country, and for a variety of occasions from spectator sports to formal evening events. Visitors could inspect couture models, designed to be custom made for individual clients, ready-to-wear clothes from the leading wholesale houses at a range of prices, and garments made under the government

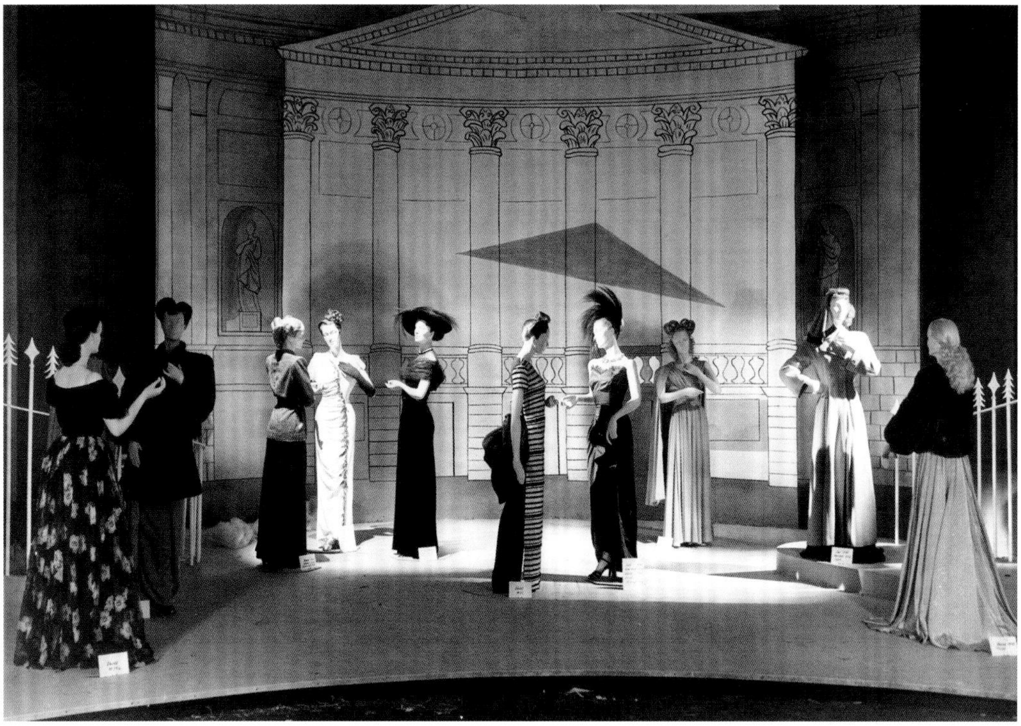

88. Couture models set against the backdrop of Geoffrey Bennison's Covent Garden mural. Bennison was an interior designer, painter and stage designer.

Utility scheme from Utility cloth. While outfits were accessorized with hats, gloves, shoes and furs, some accessories and garments such as knitwear and lingerie were displayed in groups. The corset industry was not represented in spite of their willingness to participate and the importance of foundation garments to women.[4] A special section was dedicated to rainwear and umbrellas [90]. Within the general category of textiles, waterproof fabrics and garments were considered an export with the potential for growth while undoubtedly being 'essential to comfortable life in the British Isles'.[5]

The selection panel for the women's fashions and dress fabrics comprised the Honourable Mrs Reginald Fellowes, President of the Incorporated Society of London Fashion Designers (INCSOC), which represented a small group of London's leading couturiers, five editors of fashion and textile magazines and three retail specialists. With the exception of Albert Stead, the Drapery Manager and Buyer for the Birmingham Cooperative Society and Hans Juda, the publisher and editor of *The Ambassador*, a highly respected export magazine, the panel members were all women.[6]

Given the exhibition's aim to stimulate a domestic market for good modern design by demonstrating its daily benefits to the public, and in turn to encourage manufacturers to make well-designed products that

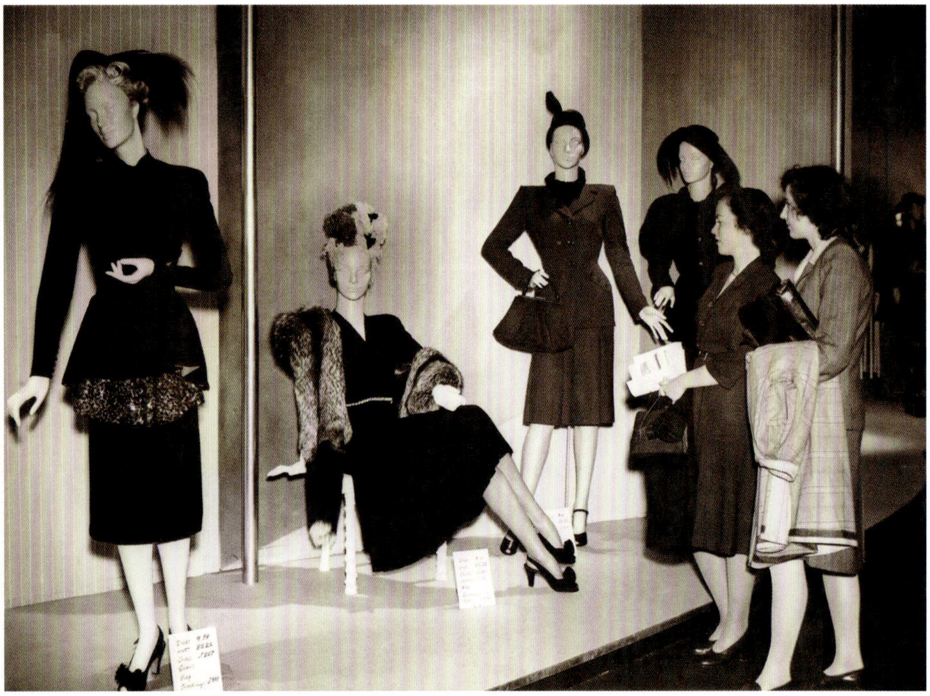

89. Women visitors viewing autumn fashions and accessories in the fashion section. The garments include black day dresses by London Model House Group members Frederick Starke Ltd and Rima (Wholesale) Ltd and a purple suit by Brenner Sports Ltd.

would also sell abroad, it made good sense to draw on the expertise of those whose jobs involved communicating with potential consumers and selecting retail stock.[7] Most of the female panel members already had well-established contacts with government. As one member, Audrey Withers, who was editor of *Vogue* from 1940 to 1960, later remembered, 'editors of magazines specially concerned with fashion were much in demand at the Board of Trade' during the war for their advice on such matters as the Utility scheme.[8] Withers and her fellow panel members, Anne Scott-James, who had recently moved from *Picture Post* to the editorship of *Harper's Bazaar*, and Phyllis Panting (known as Ann Seymour), editor of the middle market but influential monthly magazine *Women and Beauty*, were valued for the influence they could potentially exert on their readers, whose cooperation was vital to the war effort on the home front, and for their advice on women's issues. Withers also wrote a perceptive essay on 'Fashion, Dress Fabrics and Accessories' for *Design '46,* which stressed fashion's intrinsic mutability and the diverse skills the successful fashion designer needed, from understanding the capabilities of fabrics to swiftly recognizing the genesis of a new trend.[9]

Another panel member, Madge Garland, who had been fashion editor of *Vogue* from 1934 to 1940 and was currently the merchandise manager of the London department store Bourne & Hollingsworth, had

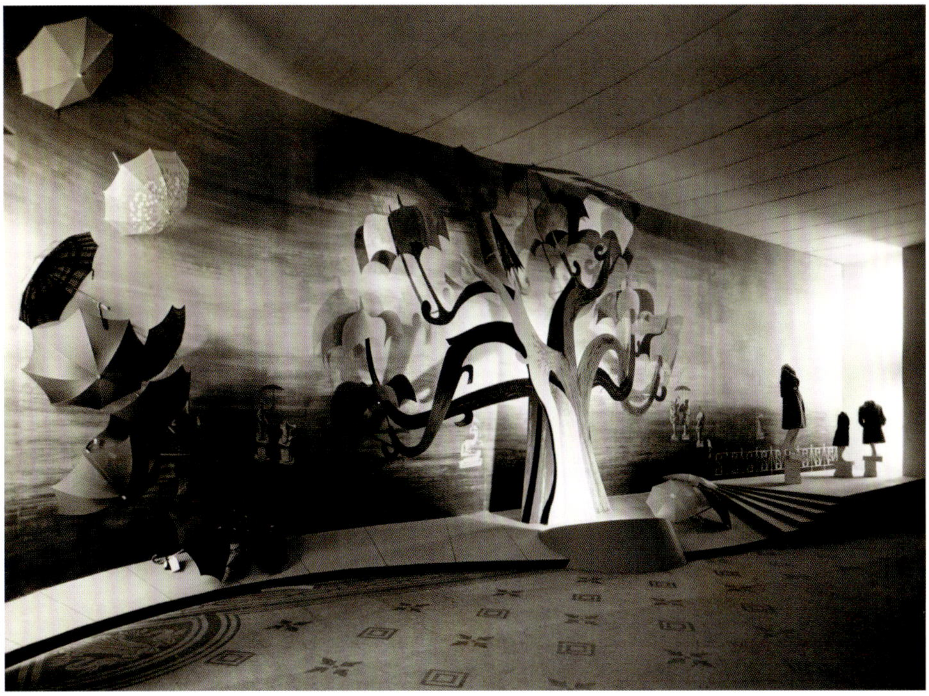

90. Mural backdrop, designed by Jan Lewitt and George Him, for the 'Wet Weather' exhibit

also advised the Board of Trade on war-time clothing regulations and had recently returned from a Board of Trade-sponsored visit to North America to research new uses of synthetic fibres and American methods of manufacturing and marketing ready-to-wear.[10] Garland, Withers, Scott-James and Panting, whose husband was the couturier Digby Morton, had a thorough understanding of the fashion industry in Britain, France and America, which was a key market for Britain's couturiers and ready-to-wear houses. Miss Gardiner, buyer for Liverpool's Bon Marché department store, represented retail in the North of England.[11]

The inclusion of so many editors on the panel guaranteed wide-ranging positive publicity for the exhibition. Hans Juda and his fellow panel member Miss Wray, the fashion editor of *Fashion and Fabrics Overseas* (which replaced *The Drapers Organiser Overseas* in 1946), represented export magazines. Although the primary target of the BCMI exhibition was the British public and British manufacturers, the government's longer-term goal was to capture a greater share of the global export market with better designed, more competitive products. *The Ambassador* flagged up the exhibition ahead of the opening in its August issue. The article, which summarized the content and layout of the exhibition, asserted that it set new standards in industrial design. It emphasized the selection of the objects by 'independent expert juries' and the quality of the displays, which transformed 'the usually sombre interior of London's Victoria & Albert Museum' into 'a spectacular contemporary showroom'.[12]

Fashion and Fabrics Overseas offered its readers a more detailed overview of the exhibition, the fashions on display, and the latest innovations in woollen dress fabrics.[13] The article stated boldly that 'it is the general policy of the Council that the name of the manufacturer and, where possible, the designer, of each article exhibited should be published'. The list of fashion exhibitors, published in the supplement to the main catalogue, includes very few designers.[14] While most of the couture houses were named after their founders, who remained at their head, none of the other houses, for instance Worth or Lachasse, and very few of the ready-to-wear houses credited their designers. Exceptions include Olive O'Neill, who designed Rose & Blairman Ltd's Dorville brand, Greta Newman at Rima (Wholesale) Ltd, and H.B. Porter for Charles Kuperstein Ltd. In the knitwear section Anny Lewinter Ltd of Newtownards, County Down, Northern Ireland, credited Charlotte Lewinter and R. Pringle & Sons Ltd of Hawick in Scotland and cited Otto Weiss (or Weisz), who had designed for them since 1934. Pringle exhibited two twinsets, a popular form of clothing which Weiss is said to have introduced to the fashionable wardrobe. Anny Lewinter was well known for its stylish knitwear and bold colour advertisements.[15]

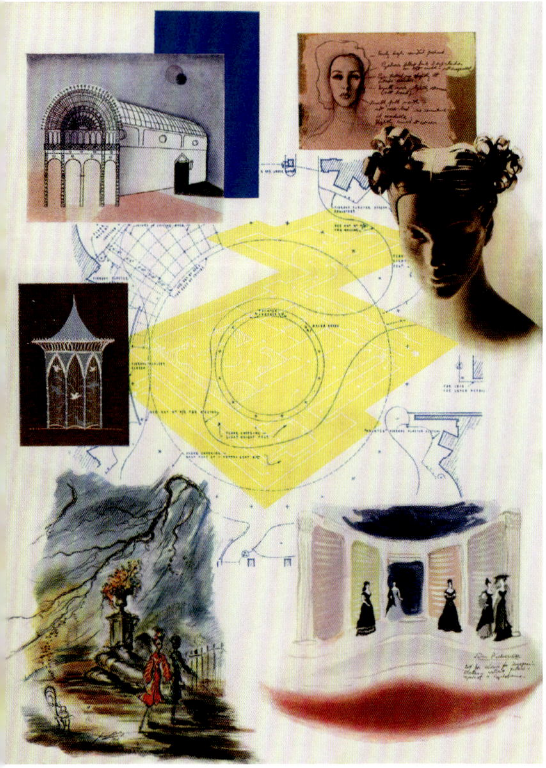

91. Working drawings for the décor of the Fashion Hall imposed on the blueprint of the displays (clockwise from top left): Geoffrey Bennison's working drawing for his Covent Garden arcade mural in the couture section; his design for a mannequin (executed by Pytrams) shown alongside a brown paper wig designed by Steiner, made by Peter Möller; John Richardson's set for inexpensive fashion; James Bailey's Hyde Park scene for the medium-priced garments; the bird cage crowning the couture's revolving stage designed by Richardson and Bailey (*Vogue*, October 1946, p. 60)

Of the three remaining magazines, *Vogue*, *Harper's Bazaar* and *Women and Beauty*, *Vogue* offered the widest coverage.[16] The article, which was illustrated with objects such as toys, ceramics and glass as well as fashion and accessories, was prefaced with an eye-catching colour-printed illustration of key features in the fashion display [**91**]. James Gardner was in overall charge of the design, the most striking element of which was a 25-foot-high revolving stage crowned with a birdcage, presumably for the peacock posed among the couture evening gowns. Gardner later said that the design of the exhibition was experimental, borrowing from 'the theatre, the pantomime and the circus, the aim being to achieve an interesting atmosphere, and to give the visitor some emotional experience, and some "fun"'.[17] The displays had a mixed reception. While some found them glamorous, the Industrial Correspondent of *The Sunday Times* thought the 'strange settings' and low-key lighting demonstrated a lack of 'practical salesmanship'.[18]

The decision to present the outfits on display models caused problems for the team dressing them. A post-exhibition report on the mannequins stated that Pytram's composition *papier maché* figures were the most successful, but a number of modifications were needed to make them more functional [**92**].[19] Praising American mannequins for their sex appeal, the author, Mrs E.M. Blackshaw criticized the composition plaster figures designed by the sculptor and commercial artist Ethelwyn Baker for their 'aggressive and unpleasant' faces and recommended that Pytrams re-model their mannequins' legs 'with more allure'.[20] The wigs, on the other hand, which were created by the London hair stylist Steiner of Grosvenor Street from a variety of traditional and more novel materials (nylon monofilament manufactured by ICI's Plastics Division, fabric, cellophane, aluminium foil, wire wool and real hair), were widely praised for their styling and technical ingenuity.

In June 1946 *Women and Beauty*, in one of three articles it published on the exhibition, featured the Swiss sculptor, ceramicist, painter and designer Raoh Schorr as their 'man of month'.[21] Schorr designed the figures of Pegasus, the flying horse, and Daphne, the goddess who metamorphosed into a tree, to support and display the dress fabrics in the exhibition. The article did not mention the exhibition but, while emphasizing Schorr's reputation as an artist and sculptor, described his varied work as an industrial designer and success across a range of materials in some detail. The article was complemented by another, on how to become an industrial designer – 'a career for the girl whose artistic abilities take a more practical form'.[22] While flagging the CoID's ambition to develop training schemes and recommending London's Central School of Arts and Crafts as the institution that offered the most wide-ranging technical course in industrial design, the author recognized that at the present time most aspiring designers would have to supplement a general course at art school with further classes at a technical school. The Council of Art and Industry's (CAI) report on 'Design and the Designer in the Dress Trade' (dated July 1938 but not published until 1945) had recognized the inadequate training opportunities for dress designers and argued that art schools teaching industrial art should offer courses in industrial dress design. While recommending that the Royal College of Art should provide training for the industrial artist wishing to become a fashion designer, the Dress Committee praised Bromley School of Craft and Industrial Design for integrating teaching on design with practical training.[23] In BCMI the couturier Angele Delanghe featured a design made with muslin and net fabric created by students at Bromley.

Women and Beauty's final article relating to the exhibition was published in August 1946 when James Cleveland Belle was presented as the 'man of the month'.[24] Belle, a member of the 1946 Exhibition Policy Committee, was in overall charge of the selection of the men's, women's and children's clothes and accessories, and the textiles. A former buyer for Liverpool's Bon Marché department store, he was then the Director of the Cotton Board's Colour, Design and Style Centre in Manchester. Belle personally liaised with INCSOC about their participation in the exhibition. The couturiers agreed to take part at their first meeting in late March 1946.[25] The exhibition design and mannequins were discussed in

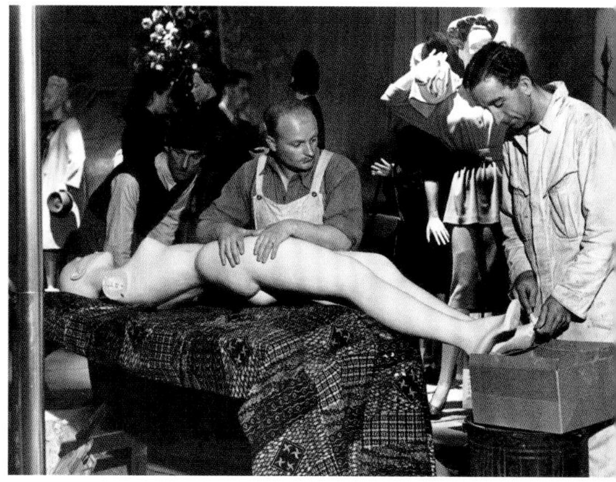

92. Assembling a mannequin for the 'Women's Dress' section

July, when it was agreed that the mannequins would be delivered to the couturiers two to three weeks ahead of the opening, so that the garments could be fitted.[26] The choice of models was left to the individual couturiers, with no involvement from Belle or the panel members, but it was agreed that the group would send Belle a summary of each outfit and, if possible, a 'snippet' of the material. The meeting closed with the couturiers commenting on the extreme difficulty of obtaining fabrics other than wool of any variety or originality. Belle replied he was confident that the wider textile industry would support the group, 'at least for this exhibition'. In the event the models included designs in rayon, nylon and cotton, although many of catalogue entries merely described the weave of the fabric and not its fibre.

If Belle worked with the couturiers it seems likely that Anne Scott-James was involved with the ready-to-wear houses. In October 1946 *Harper's Bazaar* featured a substantial article about the exhibition, which was illustrated with designs by members of the newly formed London Model House Group (LMHG), representing ten of London's leading wholesale model houses [**89**], with a second article on the exhibition wigs.[27] The selection of the fashion exhibits seems to have been necessarily pragmatic. Miss Hawes's post-exhibition report noted that 'the majority of the Fashion sections arranged their own Collecting Centre within the Trade and a member of the exhibition information staff attended the selection and as far as possible took down the necessary information for the catalogue .... However because of the nature of the material it was not possible in all instances to finalize selection until a much later date and the Selection Committee in some instances visited various firms and selected on the spot.'[28] Labour and fabric shortages affected all levels of the fashion industry.

It was left to the general press to highlight the quality of the Utility fashions. Leonora Curry of *Tailor and Cutter & Women's Wear* felt that they 'revealed a cleaner cut knowledge of design, and stronger colour sense' than many of the other garments, whose design and quality were 'not worthy of the display efforts and are not truly representative of the best we can do'.[29] This assessment of the high standard of the best Utility fashions was widespread. In December Alison Settle commented, 'Utility [clothes] ... made by leading fashion houses improve with each collection: I am no longer able to pick out which are Utility and which "non-U" models. Materials, workmanship, details are almost the same.'[30]

Even to contemplate an exhibition of the size and complexity of BCMI in 1945 was bold, if not foolhardy. However, the exhibition was generally acknowledged to have been a success. An editorial in *The Tailor and Cutter* two weeks into the exhibition's run effectively conjures up the mixed reactions it provoked. Acknowledging that the ideas behind the presentation were good, the show was described as a 'triumph of disorganisation', 'totter[ing] rather late and slightly chaotically' into the V&A:

> But the Exhibition is among us and, with that typical British success at Muddling Through – the result is one in which the manufacturers (if not the organizers) can take some pride. Anyway the Exhibition has proved that Britain can make it. Now let's see if we can prove that Britain can make lots of it – enough to allow some of the fancy products to our own folk.'[31]

NOTES

1 Audrey Withers, 'Fashion, Dress Fabrics and Accessories', *Design '46* (London 1946), p. 48.
2 Paddy Maguire, 'Designs on Reconstruction: British Business, Market Structures and the Role of Design in Post-War Recovery', *Journal of Design History* (1991), vol. 4, no. 1, p. 18; http://www.jstor.org/stable/1315862; accessed 2 January 2015.
3 'Conversion to Peace Production', advertisement for Debenham & Company, 91 Wimpole Street, London W1, *The Ambassador* (1946), no. 9, p. 22.
4 Alison Settle, 'From a Woman's Viewpoint', *The Observer* (6 October 1946), p. 7. The corset trade attributed the decision not to display corsets to male prudery. Settle emphasized the raft of difficulties faced by the trade due to war-time regulations and the particularly severe restrictions on the amount of non-Utility fabric available to it.
5 'Britain can make it', *The Ambassador* (1946), no. 8, pp. 81–82.
6 The CoID had only two women members in 1946: Margaret Allen, a prominent member of the Co-operative Movement, and Mary Harris (d. 1990) of Mary Harris Gowns, Gateshead. For an assessment of the impact of the narrow social make-up of design reform organizations in the twentieth century see Jonathan M. Woodham, 'Managing British Design Reform I: Fresh perspectives on the Early Years of the Council of Industrial Design', *Journal of Design History* (1996), vol. 9, no. 1, p. 56; http://www.jstor.org/stable/3527213; accessed 2 January 2015.
7 V&A Archive, Exhibition File: AAD/1977/4/12, Britain can make it: 'The Council of Industrial Design Summer Exhibition 1946: points to be borne in mind in connection with the organization of the Exhibition', 24 September 1945.
8 Audrey Withers, *Lifespan* (London 1994), p. 51.
9 Withers 1946, pp. 45–48. *Design '46* went to press before the garments exhibited in BCMI could be photographed. The photographs of the fashion models in the publication were supplied by the *Vogue Book of British Exports*. While representing current fashions they were not 'necessarily those which were selected for the exhibition'.
10 Lisa Cohen, *All We Know: Three Lives* (New York 2012), pp. 297–98.
11 North America (including Canada) and the Dominions were the key British markets. No attempt was made to sell in France but as French fashion led the world an understanding of current trends in Paris was essential for designers.
12 *The Ambassador* (cited note 3).
13 'Design for a Nation's Shop Window', *Fashion and Fabrics Overseas* (July–August 1946), pp. 94–96.
14 *Britain Can Make It Catalogue* (London 1946), pp. 182–87.
15 See for instance, *Fashion and Fabrics Overseas* (September–October 1946), p. 19.
16 'Britain Can Make It', *Vogue* (October 1946), pp. 60–67.
17 James Gardner, 'Exhibition Display', *Journal of the Royal Society of Art* (30 January 1948), vol. 96, no. 4761, p. 164; http://www.jstor.org/stable/41363544; accessed 2 January 2015.
18 '"Britain Can Make it" Some Criticisms', *The Sunday Times* (29 September 1946), p. 8.
19 The display mannequins made by Pytram Ltd were used for the couture garments on the revolving stage and in the Covent Garden opera house setting.
20 V&A Archive, Exhibition File: AAD/1977/4/14A, Britain can make it: Mrs E.M. Blackshaw to Miss Disher, Notes on Display Models Used in the B.C.M.I. Exhibition, 30 December 1946. This report covers all the mannequin types used in the exhibition and includes recommendations for improvements to their functionality and appearance.
21 Gordon Beckles, 'Raoh Schorr', *Women and Beauty* (June 1946), pp. 72–73, 114–15, 117; the front cover of the June issue features a colour image of a head modelled by Schorr, dressed with a wreath of flowers.
22 'How to become an industrial designer', *Women and Beauty* (June 1946), pp. 52–55, 109.
23 University of Brighton Design Archives, Alison Settle Archive B 30S 14: notes on the 'Report of the Dress Committee of the Council for Art and Industry'.
24 Gordon Beckles, 'James Cleveland Belle', *Women and Beauty* (August 1946), pp. 64–67.
25 V&A Archive, Exhibition File: AAD/2011/14/1: Minutes of a meeting held on 28 March 1946, Minute book of the Incorporated Society of London Fashion Designers (1942–1949).
26 V&A Archive, Exhibition File: AAD/2011/14/1: Minutes of a meeting held on 11 July 1946, Minute book of the Incorporated Society of London Fashion Designers (1942–1949).
27 'Britain Can Make It', *Harper's Bazaar* (October 1946), pp. 52, 90; 53–57. The members of LMHG included Brenner Sports, Dorville, Jersey Company, Koupy Models, W.&O. Marcus, Simon Massey, Rima, Silhouette de Luxe, Spectator Sports and Frederick Starke. Many of these names were brands: for instance Charles Kuperstein Ltd designed and manufactured Koupy Models.
28 V&A Archive, Exhibition File: AAD/1977/4/14A, Britain can make it: Miss Hawes, Report on the assembly and selection of the fashion exhibits.
29 'Leonora Curry's Observations on the "Britain Can Make It" exhibition', *Tailor and Cutter & Women's Wear* (4 October 1946), pp. 628–29, 637.
30 Settle 1946, p. 7.
31 'The Exhibition', *Tailor and Cutter & Women's Wear* (4 October 1946), p. 625.

DIANE BILBEY

Time for Tea

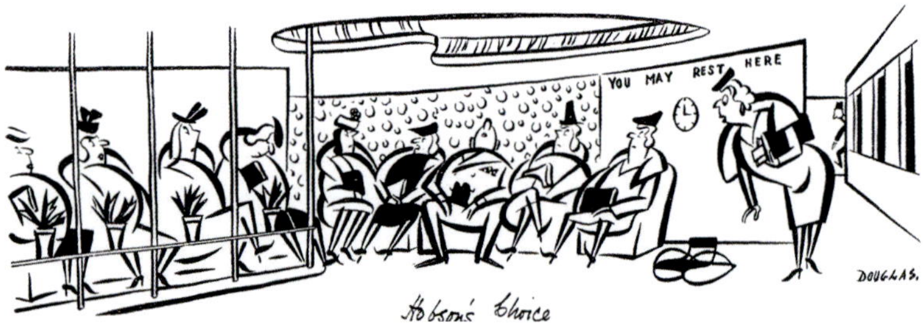

The need for rest breaks and refreshments for the vast numbers of visitors making their way around *Britain Can Make It* (BCMI) was satisfied by a series of 'rest lounges' situated at regular intervals throughout the exhibition, a dedicated 'Tea Lounge', and the Museum's own large, if utilitarian restaurant (or Refreshment Room), then located in the South Court.

A correspondent from *Punch* reported on the shared sense of weariness felt by visitors walking round the exhibition, stating that

> It was hard work, and I yearned for a seat, but the oases graciously labelled 'You May Rest Here' were always full of people taking the hint. The empty chairs were either those cute little tubular ones for children, who as any parent knows, do not sit down a great deal at exhibitions, or bigger ones marked 'For Staff Only'.[1]

The rest lounges – designed by the architect F.R.S. Yorke – were a welcome respite for visitors. Significantly, they included examples of furniture displayed in the exhibition so that visitors could try out the latest designs for themselves.[2] Wire lattice garden chairs, developed and designed by Leonard Thoday and Arthur Greenwood using techniques pioneered during the war and manufactured by Heal's Ltd, added a contemporary touch [93, 94].

93. Cartoon by Thomas Douglas from *Punch* (published 30 October 1946) entitled 'Hobson's Choice', making light of the modernistic, tubular furniture manufactured by Heal's Ltd available for the public to use in the Rest Lounges of the exhibition

The rest areas, however, proved insufficient for the unprecedented (and unexpected) number of visitors attending the exhibition. A report written by the Sales Manager of The General Electric Co. Ltd sent to

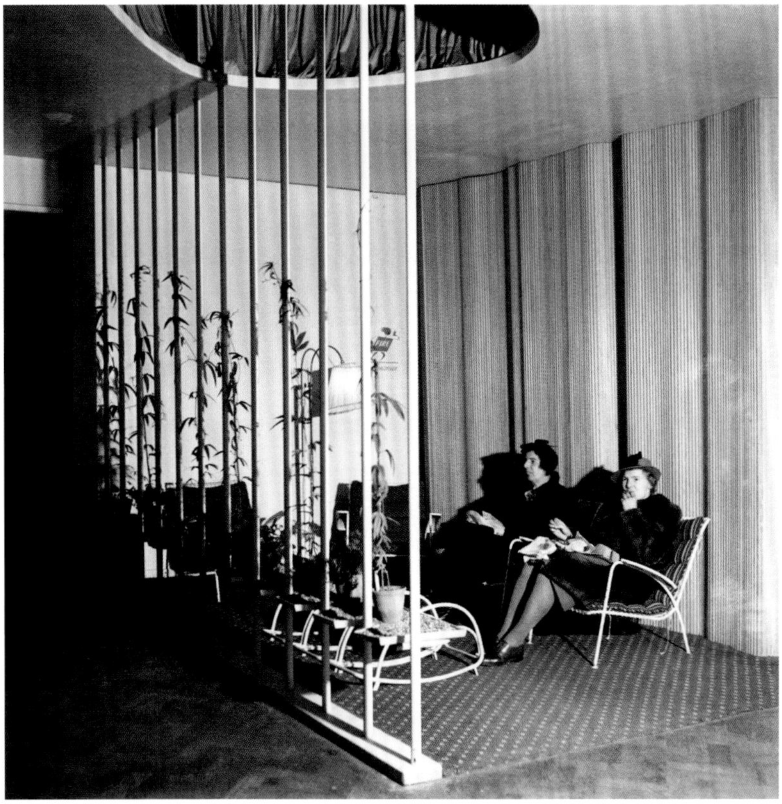

94. Visitors take a break in one of the rest lounges and try out the Leonard Thoday and Arthur Greenwood chairs designed for Heal's Ltd

the Council of Industrial Design (CoID) following his visit to BCMI highlighted the problem:

> The amenities do not appear to be very complete as it would seem that there is not enough restaurant space, nor on Monday was there apparently sufficient chairs and other rest arrangements available. It took me two hours to walk round without sitting down at all and the average person will want at least one rest in this time.[3]

Mecca Cafés

Austerity still being very much the order of the day, the catering facilities, run as a concession for the duration of the exhibition by Mecca Cafés, offered a modest, limited menu. Appropriately, the crockery used both in the main Museum restaurant and in the Tea Lounge was from the white 'Utility' range purchased direct by Mecca Cafés from the British pottery manufacturer Moorcroft. The Museum's Director, Leigh Ashton, wrote to S.C. Leslie (Director of the CoID) confirming the catering arrangements. Jovially touching on the Council's ethos of good design, Ashton commented that he felt Mecca would do an adequate job, 'though I fear

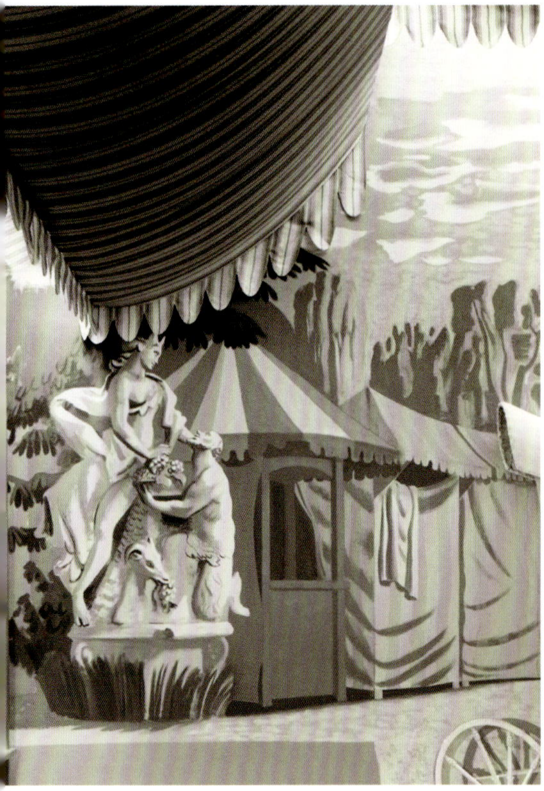

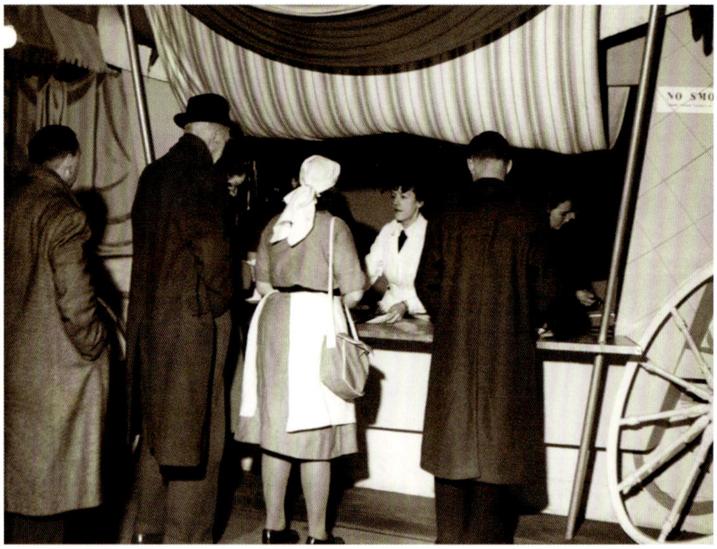

95. Part of the mural by Hans Feibusch in the Tea Lounge

96. The Tea Bar in the Tea Lounge, supposed to represent a gypsy caravan

that the shape of the cups and saucers will probably be not quite up to standard!'[4]

The menu of the Museum restaurant similarly reflected the austere post-war times of food rationing and shortages: it included paste, cheese and liver sausage sandwiches (6d, 8d and 1s respectively), baked beans on toast (8d) and macaroni cheese (9d). One visitor to the exhibition, a Mrs Thomson of Barnes, felt compelled to write to the Exhibition Director to report on the poor levels of service and unacceptable lack of cleanliness in the main Museum restaurant, and on the negative impact it must have had on foreign visitors when, she felt, Britain should be trying to impress:

> The service was 'help yourself' but the assistants behind the counter were wearing dirty overalls and one was observed through the half open service door smoking a cigarette .... What the effect of this is on foreign visitors I can imagine, it hardly enhances the exhibition. Britain may be able to make it but apparently keeping clean is beyond the management.[5]

The Tea Lounge

In stark contrast to the austere and shabby Museum restaurant, with its heavy, dark, cane furniture, the cheerful, carnival-themed Tea Lounge, with its light, sleek, modern tables and chairs must have seemed dream-like [95–97]. Designed by James Gardner, and entered from the area displaying dress fabrics and children's wear, this was an oasis of fun, with a pavilion-like structure, striped awnings, chequer-board floor and brightly painted murals on the four walls, forming the boundaries of

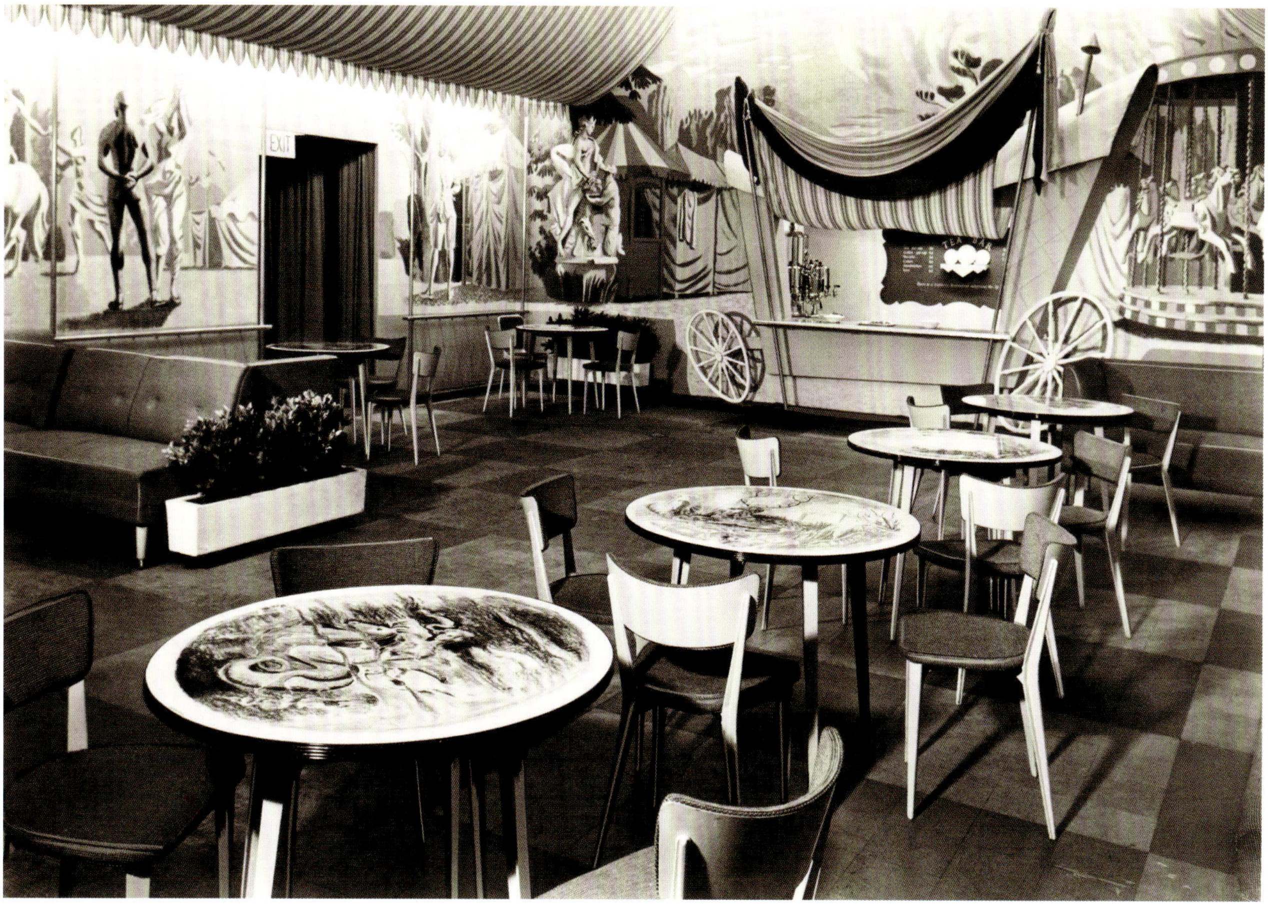

the ‘encampment’. An open-fronted gipsy caravan acted as the service counter for the Tea Bar, and cherry-coloured fabrics on the chairs and settees, added to the holiday feel. The theme for the murals – a circus and fairground, with tents, animals and Commedia dell’arte figures – was imaginatively designed and painted by the German émigré artist Hans Feibusch [95, 97–99].[6]

Feibusch was an accomplished mural artist and in 1946 wrote a treatise on the techniques, history and tradition of mural painting.[7] The decision to commission Feibusch illustrates the imagination of the organizers in using established contemporary artists, not just to decorate but to enhance the exhibition space, and visitor experience, with art. The architect Clough Williams-Ellis was one of Feibusch’s patrons, commissioning murals and metal cut-outs from him for Portmeirion. In praise of the convivial effect produced by the Tea Lounge murals, Williams-Ellis commented that ‘It was clear that tea and buns alone could not account for the prevailing atmosphere of gay relaxation’.[8]

97. The Tea Lounge, designed by James Gardner with murals by Hans Feibusch and table tops designed by Phyllis Bray

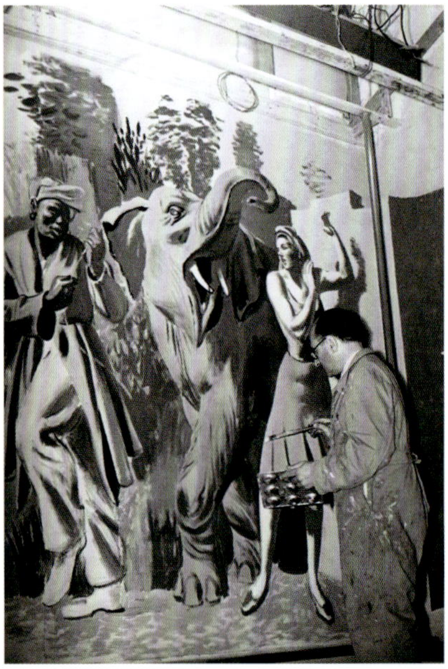

98. Hans Feibusch at work on one of the mural designs in the Tea Lounge

99. Hans Feibusch, Preliminary sketch for part of the Tea Lounge mural, showing the Tea Bar in the form of a gypsy caravan, 1946; gouache on card, 17.9 x 38.8 cm. V&A: E.138 & 139–1994

A series of twelve gouache sketches for the Tea Lounge murals were acquired by the V&A from the artist in 1994 [99]. Feibusch had written to the Museum the previous year to offer

> the small, but complete sketches for the friezes of paintings round the entrance hall of your museum and which disappeared after the closing of the 1946 'Britain can make it' Exhibition for which they had been commissioned .... I wondered whether you might care to see these designs which were part of the first exhibition held in London after the war, and, if you liked them, acquire them perhaps for your collections.[9]

Feibusch was quite right, the murals decorating the Tea Lounge, together with other murals commissioned by the CoID for the exhibition, were removed to the CoID's store for 'disposal' at the close of the exhibition.[10]

Fitting it out

Eight Warerite topped tables decorated with designs by Phyllis Bray depicting mythical beasts, unicorns, stags and horses complemented the murals and added to the theatricality of the Tea Lounge. Bray, a successful artist and member of the East London Group of painters, assisted Hans Feibusch for some 40 years [97, 100].[11]

Warerite Ltd, from Ware in England (part of the Bakelite group of companies), pioneered the production of a layered plastic laminate, a modern synthetic replacement for traditional wood veneers. Bray's highly coloured designs were printed on to fabric and sealed into the top layer of the laminate.[12] According to James Gardner, these tabletops were the first occasion when a design had been incorporated in this way.[13] Hardwearing, heat resistant, hygienic, and available in a range of colours

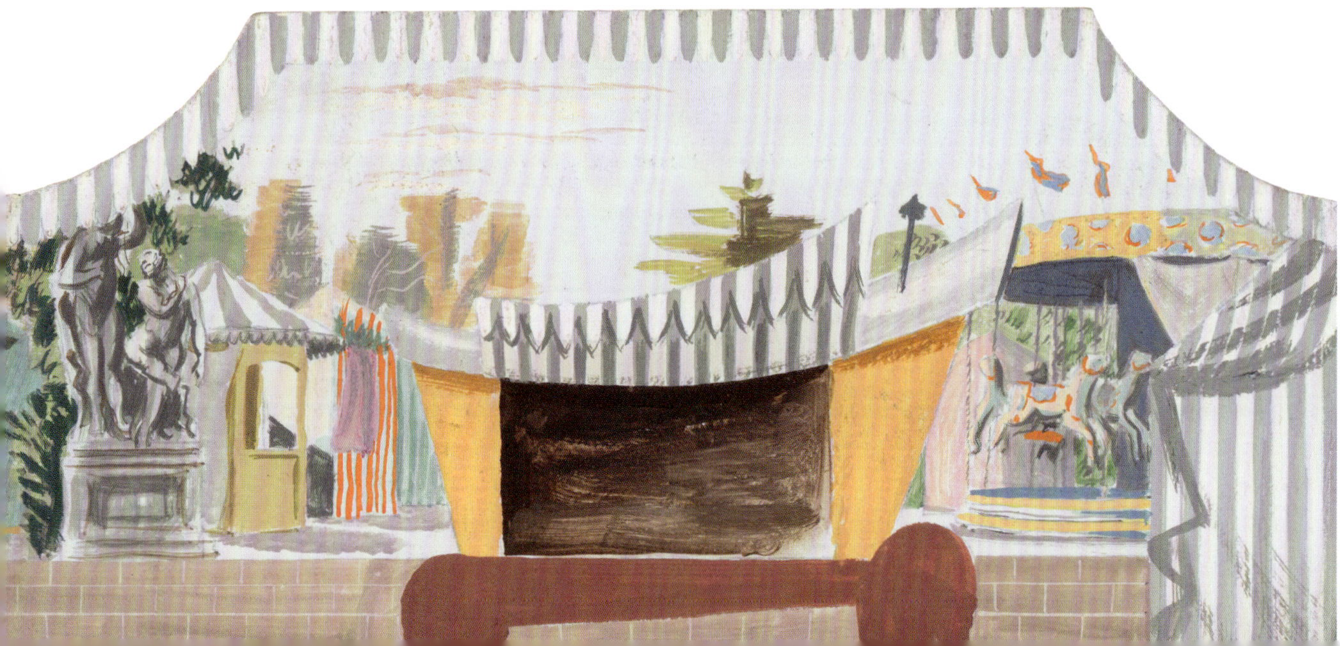

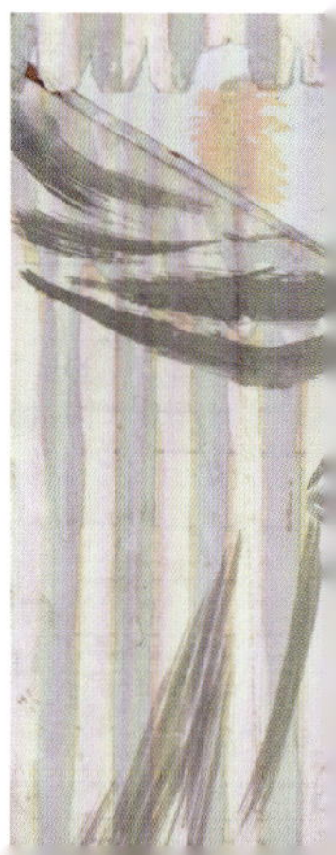

and patterns, Warerite was particularly suitable for use in tea rooms, shops, hospitals, factories, bars, ships, homes and trains.[14]

In production from the 1930s, laminates became increasingly popular and mainstream for use in the home in the 1950s, as well as being used in the interior fittings of ocean liners and 'planes.[15] Warerite Ltd also contributed to various other sections of the exhibition, including plastic sheeting on walls and tabletops in several of the 'Furnished Rooms' and on bedroom furniture. In designs by Patricia (Pat) Leith, it was used in the decorative panels designed by Peter Bicknell, for the nursery room.[16]

The decorative Warerite tabletops were fitted on to circular cast aluminium framed tables produced by Ernest Race Furniture Ltd. These were paired with cast aluminium BA 3 chairs upholstered in cherry-coloured tapestry fabric [**102**]. The chairs, also designed by Ernest Race, were prototypes, and significantly had their first showing at BCMI; they were used not only in the Tea Lounge but also in the bed-sitting room in a London apartment block designed by T.A.L. Belton for an imaginary young male BBC sports commentator [**101**]. A suitable pairing of young, middle-class professional with cutting-edge design,[17] Race's chairs were innovatory, not only in their form, but also in the methods used to manufacture them, namely the adaptation of wartime processes and advances made in the use of sand-cast aluminium.[18]

To complete the fitting out, settees around the walls of the Tea Lounge were designed by another prominent furniture manufacturer of the time, Howard Keith of H.K. Furniture Ltd. These 'unit settees' had feet in birch and aluminium, and were also covered in a cherry-coloured tapestry fabric, to match the upholstery of the Race chairs.[19]

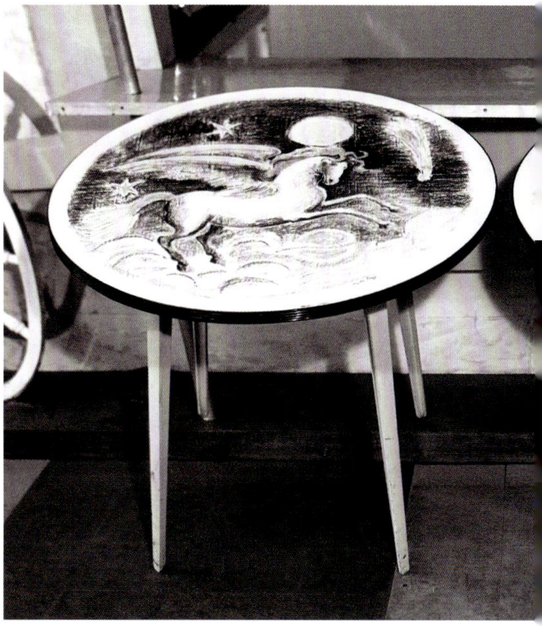

100. Phyllis Bray designed table top made by Ernest Race for the Tea Lounge

Customer complaints

Despite the cheerful appearance of the Tea Lounge it seems that the actual catering arrangements left a lot to be desired. An employee of the Public Health Laboratory Service, who had come as a visitor to BCMI (not in an official capacity), sent a letter to the CoID criticizing the

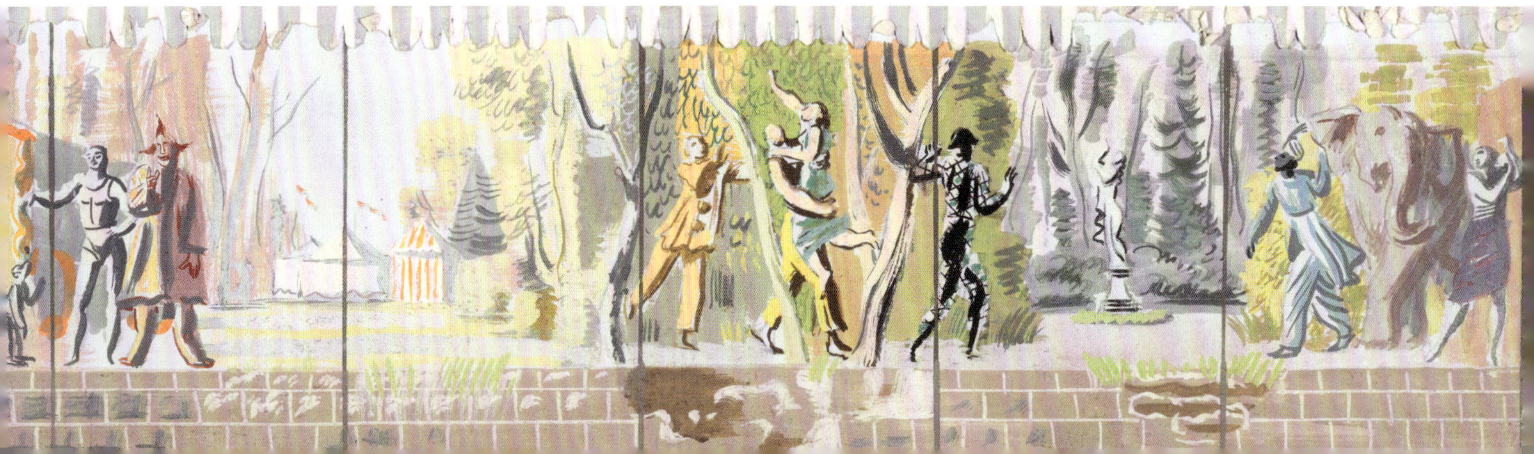

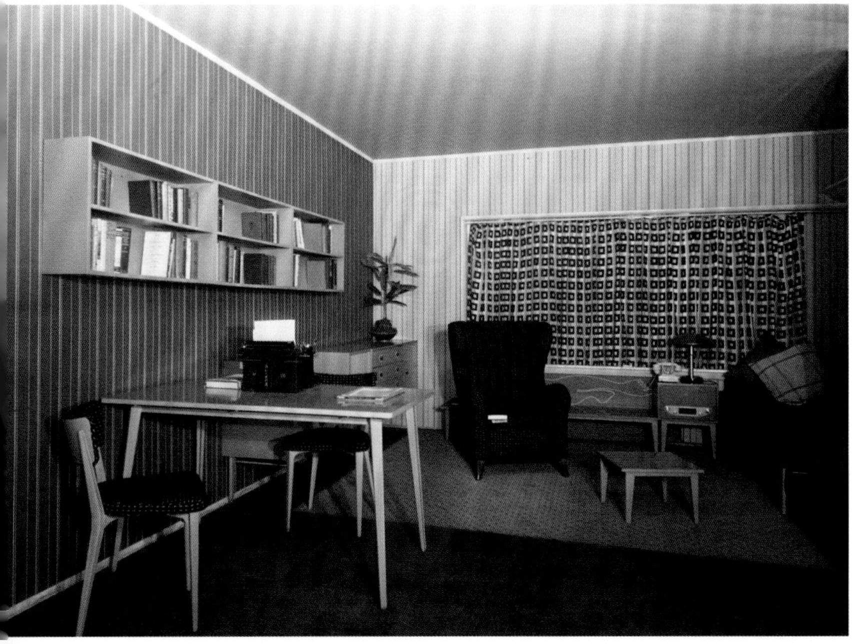

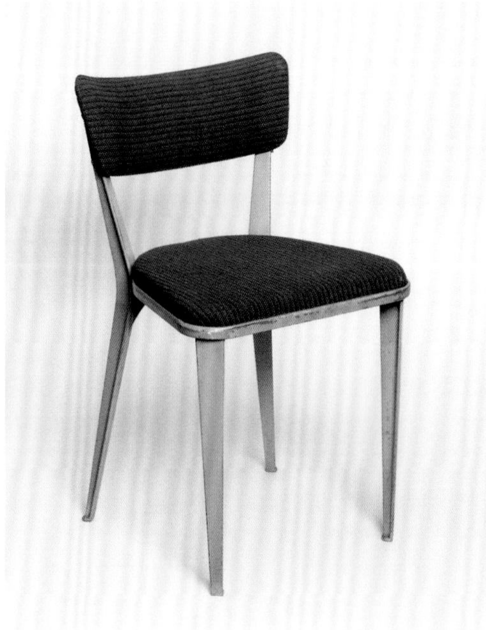

101. Bed-sitting room in a London apartment block, designed by T.A.L. Belton for the 'Furnished Rooms' section at BCMI, including an Ernest Race BA 3 table and chairs

102. Ernest Race BA 3 chair, c.1946, cast aluminium, plywood and upholstery, 77.5 x 44 x 49.5 cm. V&A: W.3–2010

standards of hygiene. Whilst apologetic in having to complain 'amid an exceedingly well arranged exhibition', he felt compelled to write:

> I refer to the service in the tea-bar. I am prepared to believe that there were good reasons for the limited accommodation available, but there can, surely, be no good reasons for the waitresses to be clad in dingy overalls and to offend all the canons, not only of food hygiene, but also of aesthetics, by pulling open each sandwich with their fingers to investigate its contents, before serving it. Such behaviour is unpleasant enough anywhere, but in an exhibition such as this it seem to me personally to be no less than scandalous.[20]

It is of course no surprise, given the context of post-war austerity, that the refreshments on offer at BCMI were basic and limited. However, with such creativity and effort put into making the catering and rest areas attractive, innovative and appealing to look at, it would be fair to suggest that – like some other aspects of the exhibition – they promised rather more than they delivered.

NOTES

1 John Basil Boothroyd, 'Britain Can Make It. A Late Look Round', *Punch* (30 October 1946), vol. CCXI, no. 5523, p. 375.

2 Gordon Russell, Council of Industrial Design, 'Furniture', *Design '46. Survey of British Industrial Design as displayed at the "Britain Can Make It" Exhibition* (London 1946), p. 98 illus; Lesley Jackson, *Modern British Furniture Design Since 1945* (London 2013), p. 60.

3 University of Brighton Design Archives: Design Council Archive (DCA) 911: "BCMI" Letters containing Criticism, Complaints and Suggestions. GEC's Sales Manager comments on BCMI; cited in P.J. Maguire and J. Woodham (eds.), *Design and Cultural Politics in Postwar Britain. The Britain Can Make It Exhibition of 1946* (London and Washington 1997), pp. 193–94.

4 V&A Archive, Exhibition File: AAD/1977/4/13, Britain can make it: Britain Can Make It Catering 3.3.47. Moorcroft's was also selected by the Council to exhibit in the Pottery section, which included ivory Utility pottery: *Britain Can Make It Exhibition Catalogue* (London 1946), p. 95, cat. nos. 570–76, also 577–83 – blue porcelain. See also '"Britain Can Make It" Exhibition', *Pottery Gazette and Glass Trade Review* (September 1946), vol. LXXI, no. 83, p. 598. V&A Archive, MA/28/75/1, Letter from Leigh Ashton to S.C. Leslie dated 1 August 1946.

5 DCA/911: "BCMI" Letters containing Criticism, Complaints and Suggestions. Letter from Mrs Thomson to Exhibition Director (R. Dudley Ryder), 25 October 1946.

6 It appears that Robert Goodden was originally asked to design a scheme for a 'garden restaurant' within the exhibition: DCA/58/Action points arising out of the third design and policy progress meeting held on Thursday 23rd May. David Coke (ed.), *Hans Feibusch. The Heat of Vision* (London 1995), pp. 8, 24, no. M5 (the murals are wrongly noted as produced in 1948).

7 For further information on Hans Feibusch see Eric Newton, 'Hans Feibusch – Mural Painter', *The Studio* (March 1949), vol. CXXXVII, no. 672, pp. 74–79; the murals are illustrated on pp. 76–77; Coke (ed.) 1995 (cited note 6).

8 Alan Powers, 'An Approach through History', in Paul Foster (ed.), *Feibusch murals: Chichester and beyond: an exploration of approach* (Chichester 1997), p. 20. For Feibusch's murals at Portmeirion, see Jan Morris et. al., *Portmeirion* (Woodbridge 2006), pp. 221, 225, 231. Feibusch was paid £315 (plus £15 for paint) for the Tea Lounge murals: V&A Archive, Exhibition File: AAD/1977/4/13, "Britain Can Make it" Exhibition. Fees of designers, display experts and mural artists, 4 February 1947.

9 The V&A paid Feibusch £220 each for five of the gouache sketches on card (a total of £1,100), with Feibusch giving the remaining six sketches so that the set could be kept together. V&A Archive, 94/18 for the Feibusch sketches; Museum nos. E.137–1994 to E.148–1994.

10 Phyllis Bray's decorated tables were also returned to the 'manufacturer'. This is likely to have been Ernest Race (who produced the table itself) rather than Warerite (who produced the tops). Together with the Feibusch murals, they are included in the list detailing items for removal at the end of the exhibition in V&A Archive, Exhibition file: AAD/1977/4/14, Britain can make it: Britain Can Make It Exhibition Dismantling. The exhibition catalogue records that a design by Henry Moore, produced by Ascher, was used for the tabletops, suggesting there was a change of plan at some stage and Bray's designs were used instead: *Britain Can Make It Exhibition Catalogue* (cited note 4), p. 46, cat.no. 111. See G. Rayner, R. Chamberlain, A. Stapleton, *Jacqueline Groag* (Woodbridge 2009), p. 66.

11 V&A Archive, Exhibition File: Britain can make it: AAD/1977/4/13, 'Britain Can Make It' Exhibition fees of designers, display experts and mural artists, 4 February 1947. Bray received a fee of £30 5s for her table-top decorations. She produced the three large murals for the People's Palace in the Mile End Road. For biographical info on Phyllis Bray see Foster (ed.) 1997 (cited note 8), 'Editor's appendix. Feibusch at work: Forms of collaboration', pp. 66–81. See also David Buckman, *Artists in Britain since 1945. Vol. 1, A to L* (Bristol 2006), p. 186; David Buckman, *From Bow to Biennale. Artists of the East London Group* (London 2012), pp. 203–06.

12 The table tops were featured along with other Warerite products from the exhibition in *British Plastics* (November 1946), vol. 18, no. 210, p. 501.

13 James Gardner, *The ARTful Designer. Ideas off the drawing board by James Gardner* (London 1993), p. 136.

14 *Warerite Plastics and You* (promotional pamphlet, copy in British Library).

15 Regarding the use of Warerite, in 1938 Warerite was used as wall panelling for a waiting room at Bakelite's HQ in Grosvenor Place, London (RIBA 72571). Warerite came into its own in the 1950s and was used in kitchen counters. Panels could also be bought to update furniture. It was particularly suitable for aircraft interior panelling as it was lightweight and fire resistant; see *Flight International* (23 January 1964), p. 137. Warerite was also used in ocean liners for wall murals (for example on SS *Chusan* and SS *Oriana* – RIBA 63454 and RIBA 52968 respectively). It was also used to fit out the interior of the Trianon Pullman refreshment car of the Golden Arrow in 1946 (RIBA 77634).

16 *Britain Can Make It Exhibition Catalogue* (cited note 4), p. 16, no. 60. Patricia (Pat) Leith also produced the toys mural in the Children's Section.

17 *Britain Can Make It Exhibition Catalogue* (cited note 4), p. 139, no. 560, 'Two dining chairs' by Ernest Race Ltd, included in the bed-sitting room in a London block of flats whose imaginary occupant was a 'Single man, sportsman and sports commentator at Broadcasting House'.

18 See Geoffrey Rayner, 'Furniture Design', *Austerity to Affluence. British art & design 1945–62* (London 1997), p. 16, F1; also Jackson 2013 (cited note 2), pp. 101–02. See Hazel Conway, *Ernest Race* (London 1982). Conway explains that a visitor to BCMI suggested adaptations to the methods used to manufacture the chairs: 'E.C. Lewin, Managing Director of a foundry in Kettering, saw the chair at the 'Britain Can Make It' exhibition ... and he suggested that he could pressure die-cast the components .... In the die-cast version the legs were stronger and more slender, and because die-casting is more accurate than sand casting the finishing was easier' (p. 27). For a discussion of the BA chair see also pp. 21–32; Jackson 2013 (cited note 2), p. 101, 102, pl. 107. For the Race items in the Tea Lounge, see *Britain Can Make It Exhibition Catalogue* (cited note 4), p. 46, nos. 110 and 111. Ernest Race furniture, including BA 3 chairs, were also used to furnish the Tea Centre cafe in Lower Regent Street, London, when it opened in 1946. For the Tea Centre see C.F.O. Clarke, 'The Tea Centre in London', *Graphis* (1947), vol. 3, no. 19, pp. 240–43; 'Tea in the air', *Display* (1947), pp. 22–23; Edmund Maxim, 'The Tea Centre', *Art & Industry* (January 1947), vol. 42, no. 247, pp. 14–17; Misha Black (ed.), *Exhibition Design* (London 1951), pp. 68–69.

19 Lesley Jackson notes that Howard Keith and HK Furniture is a furniture manufacturer largely forgotten today. For more on HK Furniture see Jackson (cited note 2), pp. 75–79. *Britain Can Make It Exhibition Catalogue* (cited note 4), p. 46, no. 112.

20 DCA/911/"BCMI" Letters containing Criticism, Complaints and Suggestions/Letter from R.E.O. Williams to The Exhibition Organizer, 24 October 1946.

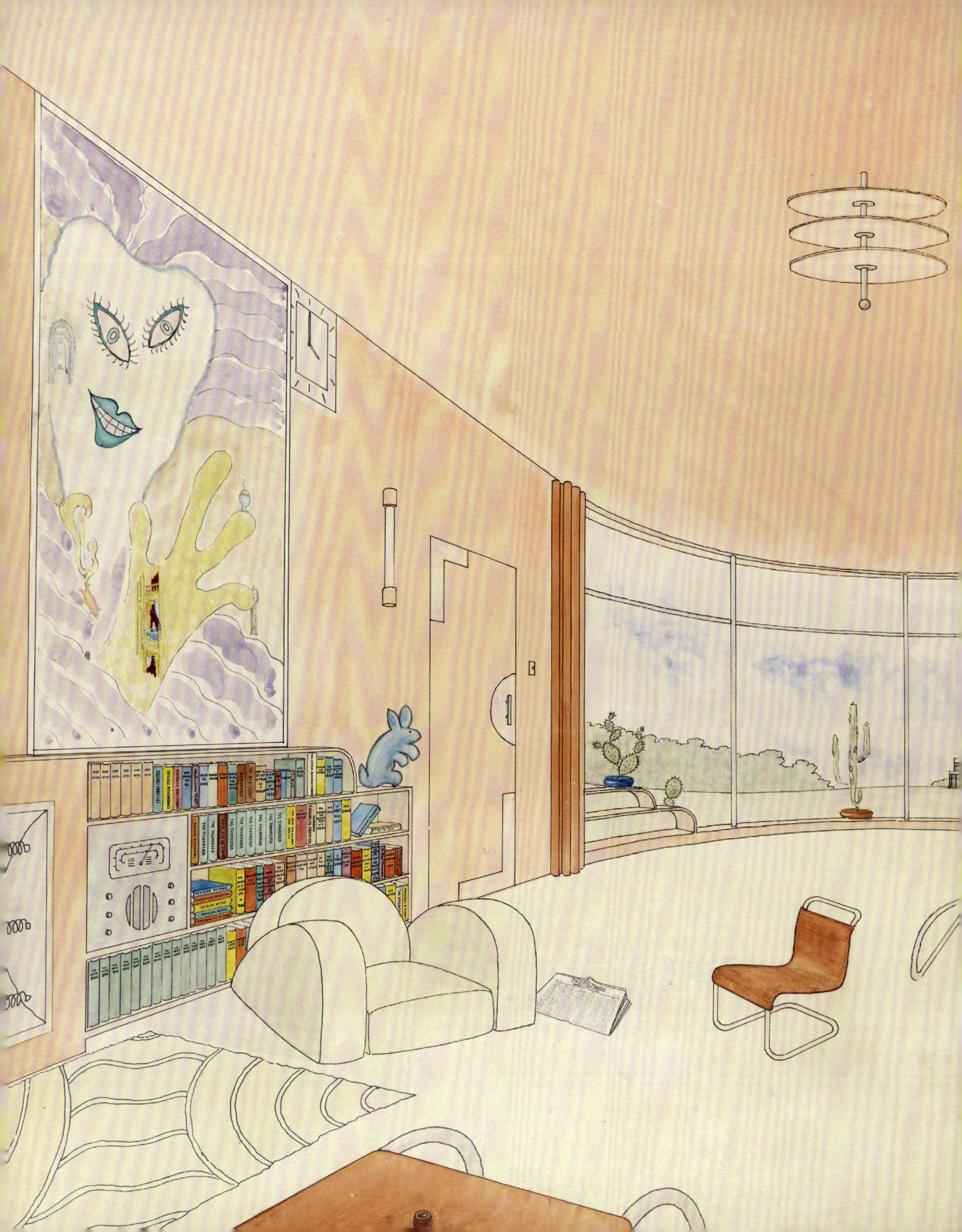

DIANE BILBEY

Introductory Story to the Furnished Rooms

> The look of people's rooms expresses the way they live.
> What would you do with four walls, a floor and a ceiling?
> The style of your home will be studied by future generations.[1]

The architect and writer Peter Fleetwood-Hesketh was commissioned by the Council of Industrial Design (CoID) to produce five illustrations depicting a history of domestic interiors and furnishings, to be shown on introductory panels in the foyer to the 'Furnished Rooms' section. The resulting coloured illustrations, representing the period from 1837 (the year Queen Victoria succeeded to the throne) to 1937, aimed to provide a potted history of interior design, and a context in which to view the furnished rooms of the present that visitors were about to see.

The decades chosen by Fleetwood-Hesketh marked milestones in the development of domestic interiors, and his room illustrations [**103**] were each accompanied by a short explanatory text, written in collaboration by Miles Tomalin (the CoID's chief copy-writer). Small line sketches for each of the five periods, depicting a traditional family group (mother, father, son and daughter) were also drawn by Fleetwood-Hesketh for the panels, 'so that people who are not too knowledgeable on architectural period might immediately date the scene by the clothes worn'.[2] The descriptions, in the form of short, punchy, sentences, summed up each period:

> **1837** The shops are full of new things to buy. Rooms are becoming more crowded.
>
> **1877** Social life is dominated by strong moral and religious views. This has had an effect on architecture, which must now conform to Christian principles.
>
> **1907** The moral severity of the Victorian age is relaxed. Pagan styles are revived, but what architecture has lost in godliness it gains in cleanliness.
>
> **1927** 'Georgian' is now a convention, no longer the monopoly of fashionable architects but part of every enterprising contractor's stock-in-trade.

103. Peter Fleetwood-Hesketh's illustration of a 1937 room – part of the 'Introductory Story' to the 'Furnished Rooms' section at BCMI

> **1937** The dawn of a new style. The death of decoration, which has all been streamlined away .... Houses are bare save for chromium plate and cacti. 'Functionalism' (fitness for purpose) is the watchword. Enthusiasts speak of something new and vital. Old-timers are shocked. Imitators and fakers debase the new style as soon as it is born.

In July 1946, James Gardner (who was responsible for designing the 'Introductory Story' to the 'Furnished Rooms') wrote to Fleetwood-Hesketh explaining the job. In his letter, a potential collaboration with John Betjeman was discussed, as well as their modest fee: 'I realise that this is not a high fee compared with commercial standards, but this is a Designers Exhibition and I hope you will join with the rest of designers in accepting a fee of this nature for the prestige of British Designers and Design'. Gardner went on to explain what he wanted to achieve: 'I suggest that we show a sequence of interiors, each of which may look very "old fashioned" to general public, but each will be of good design and good fun for this particular period.'[3]

The Betjeman connection

At the same time, Gardner also wrote to John Betjeman, who had been asked to provide the accompanying text to Fleetwood-Hesketh's illustrations. In his letter, Gardner explained the purpose of the display and the audience it was to appeal to:

> ... this section is intended to introduce the public to a series of contemporary domestic interiors. So much is taken for granted, and I wanted by this simple display to interest people in what has gone before, and put to them by indication, the problem of what they are to select for their interiors today. I think the commentary and drawings could be slightly whimsical, but the whimsies must be tough and to the point, but of course that is the sort of thing you are so good at. I don't think I need explain this any further, except that we must avoid being what is called "precious" as we are teaching people from the East End rather more than the already converted from Hampstead – we hope![4]

Although the Betjeman/Fleetwood-Hesketh collaboration was considered in various CoID meetings, it appears that Betjeman did not actually provide any text for *Britain Can Make It* (BCMI).[5] This misunderstanding clearly embarrassed Betjeman, who found himself being credited in various newspaper reports once BCMI had opened for work he had not done. Shortly after the opening of the exhibition Betjeman wrote to Peter Fleetwood-Hesketh and to S.C. Leslie (Director of the CoID). He was upset to find an article in *The Times* in which he had been wrongly credited for having supplied the commentary that accompanied the Nicolas Bentley cartoons describing the fictional inhabitants of the 'Furnished Rooms' section [104]. In his reply, Leslie explained,

> Our press people knew quite well that the captions for the section in question were not yours .... It is always possible that some junior press office, in the rush and scramble of those early days, made a false statement to a journalist based upon the original possibility that you might do the captions.[6]

Peter Fleetwood-Hesketh similarly replied sympathetically to Betjeman reassuring him that there were no hard feelings over the misattribution. His letter also provides an interesting insight into the exhibition itself:

> where the allusion to them appeared I do not know! ... The short descriptions accompanying the drawings were composed by a Mr Tomalin, based upon my notes, and considerably modified by me in collaboration with him before going into print. I also did the small figure drawings at the top of the writing, but had nothing to do with the figures or writing in the 'Furnished Rooms' section itself, which were, I believe, done, as stated in the Catalogue, by a Mr Bentley .... I thought the whole arrangement and decoration of 'Britain Can Fake It' was most attractive, spoilt only by a dismal loud-speaker giving 'explanations', though quite all right when they had music instead. I could not work up enough interest or energy, neither had I the time, to examine the exhibits themselves in great detail. There is nothing to say what anything is, except a small number, which refers to the catalogue. Even with a catalogue this is an exhasperating [sic] process, but since these had all gone before the exhibition opened, most people will have come away not much the wiser as to what it is all about! It is also a pity that there is no Chrystal Palace [sic] in which to hold such an affair, which might have spared the unfortunate thousands of visitors standing for many hours in the street in such queues as I have seldom witnessed .... Again let me assure you that if any recognition, however insignificant has accrued to you or anyone else instead of to me, I am not in the least annoyed![7]

The illustrations of the period rooms produced by Fleetwood-Hesketh were criticized in the *Architect and Building News*: 'Our feeling is that the rooms thus shown are not typical. However, these bits of frippery are virtually mere décor. The architect will see them for what they are worth, the general public may be impressed.'[8]

If the intention of Fleetwood-Hesketh's illustrations was to show the British public how interior decoration had progressed to the levels shown in the 'Furnished Rooms' section at BCMI, a Mass Observation (M-O)

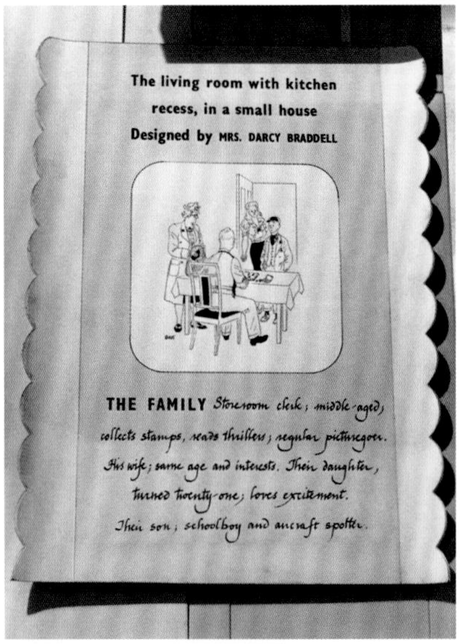

104. One of the signs that accompanied each of the furnished rooms to describe the imaginary inhabitants to visitors. This room, designed by Mrs Darcy Braddell, showed part of a living room with a kitchen recess in a small house. The occupants, a family, were described as 'Storeroom clerk; middle aged; collects stamps, reads thrillers; regular picturegoer. His wife; same age and interests. Their daughter, turned twenty-one; loves excitement. Their son; schoolboy and aircraft spotter'. The illustrations on the panels were produced by Nicolas Bentley.

report showed that, conversely, many visitors had an affinity with the old-fashioned styling of the rooms from the past. One visitor was overheard commenting on the illustration of the richly decorated living room from 1837 that 'All that stuff harboured germs and dirt, but they must have been very comfortable though'.[9]

The most dramatic change in styles demonstrated by Fleetwood-Hesketh in his illustrations was the one that took place in the decade between 1927 and 1937. The description of the 1937 room, with its reference to 'Functionalism' and 'fitness for purpose'— a slogan initiated by the Design and Industries Association but very much approved of by the CoID – has a somewhat cynical tone to it. The inclusion by Fleetwood-Hesketh of an eggcup in his illustration of the Surrealist mural must surely be an irreverent reference to the Design Research Unit display shown towards the other end of BCMI, where the creation of an eggcup is used to explain 'What Industrial Design Means' [see **154**]. The pipe laid on the table is perhaps a cheeky reference to James Gardner, a devoted pipe-smoker [**103**; see **51**].

Nicolas Bentley

The 24 furnished rooms were designed by eminent designers and architects, to suit a range of fictitious family types, and are discussed by Alexandra Goddard in the following chapter. The characters in this imaginary world were drawn by the author and illustrator Nicolas Bentley, who also provided the descriptions of these family groups in short, witty prose [see **112**]. These imagined occupants were 'specified for the guidance of the architects in order for them to create an accurate type of room settings for each'. The shilling catalogue to accompany the exhibition included Bentley's illustrations and descriptions, together with the full list of items on display in each of the rooms.[10]

Bentley's line illustrations and written descriptions were also reproduced on metal panels and attached to the entrance of each room set [**104**; see **68**]. In this way, it was hoped, visitors would immediately be able to imagine the types of people who might live there. In order to create the right atmosphere, a notice in the introductory foyer to the 'Furnished Rooms' section asked the visitor to imagine that the occupants for whom the room had been designed had just left.

A cartoon in *Punch* by a well-known popular cartoonist, Leslie Illingworth, entitled 'Available Now', makes fun of the 'Furnished Rooms', by showing the unfortunate occupants freezing cold, with no gas or electricity, a comment on the grim winter of 1946–47 and its fuel shortages. Reproduced in *Punch* opposite an article on 'How to Get Coal', this was not the kind of publicity the Board of Trade, or the CoID would have wanted [**105**].[11]

Class distinctions

Information issued to the Press by the CoID on the concept of the 'Furnished Rooms' section does not refer to class. Instead, 'Each room was visualized as belonging to a particular family of a certain size, the breadwinner being of a certain occupational status'.[12] The awkward issue of class was mainly implied, but it would not have been too difficult for visitors to recognize the type of person who might live in a 'kitchen of a cottage in a modern mining village', or those lucky enough to have a 'luxury bathroom'. The *Architect and Building News* commented on the sometimes inaccurate pairings of occupation and status with design: 'The kitchen by Edna Mosely, ARIBA, is intended for a cottage in a modern mining village. It would grace many a "middle-class" home!' [see **111**].[13]

The 'middle-class bathroom' designed by C. Entwistle was one of the displays shown to Queen Elizabeth on the tour of the exhibition following the opening ceremony, and her comments were widely reported in the Press: 'The Queen was interested in two of the bathrooms erected side by side, one described as "luxury" and the other as "middle-class". "Luxury", she suggested, was an inappropriate name for what was shown: the use of oiled teak wood in this bathroom in contrast with plastic sheeting in the middle-class bathroom was commented on by the Queen.'[14] Another article, this time in *Punch,* explains the obvious ease with which the descriptions could be deciphered (it is also one of the contemporary publications that erroneously cited John Betjeman as the writer responsible for the descriptions): 'With discrete descriptions like "young doctor newly in practice – studies social conditions; his wife likes outdoor sports and photography", Mr Betjeman neatly leaps over the barrier of social distinctions and allows us to guess, pretty exactly, the young people's income and tastes'.[15]

Some of Bentley's drawings were also later included in a CoID publication entitled *Furnishing to fit the family*. Published a year after BCMI, the booklet focused on a selection of ten of the rooms from BCMI and included photographs of the room sets, together with line sketches of the room designs drawn by Hugh Casson and Hilton Wright [**106**]. More detailed descriptions of each room, together with the types of people who might live there, were also included. All but one of the descriptions of the types of occupants remain the same as for BCMI, and do not specify class, the exception being Marjorie Holford's 'bedroom in a small house in an industrial town', which became a 'bedroom in a working-class flat' located in London.

Criticism of the accuracy of the family descriptions was touched on in an article in *The Architect and Building News*, 'We feel bound to remark that in most cases the rooms bear no relationship to the imagined families, and the class-distinction in naming the rooms appears utterly

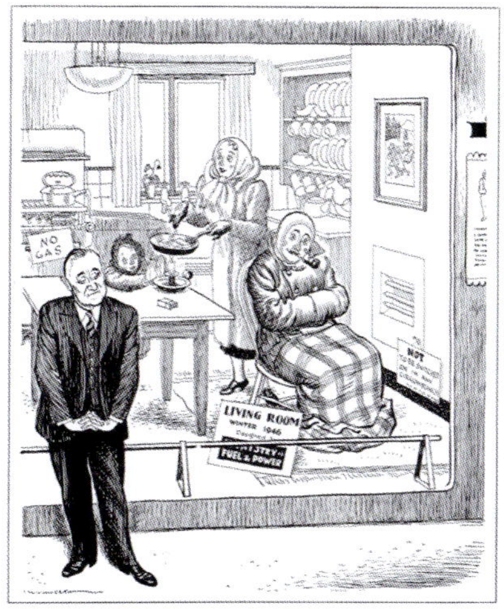

AVAILABLE NOW

105. The winter of 1946–47 was one harshest in modern times. The Minister for Fuel and Power, Emanuel (Manny) Shinwell is depicted here twiddling his thumbs in this cartoon entitled 'Available Now' by Leslie Illingworth from *Punch* (16 October 1946), who reimagines one of the idealistic room sets in the 'Furnished Rooms' section at BCMI. Shinwell was widely criticized for not averting the crisis of the coal shortage that the severe weather conditions provoked.

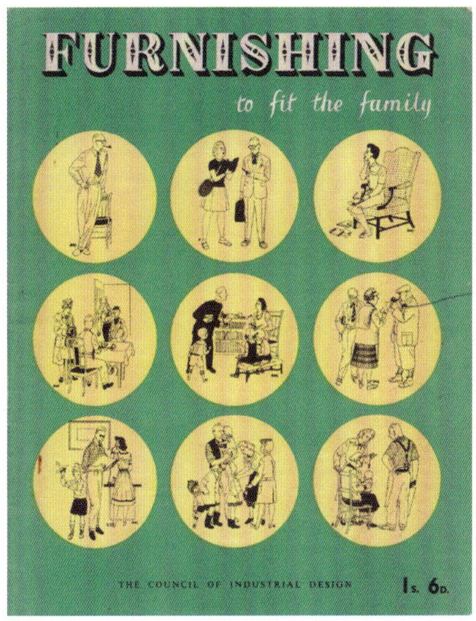

106. Cover of *Furnishing to fit the family*, a CoID publication showing Nicolas Bentley's illustrations of imagined family groups

meaningless'.[16] But discussions around class annoyed some connected with BCMI. In a review of the 'Furnished Rooms' section, Gordon Russell (the furniture designer and manufacturer, and member of the CoID) wrote, 'Many visitors will come away from these rooms with new ideas. It is unfortunate, however, that *The Times* stated that the rooms were furnished for different classes when a real attempt had been made to furnish them for people doing various jobs.'[17]

NOTES

1 Text from Peter Fleetwood-Hesketh's introductory story panels for the 'Furnished Rooms' section: RIBA Archives RIBA ON3120.

2 Miles Tomalin was a poet and veteran of the Spanish Civil War. He was the grandfather of the designer Thomas Heatherwick. The correspondence about BCMI is noted in a letter from James Gardner to John Betjeman dated 31 July 1946 in the John Betjeman fonds, University of Victoria Special Collections. I am grateful to Heather Dean for checking the John Betjeman fonds and for providing me with copies of the correspondence. I am also grateful to John Frederick for his assistance.

3 Letter from James Gardner to Peter Fleetwood-Hesketh dated 9 July 1946: John Betjeman fonds, University of Victoria Special Collections.

4 Letter from James Gardner to John Betjeman dated 9 July 1946: John Betjeman fonds, University of Victoria Special Collections. *The Times* article referred to is (Our art critic) 'Trend of Design', *The Times* (24 September 1946), p. 2. The reference to not being 'precious' almost certainly relates to a directive from Sir Stafford Cripps in a letter from him to S.C. Leslie of 10 August 1945 explaining what should not be included in the exhibition – 'no precious stuff': Design Council Archives at the University of Brighton (DCA) 312.

5 DCA/58/'The COID 1946 Exhibition. Position as at 3rd July 1946'.

6 Letter from S.C. Leslie to John Betjeman dated 3 October 1946: John Betjeman fonds, University of Victoria Special Collections.

7 Letter from Peter Fleetwood-Hesketh to John Betjeman dated 1 October 1946: John Betjeman fonds, University of Victoria Special Collections. Peter Fleetwood-Hesketh was credited in the *Supplement to the Catalogue* under Drawings and Theme in the Section introduction (*Britain Can Make It Exhibition Supplement* (London 1946), p. 237.

8 'Britain Can Make It. Exhibition at the Victoria and Albert Museum', *The Architect and Building News* (27 September 1946), p. 193.

9 DCA/903/Report by Mass-Observation on Britain Can Make It. Section C and B.

10 'Britain Can Make It Exhibition. Victoria and Albert Museum', *The Architect and Building News* (20 September 1946), p. 177.

11 The cartoon by Leslie Illingworth featured in *Punch* (16 October 1946), vol. CCXI, no. 5520, p. 303. For more on the severe weather and deepening fuel crisis see David Kynaston, *Austerity Britain 1945–51* (London 2007), pp. 193–205.

12 Mass Observation Archive, University of Sussex, SxMOA 1/2/26/1/B: "BRITAIN CAN MAKE IT" EXHIBITION. Rooms Made to Fit. Notes for the Press on the Furnished Rooms.

13 *The Architect and Building News* (cited note 8), p. 195.

14 '"Britain Can Make It" Exhibition opened by the King', *The Times* (25 September 1946), p. 2.

15 'The Great Exhibition', *Punch* (16 October 1946), p. 305.

16 *The Architect and Building News* 20 September 1946 (cited note 8), p. 193.

17 Gordon Russell, 'Furnished Rooms from the 'Britain Can Make It' Exhibition. An Assessment by Gordon Russell', *Art & Industry* (January 1947), p. 13.

ALEXANDRA GODDARD

The Furnished Rooms: Rooms Made to Fit

> The British Public will see what industry has planned for the living rooms, bedrooms, and kitchens of the post-war home: the world overseas will discover that the brains, ingenuity and taste which long gave Britain her place as a leader of world industry have not deserted her, and that her lead will be kept—*Sir Stafford Cripps, President of the Board of Trade, October 1945*[1]

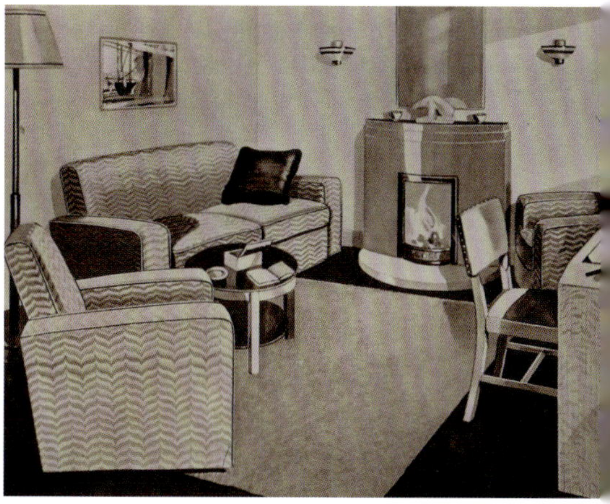

107. 'Planning a Lounge', published by the Times Furnishing Company as part of their 'Planned Home' series, c.1930s

A substantial proportion of the display space at the heart of *Britain Can Make It* (BCMI) was given over to the presentation of a series of furnished rooms. In this section, 24 designed 'rooms' offered the public a view of idealized interiors in a range of settings both inside the home and beyond it to the schoolroom and office [**110**; see also **122, 179**]. Each room allowed the public to view the furnishings from behind a low barrier, whereby they occupied the same physical space but were prevented from entering and touching what they were seeing. Given the eventual high footfall of the exhibition this was a practical approach, ensuring everything remained as intended yet at the same time the barrier lent formality to the layout befitting its location in the museum and its overall didactic purpose. Despite this, many smaller items were 'pilfered' and individual labels describing the exhibits were removed, creating problems of identification.[2]

The concept of shopping for new products arranged in suggested domestic settings would have been familiar to the middle and upper classes from department stores in major cities since the nineteenth century. From the 1920s and 1930s onwards an increasing number of illustrated retail catalogues showed idealized arrangements, as British firms began to offer hire purchase and mail order facilities to a wider public [**107**]. In 1929 the furniture store Waring and Gillow employed Serge Chermayeff (who went on to become one of Britain's foremost modernist architects) to select and arrange their new season's furniture in their shop on Tottenham Court Road. His selection for their showroom

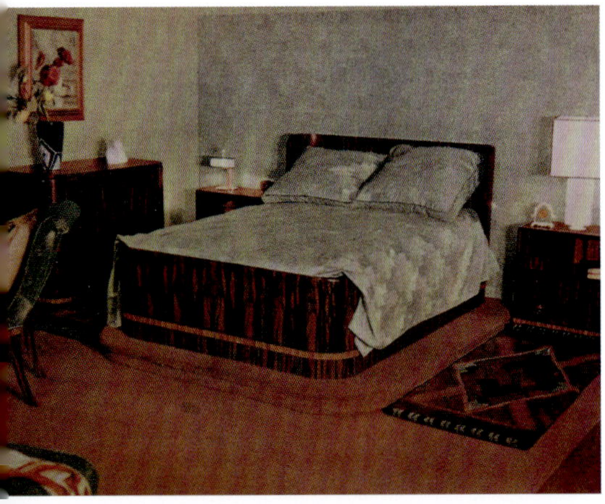

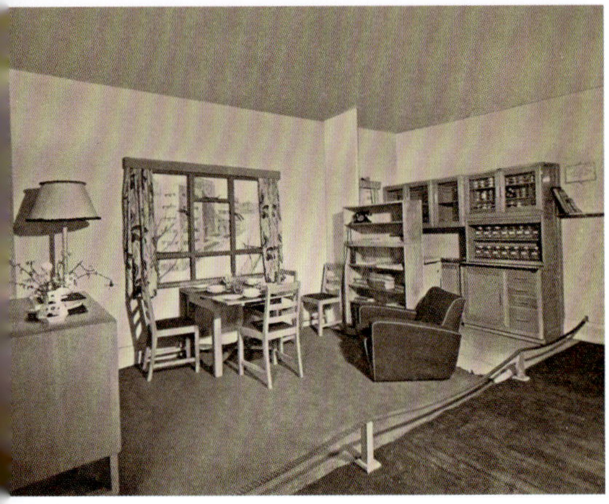

108. Bedroom designed by Serge Chermayeff for Waring and Gillow. Illustration of Waring and Gillow department store, 1929 from Martin Battersby's book, *The Decorative Twenties*, 1969

109. Gallery 12 – Modern Room, 1939, from *Handbook to the Geffrye Museum*, London, 1946

offered a more progressive choice to a wider public than the more traditional forms on offer at most of the large retailers [**108**].[3]

According to research into visitors to BCMI, 'the most widely represented class was very definitely artisan working class'. Such clear evidence of reaching a working class audience suggests that the exhibition was responsible for opening up these types of displays to a broader cross-section of the population.[4]

Since 1908 the annual *Daily Mail* Ideal Home show had been presenting the latest in labour-saving technology and home furnishings, not just in idealized room sets but in custom-built streets of houses.[5] Although the show attracted large audiences, its aspiration was not to promote so-called 'good design' in the way BCMI did. It was rather aimed at the average housewife who was looking to make her life easier and her home more up-to-date. Attempts were made in the inter-war years at public design education through exhibitions organized by the Design and Industries Association, Council for Art and Industry and others. However, none was successful in influencing public taste in the way BCMI intended.

Well before BCMI progressive educationalists were already using furnished rooms as a way of educating the public. As far back as 1917, the Bethnal Green Museum (now the V&A Museum of Childhood) had presented 'demonstration living-rooms' to its visitors, and in the late 1930s the Geffrye Museum in Shoreditch established a series of chronologically arranged period rooms to educate its visitors about the evolution of domestic life, from behind a series of low barriers. The museum was run by the London County Council (LCC), at that time a breeding ground for progressive educational ideas, and its administrator George Lowndes appointed Marjorie Quennell, co-author of the popular *A History of Everyday Things in England* as the museum's curator. Quennell's focus was on the objects themselves, their design and production, an approach with distinct sympathies with what the organizers of BCMI were trying to achieve. Quennell's successor, Molly Harrison, began using the period rooms to educate people about more than just history. She had identified the potential of mass production to bring good-quality, well-designed products to working-class homes but also saw the potential for bewilderment among the general public about what to choose. She began as early as 1941 to collaborate with furniture designer and retailer Ambrose Heal to select pieces for a 'modern room' intended to help guide people towards good choices in contemporary design [**109**].[6]

BCMI's 24 furnished rooms were precise in what they were intending to show. Of the four non-domestic rooms, three were offices and the fourth was a secondary school classroom. The other eighteen were

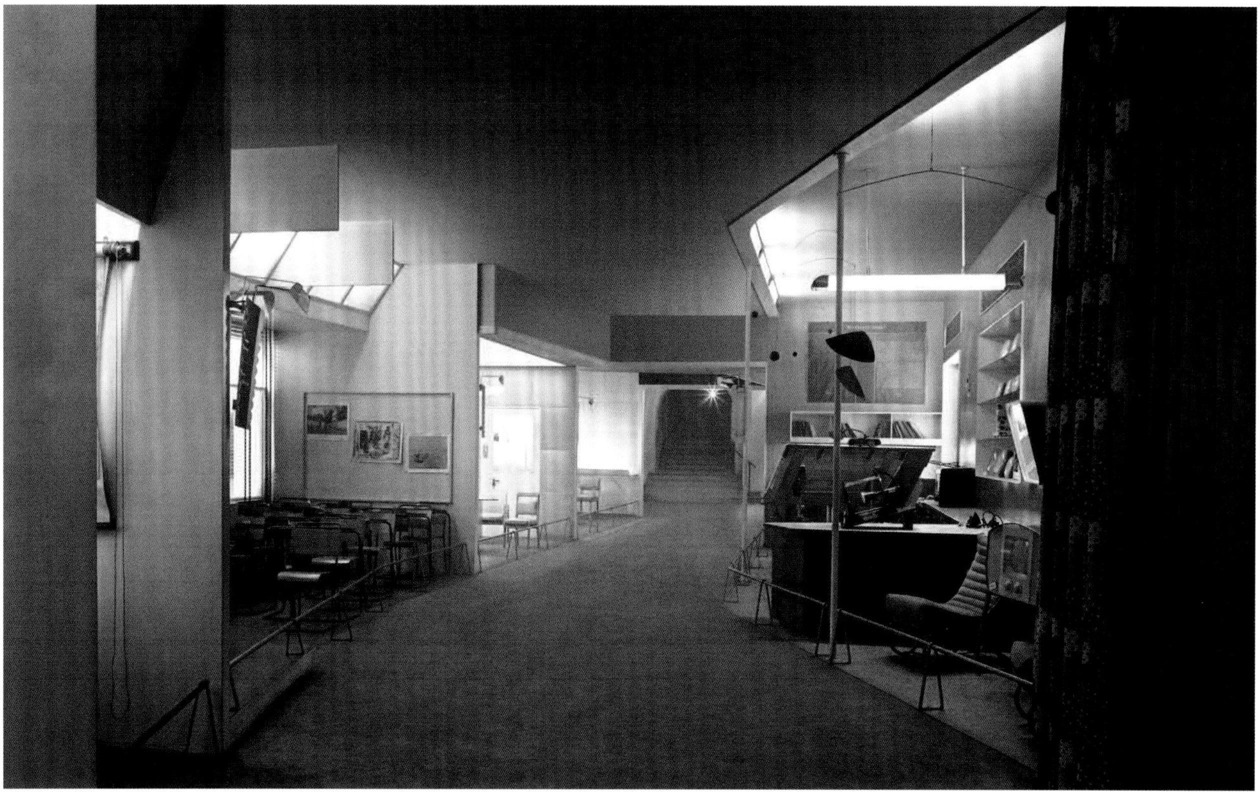

domestic: bathrooms, kitchens, living rooms, dining rooms, bedrooms, bed-sitting rooms and a nursery, referred to as 'Things in their home settings'. The intention was that

110. View of the 'Furnished Rooms' section looking back towards the introduction with Secondary Schoolroom, Manager's Office and General Administration Office on the left and the Drawing and Technical Office on the right

> Each room was visualised as belonging to a particular family of a certain size, the breadwinner being of a certain occupational status. Information about each family, with a sketch of imaginary members by Nicholas [sic] Bentley, is displayed on the facia [sic] of each room, and the designer's task was to work out an appropriate plan for the equipment, furnishing and decoration of the room.[7]

Designers were invited by the Council of Industrial Design (CoID) to submit schemes fitting a schedule of definitions. As well as covering a variety of room types, the schedule required a wide socio-economic spread. For instance the kitchens should include one 'in a working-class house where the kitchen is also used for dining' in a low price range, two medium-priced examples (one with an 'occasional dining recess' and the other a 'kitchenette for a small flat for a middle-class family in

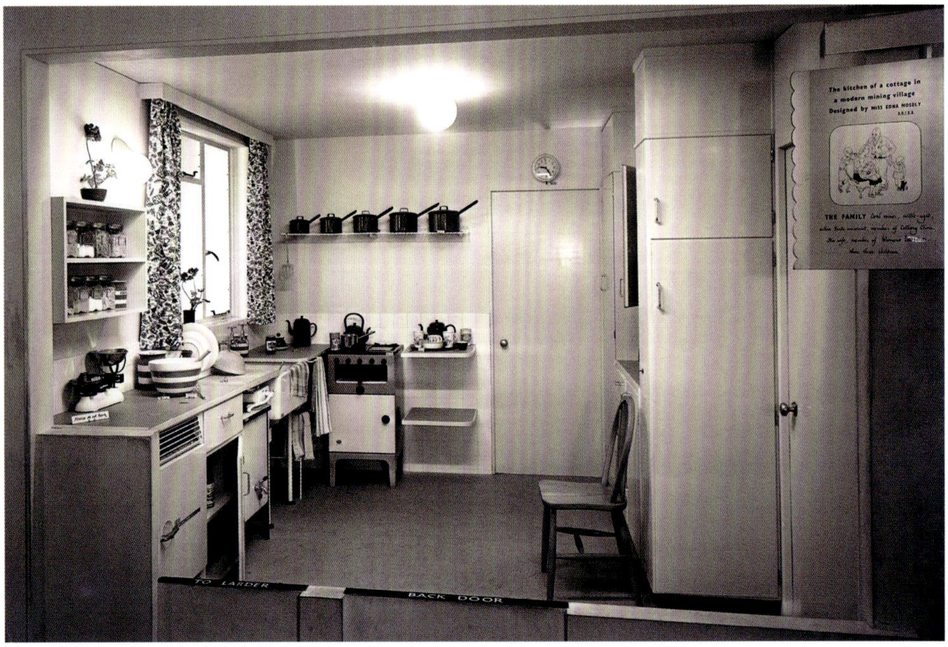

111. Kitchen of a cottage in a modern mining village, designed by Edna Mosely

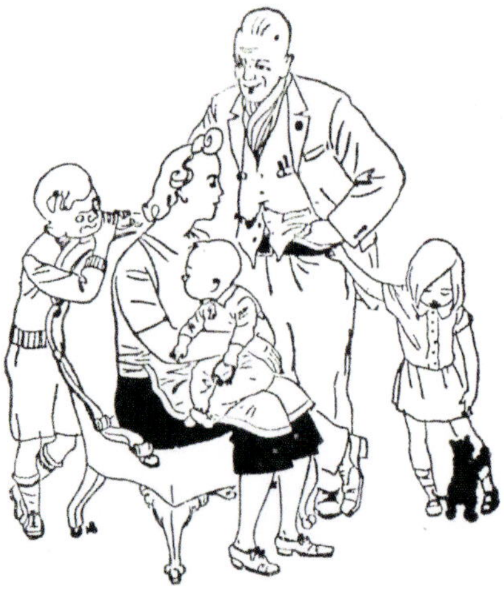

112. Line drawing of a miner's family by Nicolas Bentley from the *Britain Can Make It Catalogue*, p. 119

the London area'), and finally one 'for a large house where price is not the main consideration'.[8] These definitions are a reminder of post-war Britain's clearly defined class system and reflect a somewhat paternalistic view of what was appropriate for each sector of society.

The rooms were each presented with a scenario including the room title, a line drawing by Nicolas Bentley and the definition of the 'Family' or 'Occupant' (if it was a single person) – described in a caption that included occupation, marital status, family size and some details of hobbies and interests. For instance, the first room that visitors approached was 'The kitchen of a cottage in a modern mining village', designed by Edna Mosely [111]. Its inhabitants are described thus: 'The Family; Coal miner, middle aged, active trade unionist, member of colliery choir. His wife, a member of Women's Institute; their three children' [112]. This family is doubtless playing to certain British stereotypes of hard-working 'salt-of-the-earth' working-class people. The miner was emblematic of the British working-class throughout most of the twentieth century but the miner's home used as a touchstone for modern living standards has an interesting precedent at the Ideal Home show of 1922. A Lanarkshire miner's cottage was displayed there, not as an idealized standard but as characterizing the inefficiency which led to dissatisfaction and industrial strikes and contrasted with the ideal homes on show elsewhere in the exhibition. The importance of good housing for miners was clearly a priority.

The room set is a model of rationality and practicality. It is equipped with a refrigerator, drying cupboard and oven, all of standard design. Everything is designed and arranged for easy cleaning: cupboards are raised off the floor on legs, areas around the sink are removable, the work surfaces, tabletop and floor are covered with linoleum and fitments are finished with enamel paintwork. The layout is arranged 'in the sequence in which work is likely to be done'. It has much in common with Margarete Schütte-Lihotzky's influential 1926 Frankfurt Kitchen, the first modernist kitchen and model for all subsequent built-in kitchens in terms of efficiency and hygiene [**113**]. At BCMI this working-class family are shown what they can expect the post-war era to bring them, clearly defined against rooms intended for their working-class rural counterpart – the agricultural worker – as well as those higher up the social scale including the curate, doctor, barrister and bank manager.

The line-up of chosen designers featured some well-established names, including husband-and-wife team Maxwell Fry and Jane Drew, David Medd and Dorothy Braddell, as well as others whose influence has not endured. The 'Kitchenette in a small flat' intended for a 'single woman, dietician at a hospital; excellent cook' was designed by Frederick Gibberd, now best known for his work on Harlow New Town and Liverpool Metropolitan Cathedral [**114**]. He was the architect responsible for Pullman Court, Streatham (completed in 1935), a ground-breaking

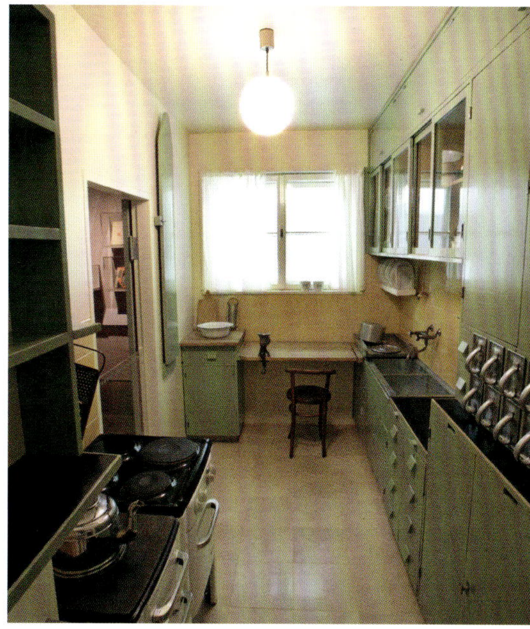

113. Frankfurt Kitchen, designed by Margarete Schütte-Lihotzky, V&A: W.15:1 to 89–2005

114. Kitchenette in a small flat, designed by Frederick Gibberd

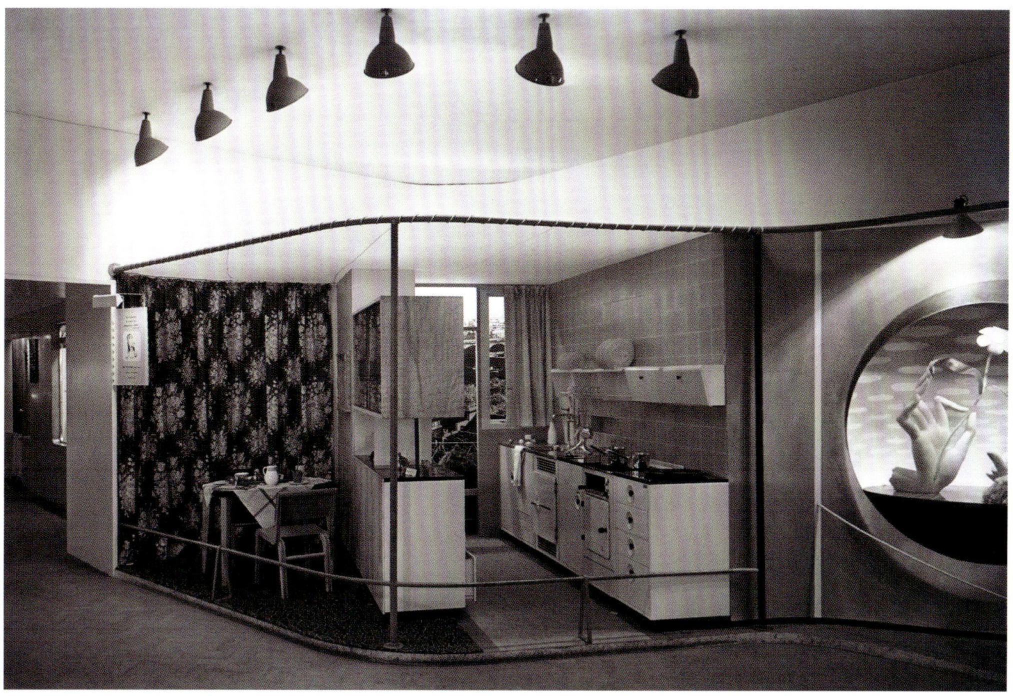

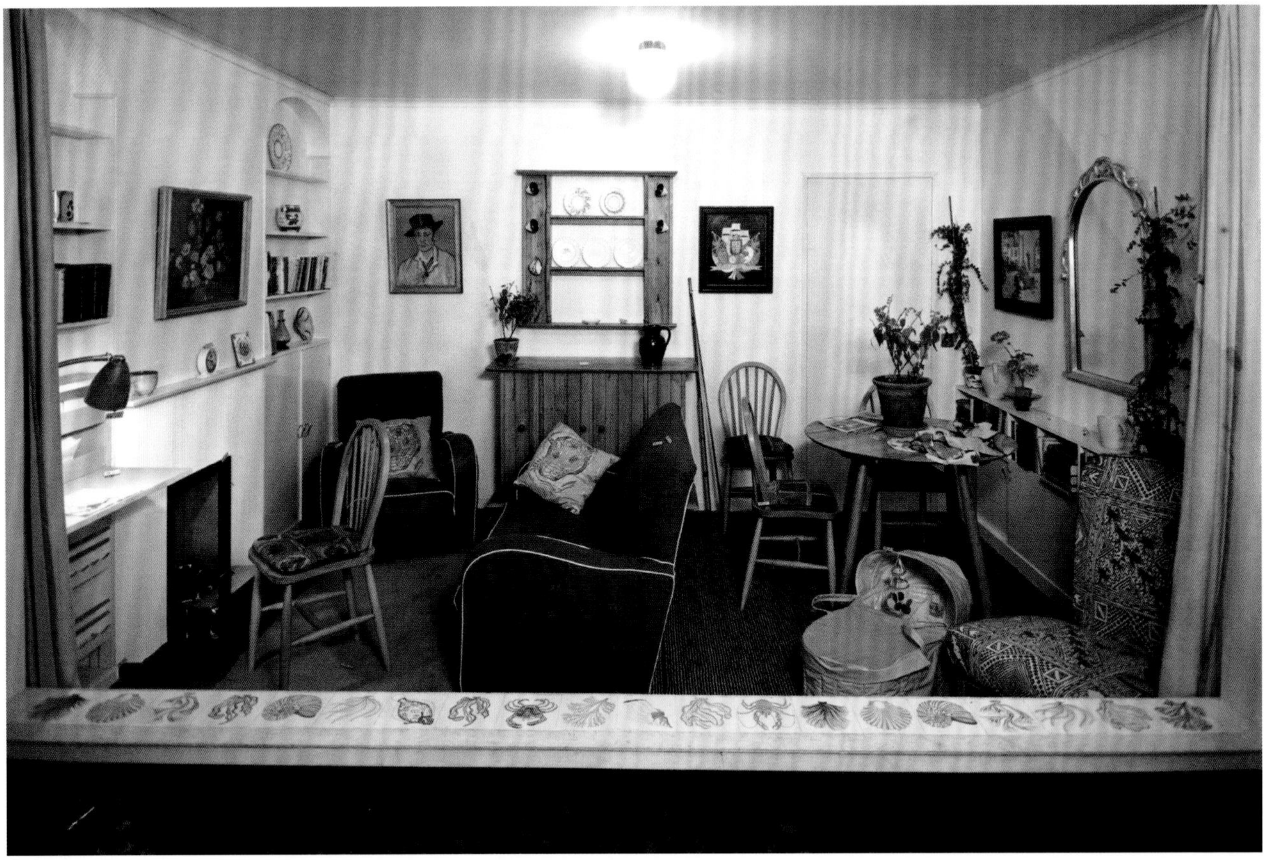

115. Living room in a small semi-detached house on a new estate, designed by Elizabeth Denby

high-density development of 218 studio, one- and two-bedroom flats built on modernist principles. As well as the flats themselves, he also designed a special range of furniture to suit the scale of the interiors and their occupants.

The 'Living room in a small semi-detached house on a new estate' was designed by urban reformer Elizabeth Denby [**115**]. The intended occupants were 'The Family: young artisan, enjoys a bit of carpentry, motor cyclist. His wife; formerly a school teacher. His mother in law; clever at needlework and embroidery. Their baby'. Denby had campaigned during the 1930s for improved housing and affordable furnishings for the urban working classes. Although not a qualified architect herself, she worked with Maxwell Fry on Sassoon House, Peckham (1933), the first modernist housing for workers in Britain, and later on Kensal House, Ladbroke Grove. Her focus was always on the needs of the occupants and she was noted for her attention to popular taste, endeavouring to give people what they wanted rather than just what was good for them.[9]

Higher up the social scale, R.D. Russell had designed the 'Living room in a large town house' for 'The Family; Barrister at law, collects books. His wife, gives musical parties' [see **116**]. Russell had designed furniture for his brother's firm Gordon Russell Ltd in the village of Broadway in the Cotswolds since 1919. Although ideologically rooted in the Arts and Crafts tradition, in the 1930s the firm moved into the production of high-quality machine-made simple modernist furniture, which strongly appealed to the upper-middle classes. Who more appropriate then to undertake the barrister's living room?

The rooms proved popular with visitors, with 24% of people citing them as 'the thing that interested them most' in the exhibition, and evidence suggesting they spent one fifth of their time looking at them, spending a proportionally longer 'dwell time' there than in other areas of the exhibition.[10] The miner's kitchen proved most popular. However, there is little evidence to suggest people were making direct comparison with their own homes, or had any intention of emulating what they saw, despite every commodity being listed in the catalogue should they wish to get hold of it. They were ultimately more likely to pick out and comment on particular fixtures or furnishings and express an opinion about them, suggesting behaviour perhaps more reminiscent of shopping than of museum visiting.

The use of furnished room sets is still very much alive today for both educational and retail purposes. Social history museums such as Den Gamle By (Denmark), Norsk Folkemuseum (Norway) and Winterthur Museum, Garden and Library (USA), as well as the Geffrye Museum of the Home in London, continue to educate their audiences by means of constructed period room sets, often with very specific criteria or scenarios which help to provide an accurate research hook. And the presentation of idealized interiors is no less successful in retail outlets than it was. Large department stores such as IKEA continue to present idealized interiors furnished with their products, encouraging their customers to recreate them at home. Whether or not the furnished rooms of BCMI were successful in providing a design education for the British public, it seems that they mark an important place in British history where design, trade, consumerism and education collided.

NOTES

1 Design Council Archive at the University of Brighton Design Archives (DCA) 312/Summer Exhibition 1946/'Preliminary Notice/The 1946 Exhibition of Design in British Industry. Statement by the President of the Board of Trade, 25 October 1945'.

2 V&A Archive, Exhibition File/AAD/1977/4/13, Britain can make it: Report by G. Kingston Stewart.

3 John Charles Rogers, *Modern English Furniture* (London 1930), gives an account of the display at Waring and Gillow's Tottenham Court Road store in 1929.

4 DCA/903/Mass-Observation report entitled 'Those who came'.

5 The Ideal Home show was suspended during the war and did not resume until 1947.

6 I am indebted to Laura Carter, University of Cambridge, for this information, which she presented at a conference at The Bowes Museum 18–20 September 2014, and which forms part of her subsequent PhD thesis, *The 'history of everyday life' and democratic culture in Britain, 1918–1968*, unpublished PhD thesis, University of Cambridge (2017), pp. 191–226.

7 Mass Observation Archive, University of Sussex, Britain Can Make It exhibition: SXMOA 1/2/26/1/1B.

8 V&A Archive, Exhibition File/AAD/1977/4/12, Britain can make it/'The Council of Industrial Design 1946 Exhibition, Schedule of Furniture Designs Required'.

9 Lynne Walker, *Golden Age or False Dawn? Women Architects in the Early 20th Century*; http://www.english-heritage.org.uk/discover/people-and-places/womens-history/visible-in-stone/resources/resources-used; accessed 12 January 2015.

10 DCA/903/"Britain Can Make It" Exhibition, Report prepared by Mass-Observation, December 1946.

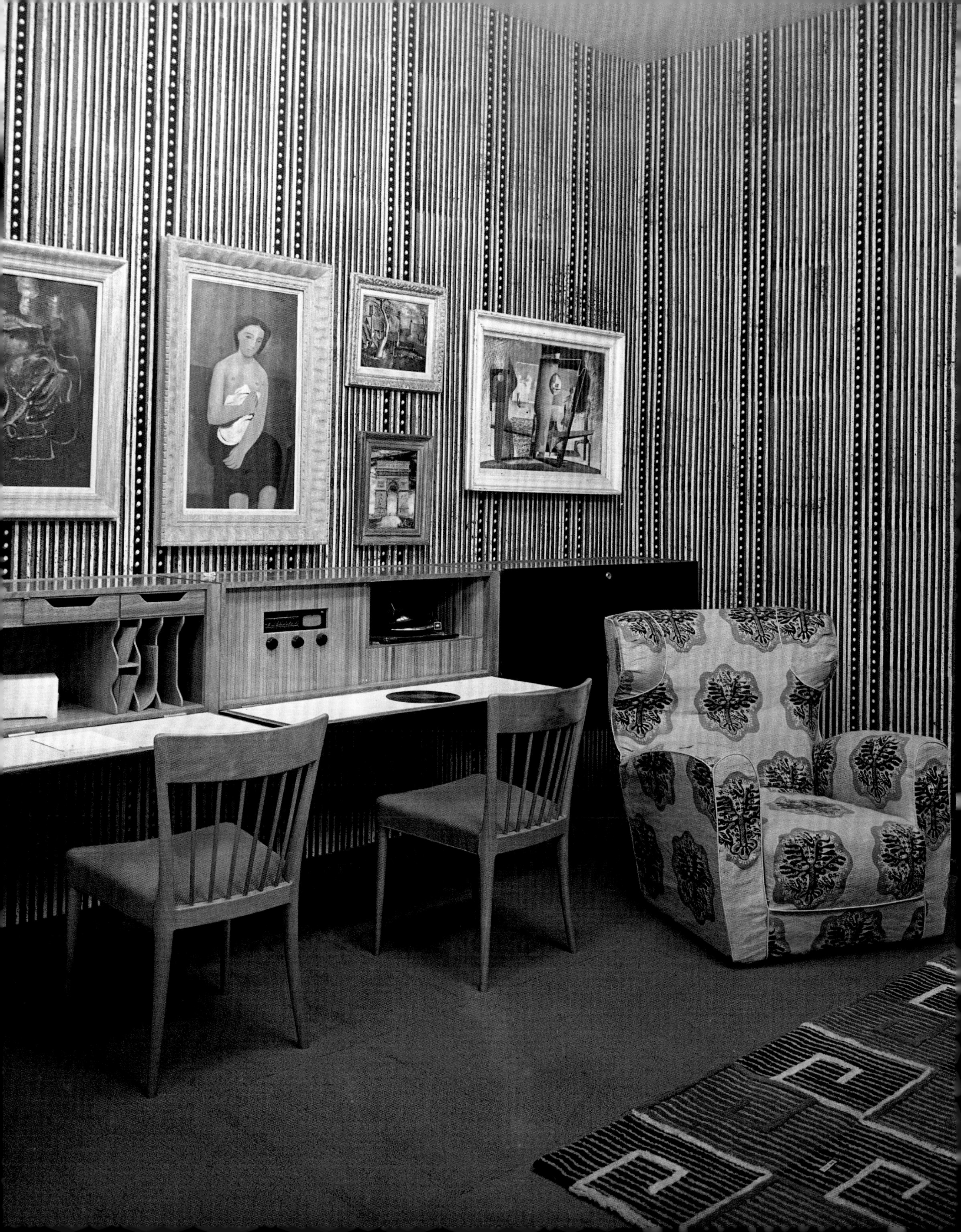

DIANE BILBEY

Displaying Art at Britain Can Make It

Unexpectedly, perhaps, for an exhibition focused on design, art in various guises was well represented at *Britain Can Make It* (BCMI). Early in the planning stages, the idea to illustrate themes throughout the various sections of the exhibition using murals was discussed by the Council of Industrial Design (CoID). In addition, paintings, prints and sculpture by leading contemporary artists were used to adorn the furnished rooms, and artists were also involved in the design of many of the products on display, including furnishing and fashion fabrics, clothing accessories, wallpapers and ceramics.

Paintings in the furnished rooms

A number of mainly Modern British paintings, prints and sculpture were displayed in the 'Furnished Rooms' section. Their inclusion was very much supplementary to the overall display, and certainly not intended to distract visitors from the primary objective of the organizers, which was to instill in the public an appreciation of good design. A review in *The Times* put this into context, 'Some good examples of modern English painting and sculpture are to be seen in the rooms and look, as they should, the most conspicuously decorative objects there.'[1]

Some of these artworks were produced in or around 1946, and so provided an unexpected insight into the contemporary art world. John Tunnard's *Painting 1944 (O.85)* was one of six paintings lent to the exhibition by the London art dealers Ernest Brown & Phillips Ltd, Leicester Galleries.[2] These were displayed in one room, the living room of a large town house designed by R.D. Russell for the imaginary family of 'a barrister at Law, who collects books, and his wife who gives musical parties'. Five of the paintings were displayed together, as a block on one wall, above the writing desk and radio cabinet by the established furniture manufacturer B. Cohen & Sons Ltd [116]. They were noted in a report by one Mass Observation (M-O) investigator who found it surprising that so few people commented on them, particularly as they

116. Wall of a living room designed by R.D. Russell for a barrister at law and his wife, with John Tunnard's painting (far right) on the wall above the radio cabinet and gramophone cabinet, manufactured by B. Cohen & Sons Ltd. The modernist wallpaper, called 'Flute', was designed by Edward Bawden.

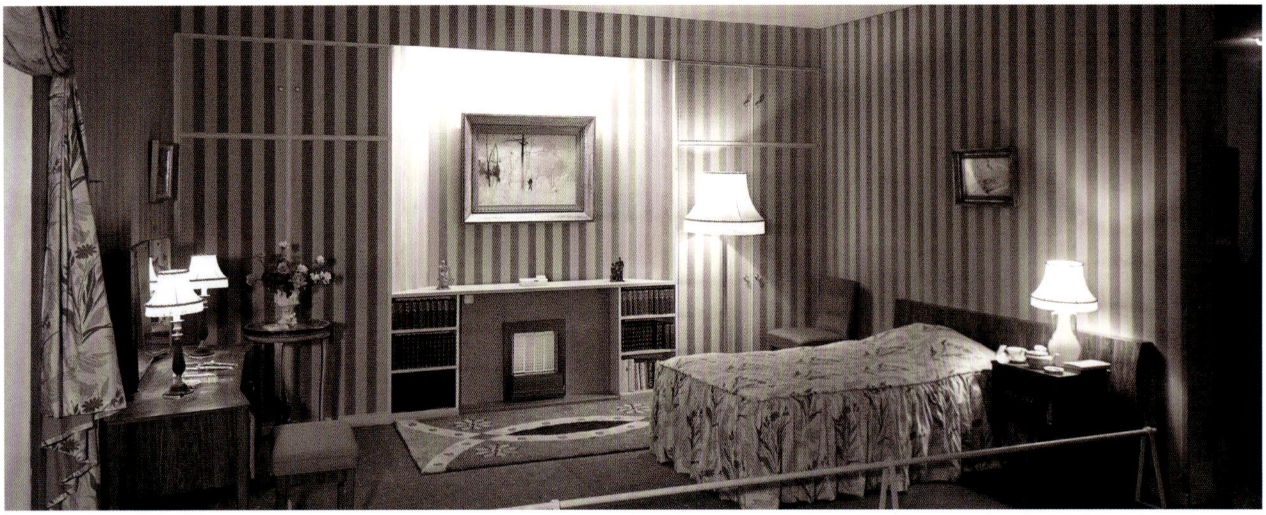

117. Bedroom designed by John Hill for a County Borough Councillor and his wife with Victor Pasmore's *The Wounded Bird* painting placed centrally above the fireplace. Two Henry Moore maquettes are on the mantlepiece and two small paintings by J.B. Yeats, *Children Playing* and *Rose in Basin*, are on the side walls.

took up most of the left-hand wall of the room. The investigator, did however, note part of an overheard conversation about the paintings between a 'middle-class woman of 20' and her friend of the same age, which implies that their dislike of the paintings adversely affected their opinion of the room as a whole: "Look at them pictures." The second woman echoed, "Look, vile, aren't they? Awful room, isn't it?'[3]

Elizabeth Denby, the social housing reformer, was asked to design a room for an imagined family living in a small house on a new housing estate. This family clearly had an eclectic taste in art, as the room included paintings of *Two Clowns* by the Austrian émigré artist Georg Eisler; *Harbour* by Susan Williams-Ellis; and a portrait said to be by Christopher Wood, though in fact this was a reproduction of Van Gogh's portrait of *Armand Roulin*. Tiles decorated with motifs of sea shells and crustacea (designed by Susan Williams-Ellis for Carter Tiles) were included as a fake windowsill at the front of the room from which visitors were encouraged to peer in. These tiles were singled out for a special mention in the CoID's publication *Furnishing to fit the family* (1947), that included photographs of some of the room sets from BCMI. Illustrating that art was for all ages, including the very young, the description reads, 'Window sills always catch dust and damp plant pots spoil paint quickly, so their row of hand-painted tiles is at once decorative and practical – besides making an open picture-book for the baby when he begins to walk' [see **115**].[4]

An appreciation of art was often one of the characteristics proposed for those who might occupy the rooms shown in the 'Furnished Rooms' section. There was an architect 'who paints in his spare time'; the wife of

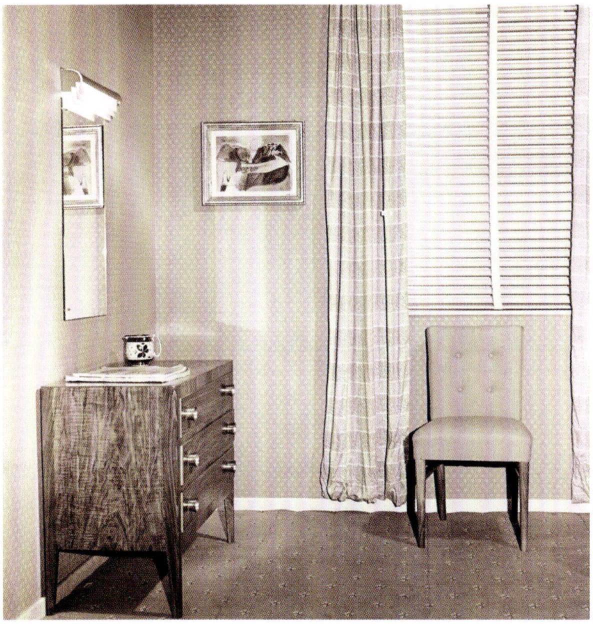

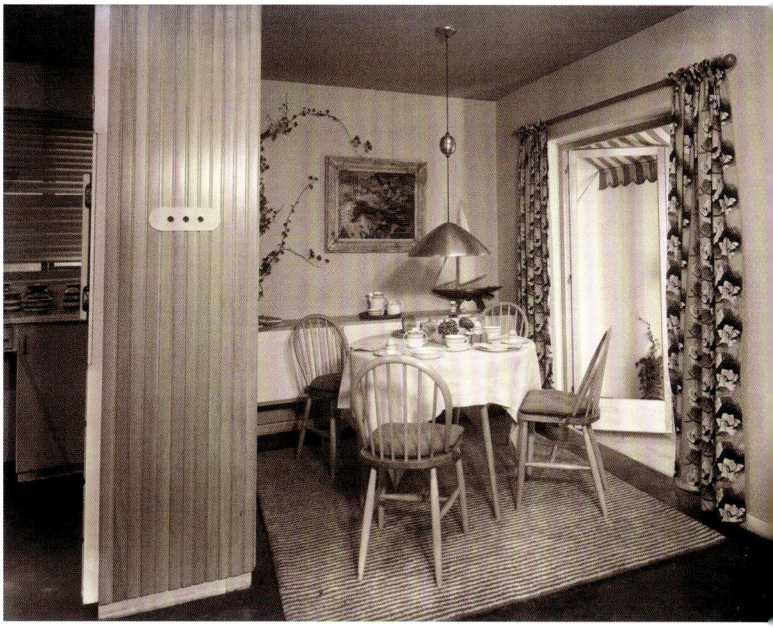

a doctor, who is interested in photography; and a well-educated husband and wife, 'interested in modern art and literature'. For one 'connoisseur', described as a 'County Borough Councillor', a bedroom with man's dressing room and bathroom was designed by John Hill. In the bedroom, two small paintings by J.B. Yeats (*Children Playing* and *Rose in Basin*) and Victor Pasmore's *The Wounded Bird* were displayed. Painted in 1944, Pasmore's *The Wounded Bird* was prominently displayed above the modern Terrazzo fireplace manufactured by Bratt Colbran [117].[5]

The Wounded Bird (and *Rose in Basin*) were part of Sir Kenneth Clark's substantial art collection, and are likely to have been in his possession at the time of the BCMI, and were presumably lent by him, although, frustratingly, no documentary evidence to support such a loan exists in the archives. Clark was on the selection committee for the 'Furnished Rooms' section for the exhibition, and almost certainly played a part in the selection of the art used to decorate the rooms. Together with Sir Charles Tennyson, Sir Francis Meynell and Leigh Ashton, Clark was invited to be part of a special supervisory committee to make a 'last minute survey of the Exhibition' just before it opened to ensure 'that the highest standards of display are maintained and that no blunder has been committed in the selection arrangement of goods'. However, there is no mention in the archives of any discussions on the art displayed at BCMI, suggesting that it was an informal, impromptu arrangement. However, by supporting the display of works by contemporary artists he admired Clark was also clearly legitimizing the value of their work, and even subtly, it

118. Dressing room in a large house designed by John Hill for a County Borough Councillor and his wife with a painting by Graham Sutherland, probably *Sun Setting Between Two Hills* (1938). A.E. Gray's lustre-ware jug sits on a chest of drawers in walnut by W.H. Russell for Gordon Russell Ltd. The curtains in yellow and terracotta are by Edinburgh Weavers. The hand-printed wallpaper – Regency button design by Cole & Son – is in terracotta with small rosette design in French grey.

119. Julian Trevelyan's *Scilly Isles* painting on the wall of a kitchen with dining recess designed by Frederick MacManus for a young architect, his wife and their son could be argued, his agenda to promote visual arts appreciation in Britain.[6]

A further two paintings, both by Graham Sutherland, entitled *Sun setting between two rocks* (1938) and *Landscape* (possibly *Landscape with Black Rock*, also from 1938), were hung in the dressing room for an imaginary 'County Borough Councillor'. These were noted as being lent by the artist himself, but it is likely that Clark acquired them from Sutherland at this time or soon after, as they too were later in his collection **[118]**.[7] Displayed on the mantelpiece, domestic sculpture in the form of Henry Moore's *Family Group* (1945) (lent by Henry Moore via John Hill, the designer of the room) and a maquette for the *Claydon Madonna* (1942) (in the possession of Kenneth Clark), completed the Modern British emphasis of the room **[117]**.[8]

Another, more modest room for a 'young architect who is an amateur painter', in a small modern house designed by the architect Frederick MacManus, appears to have had a painting by Julian Trevelyan of Tresco in the Scilly Isles on the wall **[119]**.[9]

Graham Sutherland

Graham Sutherland was an artist whose work spanned a range of media at BCMI. Not only were his paintings displayed in two rooms of the 'Furnished Rooms' section, but so too were his designs for furnishing fabrics and wallpaper. Sutherland (along with Henry Moore) was also one of the artists commissioned by Ascher Ltd to design one of their 'artist squares' – scarves, produced in silk or screen-printed rayon. Silk squares designed by Moore and Sutherland were some of the 50 items the firm showcased at BCMI.[10] The scarf by Sutherland, entitled *Black Trellis,* ties in with the series of paintings entitled *Thorn Tree* that he produced around the same time (1945).

The theme for the series of *Thorn Tree* paintings came to Sutherland as a response to working on the *Crucifixion* for St Matthew's Church, Northampton, and his intense interest in the symbolic representation of Christ on the Cross, with the crown of thorns. A version of *Thorn Tree* lent by Sutherland himself was hung above the fireplace in the room designed by F.R.S. Yorke for an imagined bank manager and his wife, and later found its way into Kenneth Clark's personal collection. *Thorn Tree*, with its vibrant green and blue colours, was very different to the monochrome *Black Trellis* scarf shown in the Ascher display, but shows Sutherland's adaptation of the same imagery for different audiences and materials. Sutherland's *Thorn Tree* painting in this room set was accompanied by a small piece of domestic sculpture by Elizabeth Spurr that was photographed sometimes displayed on the mantelpiece or on the radio gramophone **[120; see 133]**.[11]

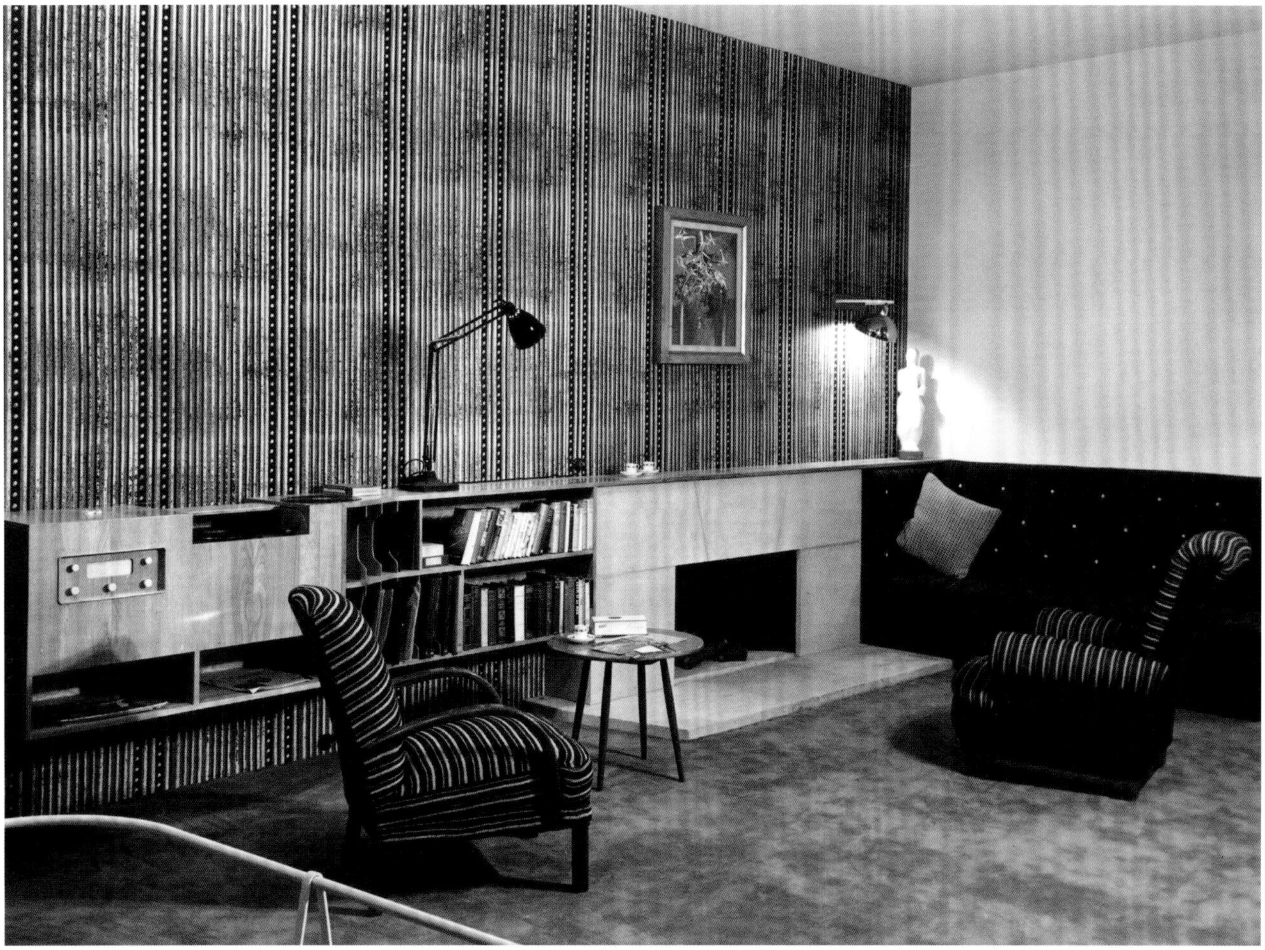

120. Graham Sutherland's *Thorn Trees* above the fireplace of the living room designed by F.R.S. Yorke for a barrister at law and his wife. A sculpture by Elizabeth Spurr is on the mantlepiece. Edward Bawden's 'Flute' wallpaper forms the backdrop.

Sutherland's connection to BCMI, as well as his difficulties in obtaining artists' materials after the war, were highlighted in his correspondence with Walter Hussey (from whom Sutherland was commissioned to produce the *Crucifixion* at St Matthew's):

> The position, as you probably know, with regard to artists (painters) is this: – nothing, with the exception of colours, brushes, or paper can be obtained without a priority permit. Canvas (a very small quantity) appears in the shops every six months and each customer is rationed by the shopkeeper to about one yard. In the war time days I had a permit through the War Artists' Advisory Committee. This has now been wound up and it has taken me a month of abortive enquiry to find out about permits, and when they can be obtained. There is a priority department in the Board of Trade, I learn. But work for churches is not, so I'm told, high priority! Canvas, anyway, is held up at the manufacturers, owing to lack of canvas workers. But I have managed to get

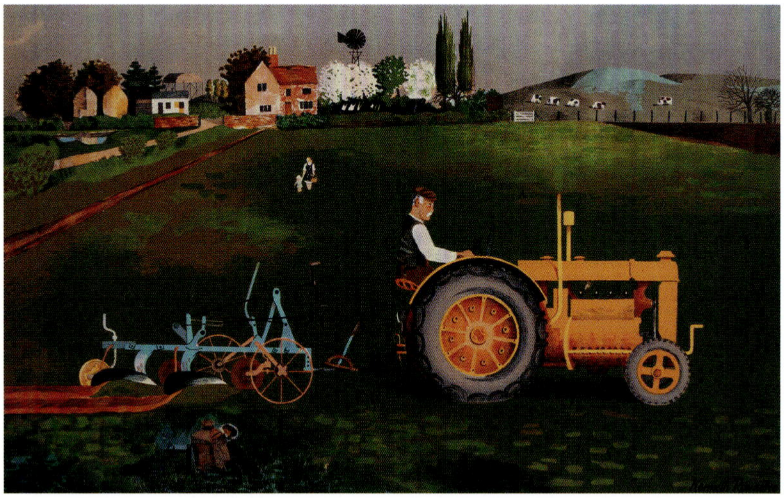

121. Kenneth Rowntree, *Tractor in Landscape*, c.1946. Produced for School Prints Ltd by The Baynard Press. Colour lithograph, 49.53 x 76.2 cm. V&A E.5233–1958

> a man in the Board of Trade, to whom I was introduced, to get me some Sundeala Board (this is the stuff on which I did the Crucifixion you saw, and it is very permanent and rigid) on the grounds that I am doing some work for the 'Britain can make it' Exhibition. Highly black market![12]

Textile design

Textile manufacturers commissioned artists including Henry Moore and Enid Marx to provide designs which were then translated into fabrics, and their designs were to be found on fashion, curtain and furnishing fabrics displayed at BCMI. Graham Sutherland's 'Sutherland Rose' fabric was used for curtains in the kitchen designed by Frederick MacManus for a young architect, as well as being part of a display for the manufacturer Helios Ltd [see **119**]. A forward-looking design in spun rayon crepe fabric, called 'Aztec', was produced by Patrick Heron for his father's company Cresta, and was displayed alongside textiles designed by Mary Duncan and Lana Mackinnon for the same firm.[13]

Class

Class stereotypes were inevitably reflected in the type of art displayed in the 'Furnished Rooms'. Cheerful, affordable colour lithographs produced in 1946 by School Prints Ltd were shown in what would have been seen as the traditional working-class home. Although originally intended to be displayed in classrooms, School Prints were also made available to the general public, and two examples were shown in domestic interiors at BCMI: one in an agricultural worker's bedroom (Kenneth Rowntree's *Tractor in Landscape)* [**121**], and another in the bedroom of a small house in an industrial town (Edwin La Dell's *Tower* *of London*).[14] Affordable reproductions such as this, framed and hung on the wall, demonstrated that there was a place for art in all homes. It also unintentionally illustrated the striking contrast and economic gulf between widely available prints for the working classes and the one-off fine art paintings hung in the rooms designed for wealthier inhabitants.

This issue of class stereotyping was highlighted by a writer for *Punch* who supposedly overheard a conversation in the room designed by D.L. Medd for an agricultural worker:

> There were bound to be a few laughs at the designers' expense of course, as people moved from room to room trying to find their own level, and I hear a farm labourer's bedroom come in for some hard words from a visitor of the farm-labouring type. Pointing to a framed lithograph on the wall he said bitingly, 'Ah, stands to reason, I s'pose, I'd have a picture of a ruddy tractor 'anging in my bedroom!'[15]

The labourer's bedroom also included (perhaps rather disingenuously) a hand-painted wallpaper designed by Graham Sutherland for Cole & Son and a bedspread and curtains by Alastair Morton of Edinburgh Weavers.[16]

A progressive choice was to hang an abstract print by John Piper on the wall of the secondary school classroom. Piper's *Abstract Composition* (Levinson 7) from 1937 was one of fifteen prints in the first series issued by Contemporary Lithographs Ltd that was aimed at schools. It is considered 'Piper's last and most impressive exercise in abstraction. It is also the only abstract print in the entire series of Contemporary Lithographs' **[122**; see also **8, 110]**.[17]

Public opinion

M-O investigators reported that, generally speaking, visitors did not really seem to notice or comment that much on the paintings displayed at BCMI. When they did, it was often critical and mocking:

> Generally unpopular were the pictures. Comment in the furnished rooms over and over again picked out the pictures for abuse, and the only picture in the furnished rooms at all subject to favourable comment – was one photography representation of a fairly uncontroversial, safe subject, of a bowl of flowers. For instance: Man and his wife leaving room in small house on new estate. He looks at the Christopher Wood and a modern painting of clowns, laughs out loud and points to them. 'Just look at those pictures up there. Come on!'[18]

The art on display was considered incidental – not practical, not functional, and certainly not essential – so perhaps the majority of visitors could see little intrinsic value in it. Although M-O did not directly

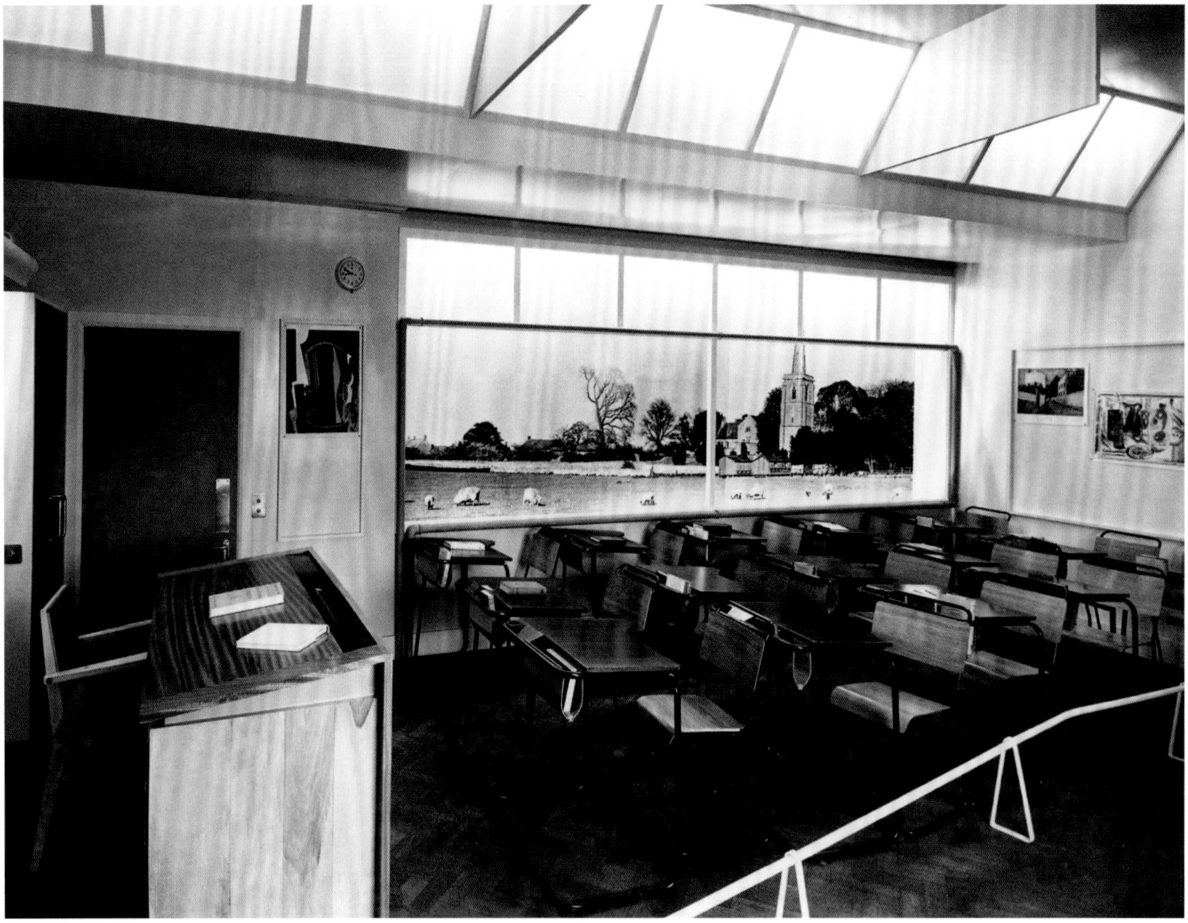

122. Secondary schoolroom, designed by D. Clarke Hall. John Piper's lithograph entitled *Abstract Composition*, produced by Contemporary Lithographs Ltd, lent to the display by Zwemmers, is on the wall to the right of the door beneath the clock.

ask visitors any questions about the art displayed, its investigators did make reference to it in their note-taking:

> No direct questions were asked about the pictures in the exhibition. Very little reaction to them was noted in overheard remarks. The few casual comments made were nearly all derogatory. The only remark of approval overhead was for the lithograph [*Tractor in Landscape*, Kenneth Rowntree] in the farm cottage bedroom. Eleven people, out of the whole of ... Sample ... spontaneously mentioned the pictures and murals as items which they disliked. No one mentioned them as things they particularly liked.[19]

The paintings by artists including Graham Sutherland and Victor Pasmore in the more affluent homes went down particularly badly in the M-O survey, as the general public generally considered them to be too radical and ahead of their time. 'That's a pretty awful painting, isn't it?'

— spoken by a middle class woman — was another comment picked up by Mass Observation, though it's not known to which painting it referred. Another recorded:

> Two middle class elderly women passing through the free design furniture section and taking eight minutes to go round it:—
> First Woman: 'What a lovely room and a beautiful fireplace'
> Second Woman: 'The chairs are very comfortable looking'
> First Woman: 'I wouldn't like pictures like that on my wall'

This ties in with some general criticism of the whole exhibition, which was derided in one newspaper article for being ostentatious by putting 'too great an emphasis on what might be termed the "Bloomsbury school of display"'.[20] Similar sentiment was expressed by an article in the journal *Display* by Stafford Bourne, then Governing Director of Bourne and Hollingsworth, a large London department store located in London's Oxford Street. He wrote in no uncertain terms:

> The 'Bloomsbury corduroy school' is in power and they don't know their job. The 'Britain Can Make It' Exhibition was an example of bad display when it should have been superb. Foreign buyers don't want to see surrealist and grotesque models and display stands in arty settings — they want to see the merchandise.

Such harsh comments appear to have astonished R. Dudley Ryder, Chief Exhibitions Officer for the CoID, who wrote in a memo to S.C. Leslie that he thought Bourne had 'missed the whole point of the "Britain Can Make It" Exhibition, namely that it was a selective Exhibition and not one in which stands were set aside for the mass display of any one firm's products'.[21]

But this inequity between the art and artistry of the display and the people for whom it was meant to appeal — the British public — was also referred to in *The Architect and Building News*:

> We fear that the ordinary visitor, who has not been in touch with the artistic trends of the past fifteen years, may feel bewildered and perplexed, rather than stimulated, by what he sees .... There is a good deal to be seen here which will have plenty of meaning to the artistically sophisticated, but may well be incomprehensible, and therefore irritating, to the ordinary man, because so far out of touch with the realities of his existence.[22]

M-O were also interested in the terminology used by visitors to describe their experience of BCMI. They deduced that 'design' was not a word the general public was very comfortable with. Often the words 'design' and 'new fashioned' were applied in a critical, derogatory sense, while 'old-fashioned' had positive, homely connotations. Older visitors seemed those most likely to use the word 'arty' in a negative, condescending manner, such as a remark made by a Doctor, 'I dislike some of the colour schemes. Far too arty in the wrong way.' A man of forty-five made a similar remark, 'Some of the stuff's too arty – as if the designers wanted to show off how clever they were instead of concentrating on the goods themselves'.

We can return to the M-O report to sum up the effect on visitors of the art displayed at BCMI:

> This dislike of pictures is partly attributable to the fact that the paintings displayed are rather farther ahead of, and less related to, public taste than any other item on show. As will be explained, utility is a vital factor in general approval of items on show, and even those assessed mainly by aesthetic considerations are assumed first to be satisfactory to their purpose. A picture has no direct utility at all in this sense. Equally, a designer must always have in mind whether the article on which he is working will function as it is intended to; the painter need have no such worry, his intention can largely be expressed without considering a consumer, and without thought of mass-production.[23]

NOTES

1 'Trend of design. Imagination and composure. From our art critic', *The Times* (24 September 1946), p. 2. Art displayed at BCMI was usually briefly noted in the accompanying catalogue, though sometimes inaccurately: Georg Eisler's *Clowns* painting was incorrectly noted as *Clouds*; Victor Pasmore's *The Wounded Bird* is described merely as *Snow Scene*; the titles for paintings lent by Graham Sutherland are not given. The paintings shown in the Scottish living room designed by R. Mervyn Noad are not mentioned in the catalogue, but are likely to have been by Scottish artists (see RIBA 52696, John Maltby photograph, and Design Council Archive at the University of Brighton Design Archives [DCA] 0832). Council of Industrial Design, *Britain Can Make It Exhibition Catalogue* (London 1946), Eisler p. 140; Pasmore p. 132; cat. no. 414, and Council of Industrial Design, *Britain Can Make It Exhibition Supplement* (London 1946), p. 252; Sutherland p. 144; Scottish living room, pp. 144–45.

2 Tunnard's painting had been exhibited at an exhibition of Contemporary British Painting in Buffalo, USA, early in 1946, but returned to the Leicester Galleries in May 1946. It was one of a number of paintings noted as being lent by Leicester Galleries (Ernest Brown & Phillips Ltd) in the exhibition catalogue: Council of Industrial Design, *Britain Can Make It Exhibition Catalogue* (London 1946), p. 143. The painting is shown with others lent by Leicester Galleries in a photograph included in the article by Noel Carrington, 'Britain Can Make It', *Country Life* (4 October 1946), p. 518. See also A. Peat and B.A. Whitton, *John Tunnard. His Life and Work* (Aldershot 1996), p. 169, no. 390. It was later bought for the Toledo Museum of Art. It was sold at Sotheby's, London, on 4 June 2003, no. 28. Tunnard, who trained in textile design, also designed the fabric (manufactured by Sir Thomas and Arthur Wardle Ltd) used in the curtains hung behind settees in the entrance hall and foyer of the exhibition. See *Britain Can Make It Exhibition Supplement* (London 1946), p. 222.

3 DCA/903/Report by Mass-Observation on Britain Can Make It. Section B, p. 5 (December 1946).

4 Otto Breicha, *Georg Eisler* (Vienna and Munich 1970), p. 123, no. 18, painted in 1946. Noted as being in the possession of Elizabeth Denby, the designer of the room, so presumably acquired by her from The Fine Art Trade Guild following the exhibition. *Furnishing to fit the family* with drawings by Nicolas Bentley, Hugh Casson and Hilton Wright (London 1947), p. 11. For the Susan Williams-Ellis tiles produced by Carter see Mark Eastment in Jan Morris et. al, *Portmeirion* (Woodbridge 2006), 'Portmeirion Pottery', pp. 133-34.

5 For *The Wounded Bird,* see Alan Bowness and Luigi Lambertini, *Victor Pasmore, with a Catalogue Raisonne of the Paintings, Constructions and Graphics 1926–1979* (London 1980), no. 69, pl. p. 58; and Chris

Stephens and J.P. Stonard (eds.), *Kenneth Clark. Looking for Civilisation* (London 2014), p. 33, pl. 3. For the fireplace see *Britain Can Make It Exhibition Catalogue* (London 1946), p. 132, no. 409.

6 For Kenneth Clark's patronage of British art, see Brian Foss, *War paint: art, war, state and identity in Britain, 1939–45* (London and New Haven 2007). Clark's enduring friendship and high regard for Pasmore as an artist was noted in Clark's biography, *Another Part of the Wood. A self-portrait* (London 1974), p. 251. During the war, Clark was Chairman of the Ministry of Information, War Artists' Advisory Committee. He was Chairman of the Art panel at the CEMA *Design at Home Exhibition of Furniture and Furnishings* (1945), arranged by the Arts Council by designers that were to be involved in BCMI, including Milner Gray. See Sue Breakell, 'Exhibiting "the taste of everyday things": Kenneth Clark and CEMA's wartime exhibitions of design', in Liz Farrelly and Joanna Weddell (eds.), *Design Objects and the Museum* (London 2016). Clark was on the Design Committee of the CoID until his resignation in May 1945. He was also a member of the Committee of the Council of Industrial Design, and on the Art School Exhibits Selection Panel. See Sue Breakell, '"The Exercise of a Peculiar Art-Skill": Kenneth Clark's Design Advocacy and the Council of Industrial Design, *Visual Culture in Britain* (16, no. 1, 2015), pp. 42–66. Clark purchased *The Wounded Bird* directly from the artist. It was exhibited in the recent Tate Britain Kenneth Clark exhibition: see Stephens and Stonard 2014 (cited note 5), p. 148, no. 124, and pl. 3 on p. 33. For its previous exhibition history see *Victor Pasmore. Retrospective exhibition 1925–65*, exh. cat. Tate (London 1965), no. 37. Tate Archives TGA 8812.1.4.126, Kenneth Clark archive: Letter to Clark from S.C. Leslie dated 18 July 1946.

7 *Britain Can Make It Exhibition Catalogue* (London 1946), p. 132, no. 418. *Britain Can Make It Exhibition Supplement* (London 1946), p. 252, nos. 414, 418. Stephens and Stonard 2014 (cited note 5), p. 93, no. 156, fig. 29 p. 149 (*Sun Setting between Two Hills*); p. 149, no. 155 (*Landscape with Black Rock*).

8 *Britain Can Make It Exhibition Supplement* (London 1946), p. 252, no. 415. The maquette for the Claydon Madonna was one of the trial pieces shown to Kenneth Clark by Henry Moore in order for him to decide on the Northampton *Madonna and Child*. Although it retains its original title, the Claydon Madonna is no longer at Claydon. It was moved to St Mary's, Barham, Suffolk in 1978. I am grateful to Michael Phipps for his help with identifying the Henry Moore sculpture and for supplying this information. See David Sylvester (ed.), *Henry Moore, Vol 1. Complete Sculpture 1921–48* (London 1990), p. 13, no. 222, pl. 138, for LH 222 (Claydon Madonna maquette) and p. 15 for LH 238 (Family Group). The maquette was damaged at the end of the exhibition and Clark asked the CoID for reimbursement to cover repairs and loss in value. Clark estimated the value at £30, and requested 30 shillings to cover repairs, and £6 for the reduction in value. Correspondence in Tate Archives TGA 8812.1.2.1590–3, part of the Kenneth Clark archive. The maquette formerly in the possession of Clark (LH 222) was sold at Christie's, New York, 1 May 1996, lot 375.

9 What appears to be an original watercolour sketch for this painting (pen and ink, with grey, black and pink colours) is held in a sketchbook by Julian Trevelyan held in Tate Archives TGA 892/1/11 on page 29. Another closely related sketchbook drawing by Julian Trevelyan from the Scilly Isles, c.1944–45, also in Tate Archives, is illustrated in Philip Trevelyan, *Julian Trevelyan. Picture Language* (Farnham 2013), p. 155.

10 V. Mendes and F.M. Hinchcliffe, *Ascher. Zika and Lida Ascher. Fabric, art, fashion* (London 1987), p. 25, illustration of Sutherland working on *Black Trellis* scarf; p. 49; pp. 18, 101–02 for Ascher at BCMI. *Britain Can Make It Exhibition Supplement* (London 1946), pp. 176–77, lists silk squares designed for Ascher by Moore and Sutherland.

11 The version of *Thorn Trees* formerly in the possession of Kenneth Clark was recently included in the exhibition *Kenneth Clark: Looking for Civilisation* at Tate Britain in 2014, see Stephens and Stonard 2014 (cited note 5), p. 151, no. 222. See also Ronald Alley, *Graham Sutherland* (London 1982), p. 108, no. 124. Such was the unique appeal of the Ascher artists' squares that they were also collected as pieces of art in their own right, framed, and hung on a wall. For reference to the Elizabeth Spurr sculpture, see *Britain Can Make It Exhibition Catalogue* (London 1946), p. 144, no. 667.

12 Walter Hussey, *Patron of Art: The Revival of a Great Tradition Among Modern Artists* (London 1985), pp. 55–56. Sutherland's name appears on an early list relating to the mural artists involved in BCMI, suggesting he was asked at some point to be involved in the 'From War to Peace' mural. V&A Archive, MA/28/75/9 Letter from F. Graeme Macleod (CoID) to Mr Haigh (V&A) 7 August 1946 concerning obtaining passes for mural artists working in the Museum. This commission was to given to Morris Kestelman.

13 For Patrick Heron's design, see Council of Industrial Design, *Britain Can Make It Exhibition Catalogue* (London 1946), p. 26, no. 55. Geoffrey Rayner et.al., *Artists' textiles: Artist designed textiles 1940–1976* (Woodbridge 2012), p. 34, illus. p. 36. For Lana Mackinnon see Caroline Wheater, 'Lost and Found', *Homes & Interiors Scotland* (November 2012), pp. 73–76.

14 Michael Twyman, *A history of chromolithography. Printed colour for all* (London 2013), p. 296.

15 'Britain Can Make It. A Late Look Round', *Punch* (30 October 1946), p. 375. For further reading on School Prints see Ruth Artmonsky, *The School Prints. A Romantic Project* (Woodbridge 2010).

16 *Britain Can Make It Exhibition Catalogue* (London 1946), p. 133, nos. 426, 437 438 and 443.

17 Contemporary Lithographs was established by Robert Wellington and John Piper in 1936. Wellington ran the Zwemmer Gallery and the print by John Piper was lent by Zwemmers: see *Britain Can Make It Catalogue*, p. 119, no. 96. For Piper's print see Frances Carey and Antony Griffiths, *Avant-Garde British Printmaking 1914–1960* (London 1990), p. 140, no. 122, and illus.

18 DCA/903/Report by Mass-Observation on Britain Can Make It. Section B, p. 5 (December 1946). The supposed Christopher Wood painting was in fact a reproduction of the 1888 portrait of Armand Roulin by Van Gogh now in the Museum Folkwang, Essen, inv. G63. The painting of *Two Clowns* was by Georg Eisler.

19 Mass Observation Archive, University of Sussex: FR SxMOA1/2/26/1/A 'Britain Can Make It. 1st Rough for Interim Report', 29 October 1946, no page numbers.

20 "Britain Can Make It'. Some Criticisms. By our Industrial Correspondent', *Sunday Times* (29 September 1946), p. 8.

21 Stafford Bourne, 'Better Display? The Stores Reply. In Defence of Mass Display', *Display, Design & Presentation* (April 1947), p. 22. DCA/911/Britain Can Make It: Letters of criticism, complaint and suggestions. Memo from R. Dudley Ryder to S.C. Leslie, 15 April 1947.

22 'The Box is Opened', *The Architect and Building News* (27 September 1946), p. 186.

23 DCA/903/Report by Mass-Observation on Britain Can Make It. Section B, pp. 13–14 and p. 6 (December 1946).

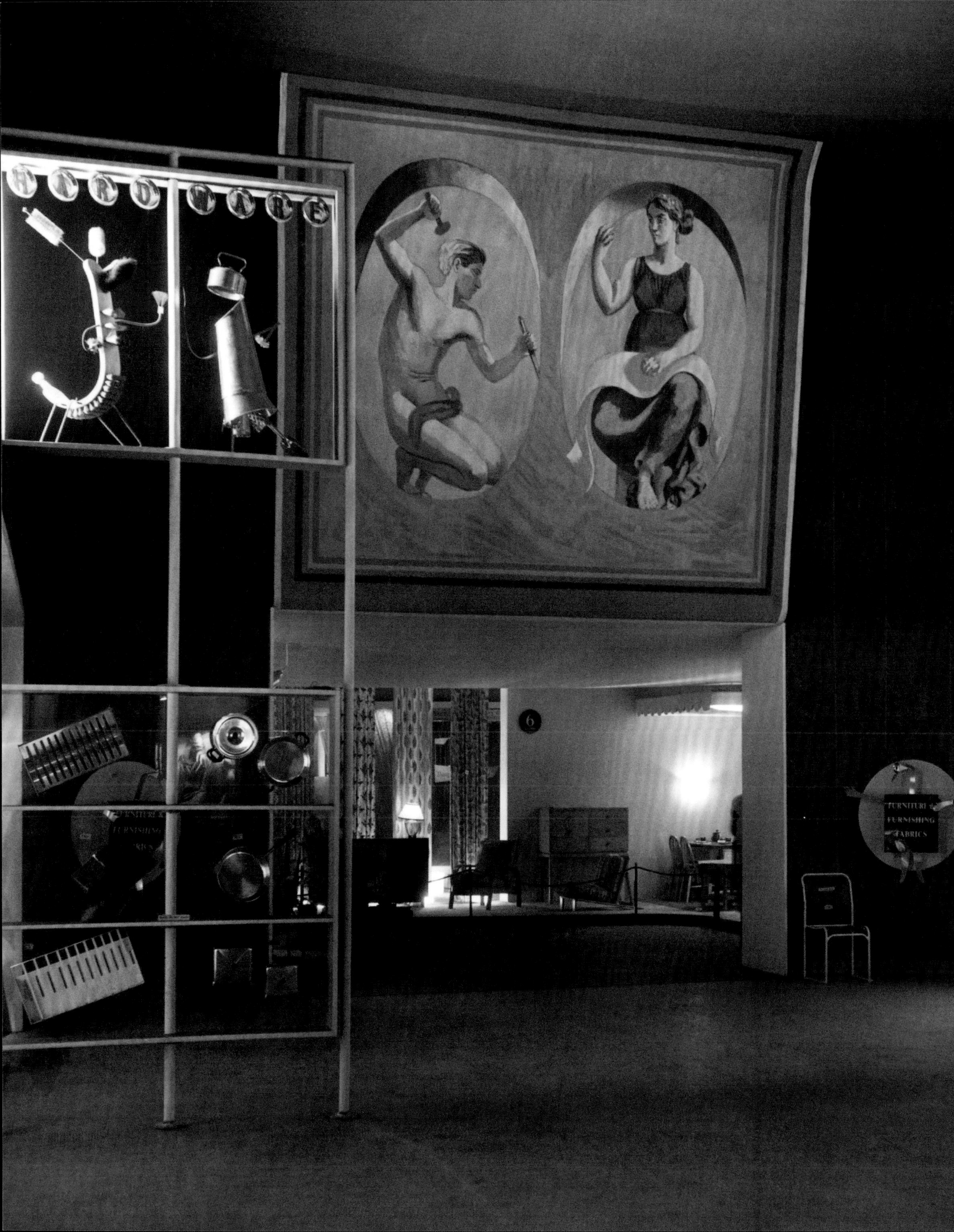
HARDWARE
6
FURNISHING
FABRICS

DIANE BILBEY

The Murals

Mural commissions provided much needed commercial (relatively well-paid) work for artists in the mid twentieth century, and they featured in many of the post-war exhibitions. *Britain Can Make It* (BCMI) was no exception. Murals (covering 3,000 square feet) – commissioned from some of the leading artists of the day – provide an interesting insight into the genre in the mid 1940s, and it is reasonable to ask what they contributed to the BCMI experience for visitors.[1]

A remark from James Holland, later joint Chief Designer at the Festival of Britain, helps place murals produced for exhibitions into some context: 'Mural decoration serves a variety of purposes in exhibitions. It can help to create the mood of the show, it can supply the background and setting for the story, or it can tell essential parts of the story itself.'[2] All three of these criteria are relevant to the murals at BCMI. With a combination of figurative and abstract compositions by, among others, Duncan Grant and Vanessa Bell [**56, 123**], William Crosbie [**54, 72, 130**], John Hutton [**172, 184**], Barbara Jones [**127**], Morris Kestelman [**37**] and Laurence Scarfe [**164, 165**], murals were to be found in almost every section of the exhibition – including a series of narrative murals by Hans Feibusch covering the walls of the Tea Lounge [see **95, 97–99**]. Many of the same artists would go on to carry out mural commissions for the exhibition held at London's South Bank as the centerpiece for the Festival of Britain celebrations of 1951.

R. Dudley Ryder, Chief Exhibitions Officer for the CoID, made particular reference to the importance of murals at the BCMI exhibition: 'During the war we were starved of colour and decoration, and in this Exhibition light and colour were used to a large extent and mural artists engaged when decoration was required, instead of covering the wall surfaces with enlarged black and white photographs'.[3] The designer and publisher Noel Carrington also commented on the increased role of the mural artist at BCMI, suggesting that 'the decorative artists have been given more scope on this occasion, at the expense of the photographer

123. Mural depicting female and male figures (probably representing Art and Industry) by Duncan Grant and Vanessa Bell at the exit to 'Shopwindow Street'.

These two figures were positioned above the entrance to the Furnishing section in 'Shopwindow Street'. Notes on the 40th progress meeting on 7 August 1946 recorded that James Gardner had been provided with two drawings by Duncan Grant and Vanessa Bell. The minutes record: 'He stated that he had asked these two people to do a joint but single mural, which will appear at the end of the Main Hall. Mr Gardner was asked to speak to Mr Duncan Grant as soon as possible and remind him that he and Vanessa Bell had been asked to do one mural, and to point out that the two drawings submitted would, if carried out in the space allotted, destroy the architectural unity of the Main Hall. If Mr Gardner experiences any difficulties in persuading Mr Duncan Grant and Mrs Vanessa Bell to modify their designs so that they can be treated as one unit, he is to report back to the Committee for further guidance' (University of Brighton Design Archives: Design Council Archive, 58). Despite this, it appears that the murals produced by Grant and Bell were individual murals. It is likely the female figure was produced by Grant and the male figure by Bell. Two preparatory sketches (identified by Darren Clarke as for the murals) are held in the collections at Charleston.

and the statistician'.[4] An early draft of 'design tasks', produced over a year before the exhibition took place, shows that James Gardner (Chief Designer at BCMI) had originally considered commissioning a number of other artists and graphic designers for the various sections of the exhibition. He had in mind Abram Games, with Jan Le Witt and George Him, for a proposed introductory section illustrating the meaning of Design for Industry, and Roland Penrose for an 'exciting and rather surrealist display' relating to materials.[5] In the end, within this group, only Le Witt and Him would make it to the exhibition, producing a mural mainly consisting of umbrellas, for the 'Wet Weather' area of the Leisure section, and another luggage-themed creation for the section on 'Travel Goods' [**126**; see **90**].[6]

Correspondence relating to the organization of the exhibition shows the different terminology used to describe the murals. Sometimes described as 'background', 'wall-painting' and 'mural', these somewhat ambiguous definitions, used interchangeably, highlight the tensions that existed between the 'creative' and 'scenic' artist.[7] Documents show that this came to a head in the later stages of the installation at BCMI, where a union dispute nearly put an end to the 'creative' artists' involvement.

The employment of creative artists and their assistants – who were not members of a recognized trade union – caused a flare-up in trade relations at the Museum a month before the exhibition was due to open. The Sign and Display Trades Union and the Society of Woodworkers called a meeting. This was a serious issue for the exhibition's organizers so close to the opening of BCMI, and senior representatives from the CoID including R. Dudley Ryder, as well as mural artists Hans Feibusch, John Hutton, Barbara Jones,[8] Clifford Ellis and William Crosbie attended. Membership of the Society of Mural Painters on the part of many of the artists did not count as union membership.

Following a tour of the exhibition space, the two union officials Mr Torode and Mr Brown put forward their concerns, that the employment of non-union affiliated artists was effectively a 'closed shop'. If no agreement could be reached shop stewards would 'immediately call for a general stoppage of work of all Trades Union members throughout the Exhibition'. R. Dudley Ryder reminded Mr Torode that prior to the meeting he had 'agreed that there was a dividing line drawn between creative artists, that is to say, artists who carried out work entirely from their own designs and with their own hands, and production artists, to whom such creative artists handed their cartoons for reproduction'. The threatened strike was averted, and a compromise agreed, that artists should become temporary members of the Sign and Display Trades Union for the duration of their work at BCMI. This did not go down well with the artists. John Hutton asked that the matter should be referred to

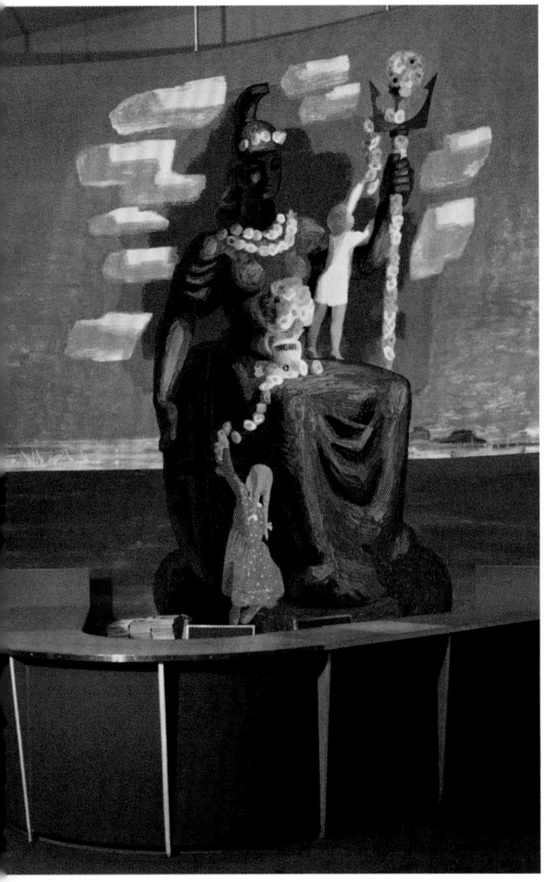

124. *Britannia* mural by Rosemary and Clifford Ellis in the entrance foyer to the exhibition. The first mural to greet visitors as they came through the Exhibition Road entrance of the V&A into the foyer of the exhibition was an over life-size painting depicting Britannia. This was an early symbolic indication of the fundamental 'Britishness' of the exhibition. Positioned unceremoniously behind an information desk, the mural depicted a seated Britannia being handed flowers by children (symbolizing a hoped-for peacetime future) whilst ships – to illustrate Britain's exporting prowess – can be seen sailing into the distance.

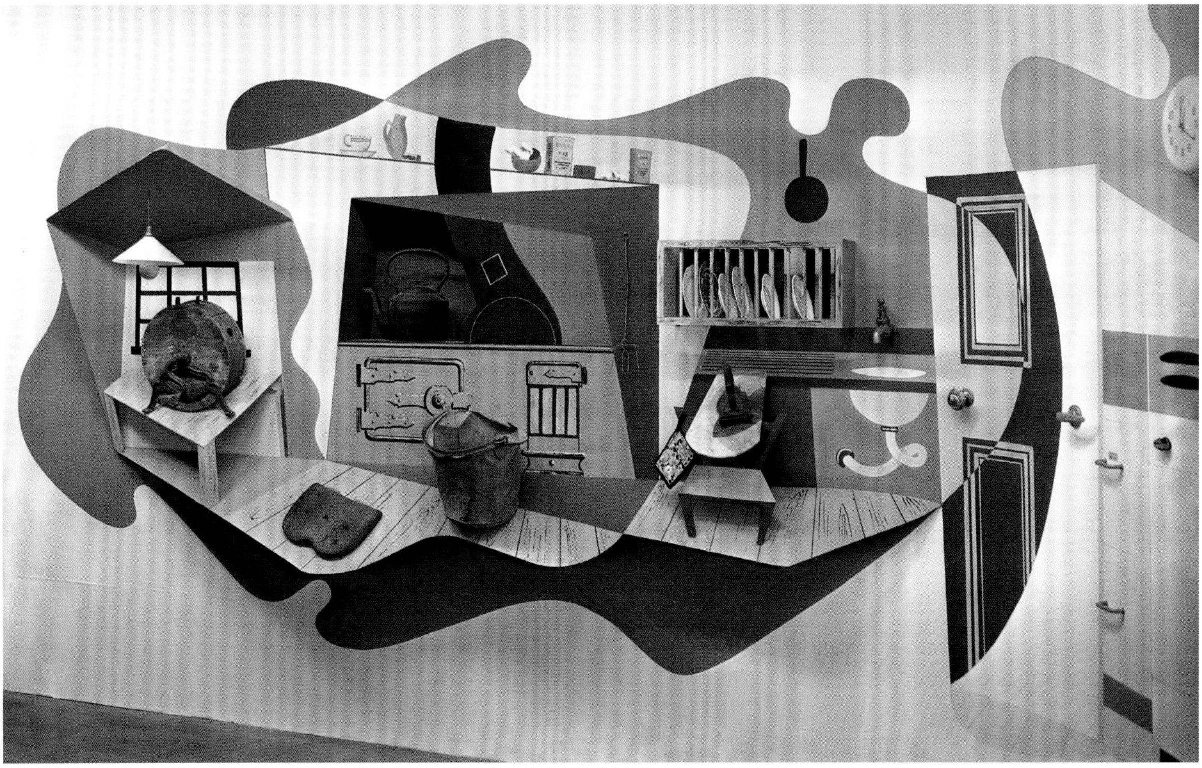

125. John Bainbridge, an Australian artist and designer, mural montage and 3-D display (Section 7) of the 'Furnished Rooms' display, showing the development of the Victorian kitchen

the Arts Council of Great Britain and the Council of Industrial Design, and was keen to reiterate that this arrangement was only for the purposes of work on the premises of the V&A for BCMI, and for no other work.[9]

James Gardner reported on the impact that industrial action would have had on the preparations for BCMI in his autobiography, *The ARTful Designer*. He wrote:

> Barbara Jones, a mural painter, was perched on a trestle, when a tubby man in a raincoat twitched at her skirt and muttered something about 'cards'. Now Barbara was an avid collector of interesting postcards, but not *that* kind, so she said, "Go away you nasty little man." The nasty little man turned out to be the Union District Organizer, so the fat was in the fire. We had a meeting. The artists were told they had to join the Union. But as artists don't get a weekly wage packet, and go to the loo in their own time, they took a poor view of being press-ganged. The Union bloke said they had members who could paint every bit as well as that bunch. He obviously could not differentiate between putting on paint and painting. A portrait artist who painted ladies and gentlemen would, I guessed, be the same to him as a man who lettered those words on the doors of public loos. Eventually, the artists joined. Had they not done so it was obvious Britain Would Not Make It.[10]

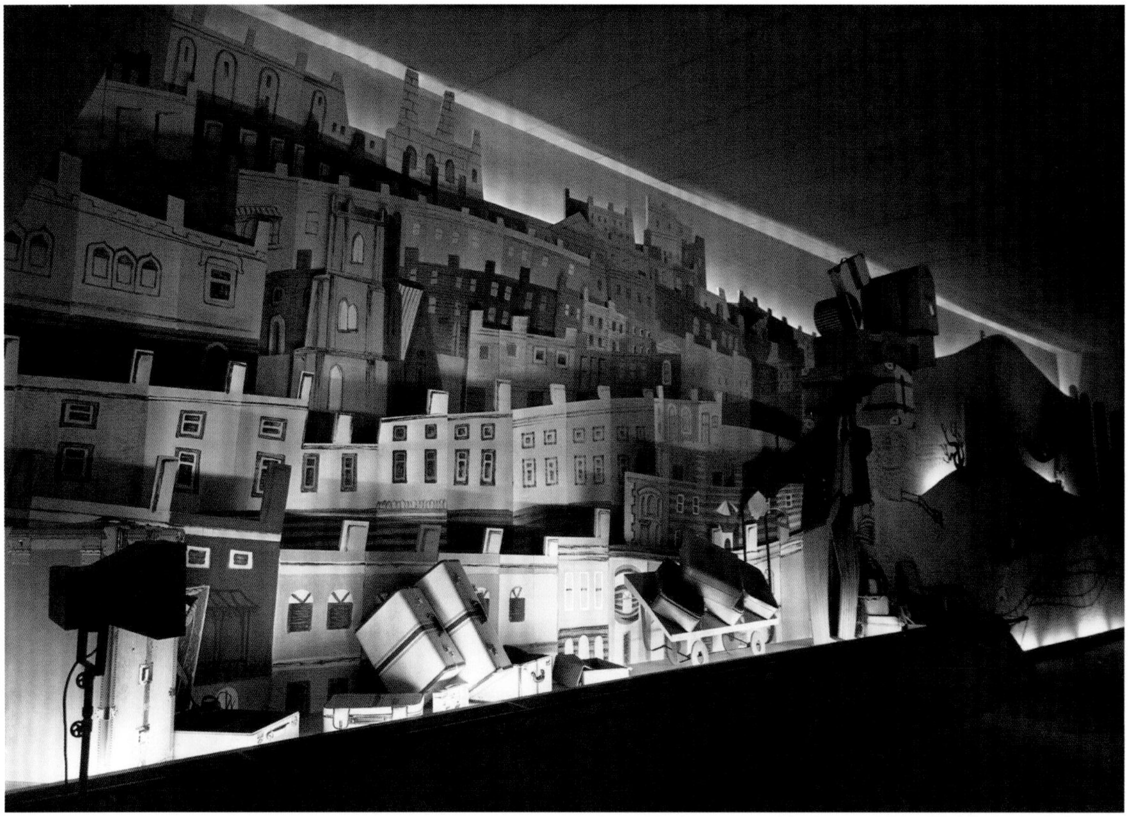

126. One of two murals produced by Jan Le Witt and George Him incorporating three-dimensional elements. The 'Travel Goods' (Section 18) included a life-size porter in relief, holding aloft a tower of luggage, against a backdrop of buildings and a railway line. Examples of the very latest in luggage design are included at the front of the display. The scene is reminiscent of the illustrations of a railway guard, luggage and rooftops produced by Le Witt and Him for a book entitled *The Little Red Engine Gets a Name*, published by Faber & Faber in 1942.

In his book *Display Presentation*, Beverley Pick (designer of the 'From War to Peace' section of BCMI) provides some insight into the frustrations of the exhibition mural artist: 'The exhibition mural presents its creator with many technical problems not encountered in the more orthodox permanent work'. The issue was primarily the speed at which artists were expected to work on murals 'undertaken in days where weeks would be desirable'.[11]

Artists appear to have negotiated or been paid different rates for the murals they produced and also handed over the rights to the work to the CoID.[12] Although all the products that had been displayed in the exhibition were returned to the manufacturers at its close, the CoID disposed of many, if not all, the murals, along with other items that formed part of the general make-up of the displays. Although the mural paintings were the property of the Council, 'the background on which they were painted belonged to the contractors', so there is a possibility that some did survive. It appears that only one artist, Robert Scanlon, arranged to remove his work, attached to columns, perhaps because they were more portable than most of the others [129].[13]

An interview broadcast by the BBC Home Service reported on the final hours of BCMI. Mr MacLeod, Assistant Exhibition Manager, was interviewed by Audrey Russell, and the impending fate of the murals was raised. She asked, 'But what's going to happen to all the other stuff – not the exhibits – all the decorations – the plaster board, the murals, all that?' Mr MacLeod replied, 'Well, the decorations, the Murals, and special features and things, some of which belong to the Council of Industrial Design, will be retained by the Council and disposed of at a later date'.[14]

As was the norm in exhibition design at the time, the murals at BCMI seem to have been regarded as ephemeral decorative installations, of no future value or intrinsic worth.

In light of this, it is not surprising that there was not a great deal of reporting or comment on the mural decoration at BCMI. The murals were installed merely as a backdrop to the exhibition, and never intended to detract from the focus of the exhibition, which was to showcase British design and manufacturing. The journalist Sylvia Pollak did discuss the murals in her review of the exhibition, pointedly suggesting that some of the work was not up to scratch, and that some of the artists could do better:

> Murals used throughout are further evidence of a range of artistic talent rounded up to form part of the Council's team. This is an admirable display. But I would suggest that the collection is not yet up to standard for the world tour advocated by some of the daily Press. Give manufacturers and artists alike a couple of years to regain and retrain skill, vitality and inspiration, to develop new materials and to experiment together, then the goods will, I hope, be up to the standard of the 'shop window' here provided, and ready to tell the world 'Britain Can Make It'.[15]

A review in *The Times* also commented on the murals, picking out for special mention those in the 'Shopwindow Street' section of the exhibition, which included the work of William Crosbie, Duncan Grant and Vanessa Bell, commenting, 'many of which are of great merit'.[16]

Despite the apparent lack of public interest in the murals, guides to the exhibition were provided with information should they be asked about them. The organizers clearly felt that they might generate interest, though it seems the public either did not like or did not even notice them.[17]

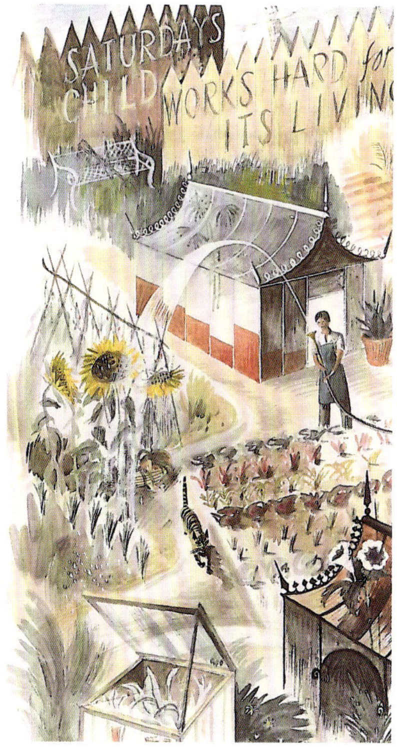

127. Barbara Jones was commissioned to produce a mural based on the nursery rhyme 'Monday's child is fair of face' for 'Things for Children' (Section 12). This is the preliminary sketch for the 'Saturday's Child' part of the mural.

Barbara Jones was one of the artists interviewed by the BBC Home Service on the opening of the exhibition. When asked how long she had to do the job, she replied, 'Well I was very wicked; at the beginning I rushed away to Ireland ... and so I had to do all the designs from drawings the size of postage stamps, but at the end I finished very calmly with a fortnight to spare' (BBC Written Archive 'Opening of Britain Can Make It exhibition', broadcast 24 September 1946, Film T126).

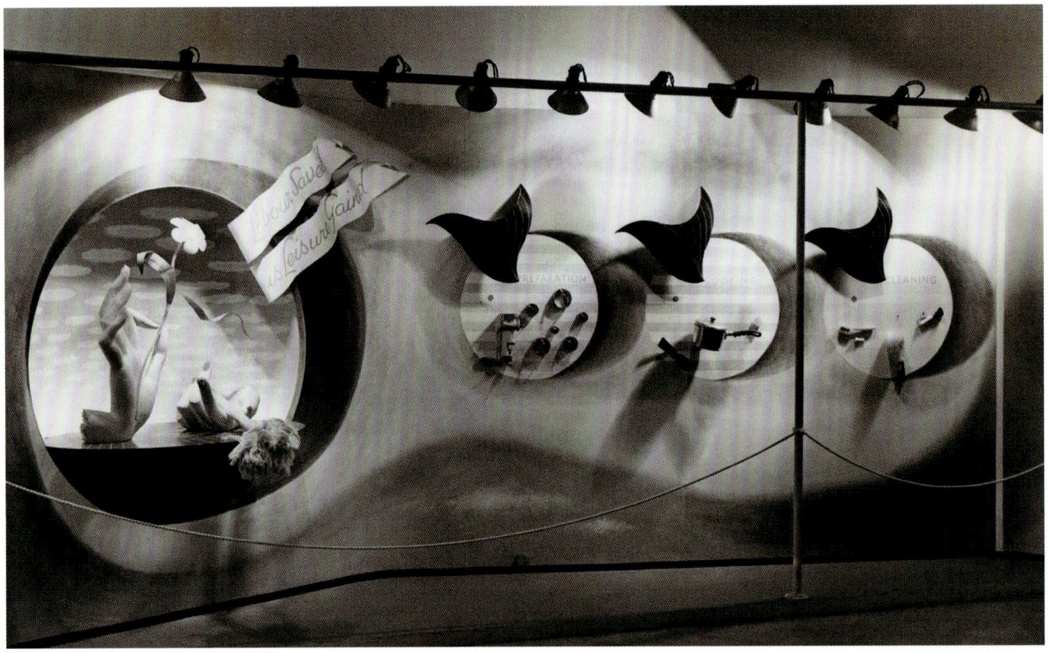

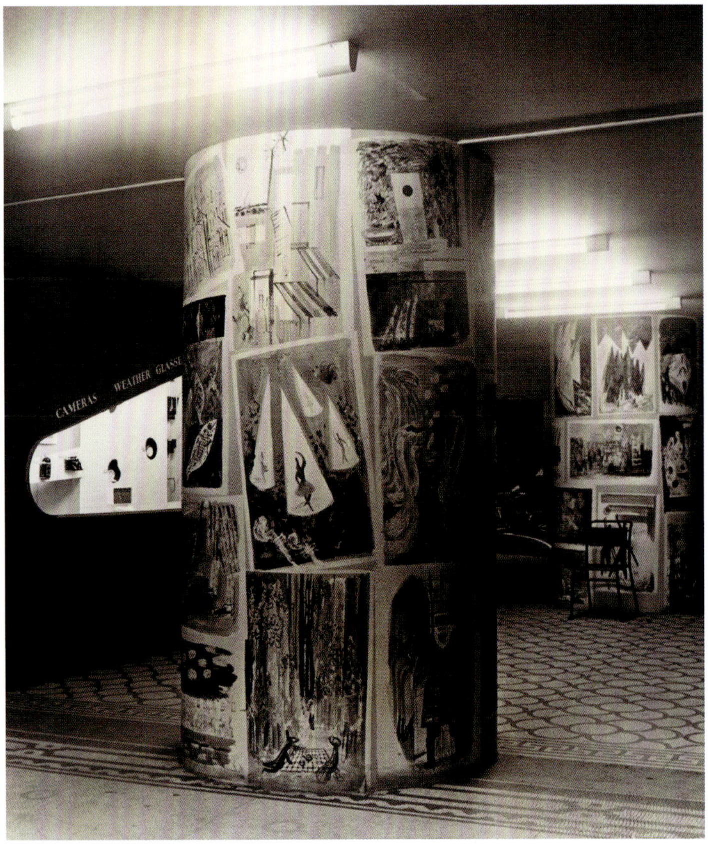

128. John Bainbridge produced this Surrealist mural montage, entitled *Labour Saved Is Leisure Gained*, for the introduction to the area devoted to kitchens in the 'Furnished Rooms' (Section 7). It was a symbolic indication of the labour-saving products that the visitor would see.

129. Robert Scanlon's Weekend murals, attached to columns at the entrance to the 'Weekend' section of the exhibition (Section 20), depicted themes of sport, leisure, entertainment and travel. The style of display is reminiscent of the outdoor advertising pillars, seen throughout France, known as Morris columns after the printer Gabriel Morris. Similar columns, established by the printer Ernst Litfaß, were also to be found in Germany. Scanlon's display would therefore have been familiar imagery for returning troops. The murals by Scanlon caught the eye of a reviewer in *The Architects' Journal* (3 October 1946, p. 239): '... two circular hoardings of high-spirited posters by Bob Scanlon – the liveliest pieces of publicity design I've seen for years'.

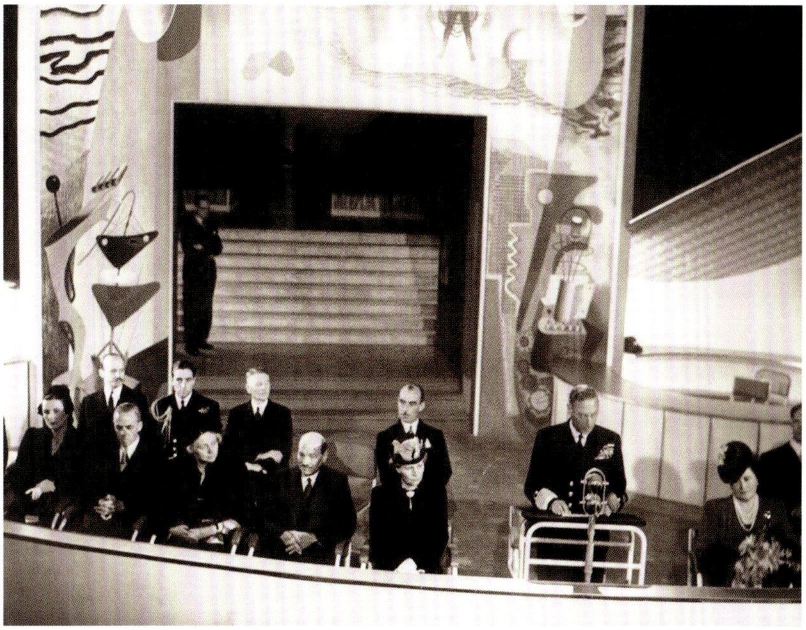

130. William Crosbie's *Voices of the Underground* mural surrounding the doorway at the entrance to the main exhibition space at BCMI, called 'Shopwindow Street' (Section 3). Crosbie's abstract painting represented voices of the Underground movement during the war.

NOTES

1 V&A Archive, Exhibition File: AAD/1977/4/13, Britain can make it: Materials used in the exhibition. For further information on mural painting see Alan Powers, 'Murals & Decorative Painting. The Mural Problem', *British Murals and Decorative Painting 1920–1960. Rediscoveries and new interpretations* (Bristol 2013); *British Murals and Decorative Paintings, 1910–1970,* exh.cat., Fine Art Society, London 2013; Ruth Artmonsky, *Exhibiting Ourselves. Fifty years of British Exhibition Design* (London 2014), pp. 103–11.

2 James Holland, 'Mural Decoration and Sculpture', *Exhibition Design* (London 1951), p. 169.

3 R. Dudley Ryder, 'The "Britain Can Make It" Exhibition', *Design '46* (London 1946), p. 12.

4 Noel Carrington, 'General Notes. Britain Can Make It', *Journal of the Royal Society of Arts* (27 September 1946), vol. XCIV, no. 4727, p. 664. In November 1945, Otto Neurath (who had devised a visual language for displaying pictorial statistics) was invited by the CoID to be a consultant for BCMI on the section explaining the 'Fundamentals of Design' (that would later become 'What Industrial Design Means'). Neurath died in December 1945, before he was able to have any involvement in BCMI, and Misha Black and the Design Research Unit were tasked with designing this display, as discussed by Lesley Whitworth in her essay. Correspondence between S.C. Leslie and Otto Neurath is held in the Otto and Marie Neurath Isotype Collection, Department of Typography & Graphic Communication, University of Reading: 1/12–13.

5 V&A Archive, Exhibition File: AAD/1977/4/16, Britain can make it: Draft List of Design Tasks Nov 22nd 1945 (drawn up by J. Gardner).

6 See Ruth Artmonstky, *Jan Le Witt and George Him. Lewitt-Him* (Woodbridge 2008); Peter Ray (ed.), *Designers in Britain* 1947. *Society of Industrial Artists* (London 1947), p. 115, illus. 9 and 10.

7 The distinction between murals and decorative painting is discussed by Robert Upstone, 'Modern British Murals', in *British Murals and Decorative Paintings, 1910–1970* (London 2013), p. 7.

8 The watercolour study for *Saturday's Child Works Hard for its Living* (illustrated) was included in the Katharine House Gallery exhibition *Barbara Jones. Paintings & Drawings,* 1999, cat. no. 36. See also Ruth Artmonsky, *A Snapper of Unconsidered Trifles. A Tribute to Barbara Jones* (London 2008), pp. 90–91. I am grateful to Neil Jennings for his assistance.

9 V&A Archive, Exhibition File: AAD/1977/4/13, Britain can make it: Notes on a meeting held at the V&A on 14 August 1946 at the request of Mr A.C. Torode, General Secretary, Sign and Display Trades Union, and Mr T. Brown, District Organizer, Amalgamated Society of Woodworkers.

10 James Gardner, *The ARTful Designer, Ideas off the drawing board by James Gardner* (London 1993), pp. 132–33.

11 Beverley Pick, *Display Presentation. Exhibitions, window and outdoor displays* (London 1957), p. 110.

12 V&A Archive, Exhibition File: AAD/1977/4/13, Britain can make it: "Britain Can Make It" Exhibition. Fees of designers, display experts and mural artists, 4.2.47. Various documents detailing the mural artists are included in the archives held at AAD, notably AAD1977/4/14A.

13 V&A Archive, Exhibition File: AAD/1977/4/14, Britain can make it: 'Britain Can Make It' Exhibition dismantling. This file contains lists detailing items for disposal and destruction.

14 BBC Written Archives Centre: Audrey Russell, 'Recordings at the Britain Can Make It Exhibition', broadcast on the Home Service 31 December 1946, Film T450.

15 Sylvia Pollak, 'Round the exhibition', *Architectural Design & Construction* (October 1946), p. 273.

16 'Home News. Britain Can Make It', *The Times* (24 September 1946), p. 2.

17 In 'British Murals and Decorative Paintings, 1910–1970. A Lost heritage?', a foreword to his book *British Murals and Decorative Paintings, 1910–1970* (London 2013), Paul Liss notes that mural commissions have often been 'written out of the accounts of the lives of the artists who created them', despite being 'amongst the most ambitious projects they ever undertook' and that they have become a 'lost heritage' of twentieth-century British art.

GILL SAUNDERS

Wallpaper

> ... when it became fashionable to dispense with wall decoration, and indeed, even with pictures, and to clear the room of almost all forms of decoration, the Wallpaper industry was, perhaps, the worst sufferer; but there is every indication of a return to patterned walls.[1]

Wallpaper reached phenomenal levels of popularity in Britain during the nineteenth century, with sales of five million rolls in 1851 rocketing to 32 million in 1874. By 1890 a two-colour machine print might cost as little as two pence a roll, although the elite products of firms such as Jeffrey & Co, designed by Walter Crane, Lewis F. Day and Voysey, sold for as much as 25 shillings a roll, 150 times as much. But from the early 1900s a prosperous middle class rejected dark dense pattern and opulent decoration in favour of lighter, less cluttered interiors with distempered walls or plain papers – of the kind which the architect Norman Shaw described approvingly as 'the "tone" wallpaper.'[2] The trend was continued when production was halted by shortages of labour and materials during the First World War. The market in the inter-war period was dominated by self-coloured and textured papers (often characterized as 'porridge' papers, an apt but damning description). Popular taste, however, eschewed this 'unnatural puritanism'[3] and continued to prefer wallpaper to distemper and patterned paper to plain, and in 1935 more than 100 million rolls were sold in Britain. But commercial success was not matched by critical reception; standards of pattern design were generally deplored by the critics – H.S. Goodhart-Rendel noted with regret "the large amount of rubbish ... joyfully accepted" by an undiscriminating public with a taste for the ornamental.[4] This decline in standards might be linked to the fact that few significant artists or designers were associated with wallpaper in this period.

Wallpaper production in Britain was suspended from 1940 until the end of the war, and in these circumstances the promotional opportunities afforded by post-war exhibitions such as BCMI were warmly welcomed by the wallpaper trade; they acted as a vital stimulus to the commission and production of new designs – though at first mostly for export. However, the BCMI initiative was not the first post-war effort of its kind, for wallpaper had already been the subject of a previous London exhibition, organized by the Central Institute of Art & Design and its chairman Thomas Acland Fennemore 'for the British Wallpaper Industry'.[5] This was held at the Suffolk Galleries, Pall Mall, and opened on V.E. Day in May 1945. Called 'An Exhibition of Historical & British Wallpapers', it offered a plethora of new designs alongside examples of the best (and a few of the worst) wallpapers representing the history of the medium from its origins in the sixteenth century to the highlights of the late nineteenth and early twentieth centuries. Many of the exhibits were lent by the V&A; others came from the archives of Cole & Son, Arthur Sanderson & Sons, and The Wall Paper Manufacturers Ltd. Six 'specimen' rooms showed wallpapers in context. These included 'A Very Small Combined Sitting-Room-Dining-Room in a Flat' designed by Lady Clark, a 'Cottage Dining Room' designed by Isabel Gallagos of *Vogue*, and a study designed by Cecil Beaton. Only one of the six rooms had its chosen wallpaper credited to a named designer, and that was Lady Clark's, which was hung with Graham Sutherland's small-scale white-on-yellow 'Abstract' pattern, produced by Cole & Son. Sutherland was also the only contemporary designer mentioned by name in Sacheverell Sitwell's introduction to the catalogue, where he applauded the wisdom of commissioning 'one of the greater painters of our time' **[131]**.[6]

Though Sutherland's designs seem to have attracted much of the critical attention, there were 180 patterns by 98 contemporary designers on display, with hundreds more available to view in portfolios. The designers – established names and newcomers – were drawn from a variety of disciplines, including painting, illustration and textile design. The exhibition was widely reviewed in the architectural press, and interior designers and critics enthused about the new designs, but only a handful ever went into production. However, a number of the featured designers went on to work with the wallpaper industry, and in the 1950s their contributions undoubtedly refreshed a rather staid and conservative business. In the short term the new designs did not find an entirely receptive audience amongst the wider public, for the British market for wallpapers was still shrinking, as consumers increasingly opted for cleaner lines and less 'busy-ness' in their homes. In an article discussing 'The Future of Wallpapers' (July 1945), the architect and designer Robert Goodden recounted a conversation with a friend in a pub. Their discussion of houses and interior decoration was overheard by the barmaid, who joined in to dismiss wallpaper out-of-hand: 'It's old-fashioned. Nothing new about it, nothing stream-lined'.[7]

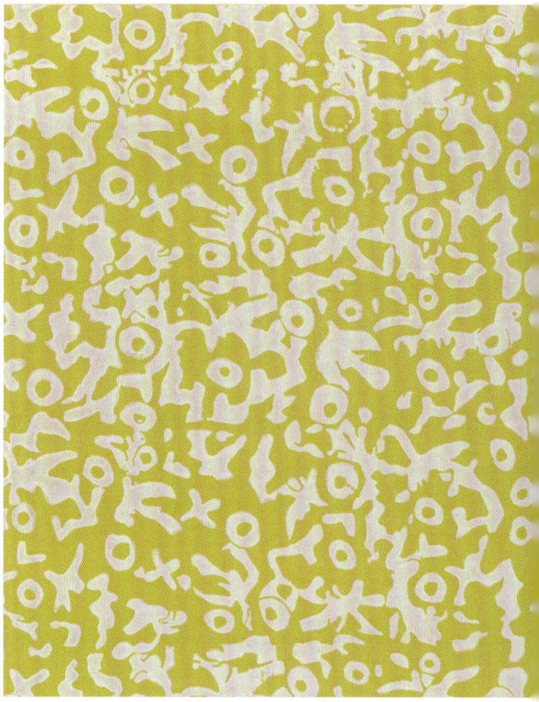

131. Graham Sutherland, *Abstract,* 1945, colour print from wood blocks, produced by Cole & Son Ltd. V&A: E.1224–1974

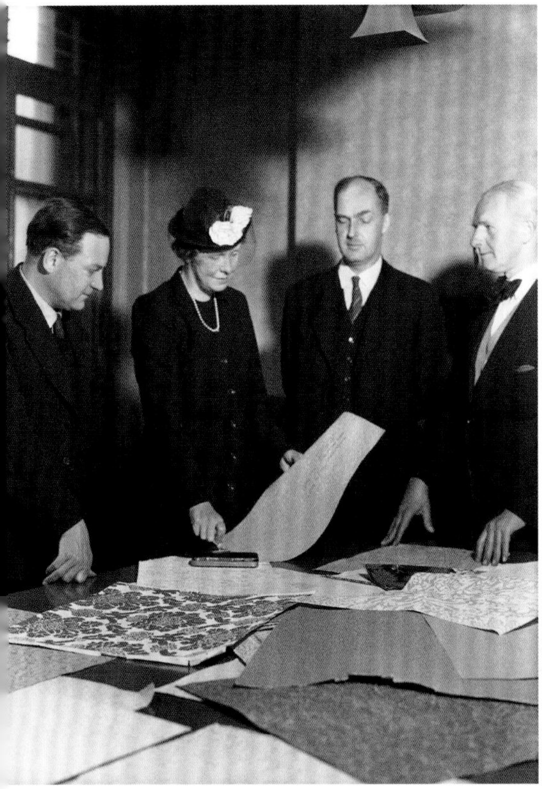

132. Wallpaper selection committee at BCMI

Nevertheless, wallpaper had a strong presence in *Britain Can Make It* (BCMI). Several of the patterns featured in the 1945 exhibition appeared again, with most of the same firms listed as exhibitors, reinforcing the claim that 'Britain leads the world in this industry'.[8] Fennemore, the organizer of the 1945 exhibition, was also on the selection panel for wallpapers (and carpets) for BCMI. Wallpapers were used throughout the exhibition – not only in the furnished rooms, but also as backgrounds to displays in Hall E ('Shopwindow Street'), and to cover decorative columns in various parts of the exhibition. The Wallpapers section of the Information for Guides explained that the main aim of the selection committee was 'to select those designs which have a freshness and gaiety and which fit in with the more simple tastes of present-day furnishing'. The committee was also keen to expand the market for patterned walls, and to this end they included machine-printed papers as well as the more expensive hand-printed papers [**132**]. Indeed, they were at pains to point out that 'the difference in price need not be as excessive as one might normally imagine'.[9] In discussions on the scope of the exhibition it had been agreed that the exhibits could not be limited to those things designed for the post-war period but must also include 'the best of what was done before the War'.[10] As a result, the wallpaper exhibits included several of the so-called 'Bardfield' wallpapers produced commercially by Cole & Son in 1938–39. These papers were the work of the artists Edward Bawden and John Aldridge, friends and neighbours living in the Essex village of Great Bardfield.[11]

The majority of the Bardfield designs are characterized by charming pastiche or quirky nostalgia, with patterns based on trellis structures, or using floral and foliage motifs. Both artists were clearly influenced by historic wallpaper patterns – the Victoriana of ivy trellises, flower sprigs and lace, as well as the architectural papers inspired by the Gothic revival of the 1830s and 1840s, and papers printed *en grisaille* to imitate moulded plaster work. In BCMI Bawden's uncharacteristically modern abstract design 'Flute' was chosen for 'The living room in a large town house' and again for 'A living room with dining recess in a house near a small country town' [**133**; see **116, 120**]. Raymond Mortimer, in a review for the *New Statesman and Nation*, noticed 'A few strikingly good wallpapers' and singled out the Bardfield papers for special praise ('bravo, Mr. Bawden') but made an explicit comparison between this and the 1945 event, lamenting that 'hardly any of the best designs shown at the wallpaper exhibition seem to have been taken up by the manufacturers'.[12]

On the evidence of the catalogue, the majority of the wallpapers used in the furnished rooms were uncredited studio designs from the major manufacturers. Of these a number were reprints of historic

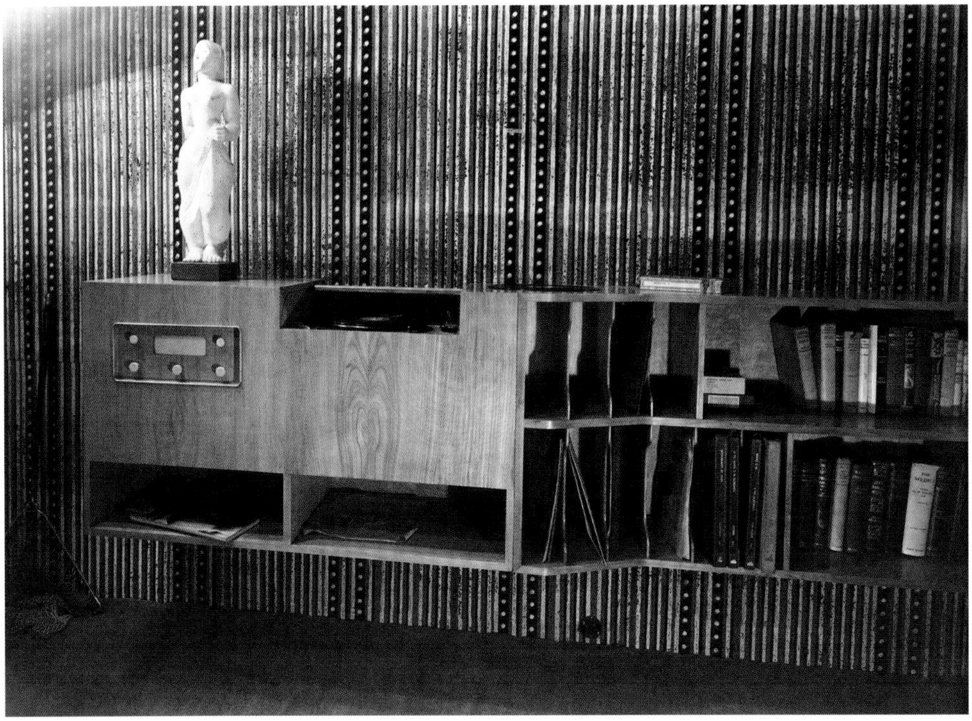

133. Bawden's Flute wallpaper in situ in the room designed by F.R.S. Yorke; on the radio cabinet is a sculpture lent by Elizabeth Spurr

patterns – such as the 'single print white star on grey ground, traditional Victorian design' used in 'A kitchen with dining recess in a small modern home', the 'Edwardian design, adapted, pale yellow wreath on yellow ground' in 'The dining room in a small suburban villa', and the 'hand-made, French grey silk flock stripe on grey ground, traditional Regency design' hung in 'A bedroom with a man's dressing room and bathroom, in a large house' [see **119, 68, 118**]. All were from Cole & Son. Though the patterns were conventional they were employed in sometimes innovative decorative schemes. For example, the grey flock stripe was used by the room's designer, John Hill, to cover the doors of the fitted wardrobes to produce a seamless wall covering [see **117**].

Even the new designs on display, which came mostly from John Line & Sons, were hardly adventurous in terms of either pattern or colour. 'Sprig', designed by Philip Pompa, was a tiny leaf motif in silver, arranged in a simple 'brick' repeat on a pale green 'watered silk' ground [**134**]; 'Walbrook' by Nat Hart was a conventional vertical stripe with stylized lines of foliage [**135**]; and Henry Skeen's 'Vanessa' was a diaper pattern with white leaf motifs on a tobacco-coloured ground, shown in 'A bedroom in a small house in an industrial town' [**136**]. All three papers were later included in a pattern book issued by the company for the years 1952–53.[13]

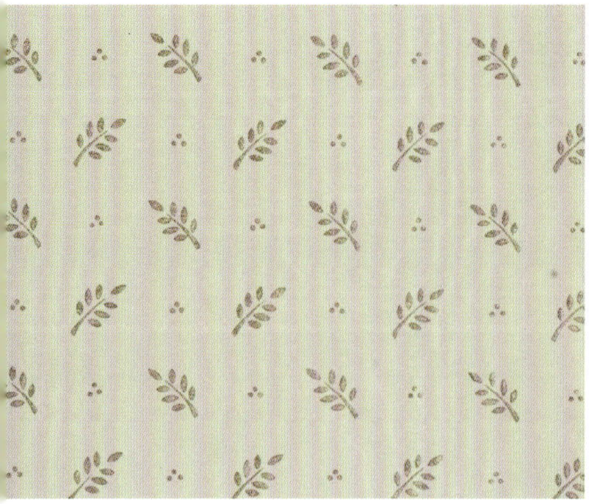

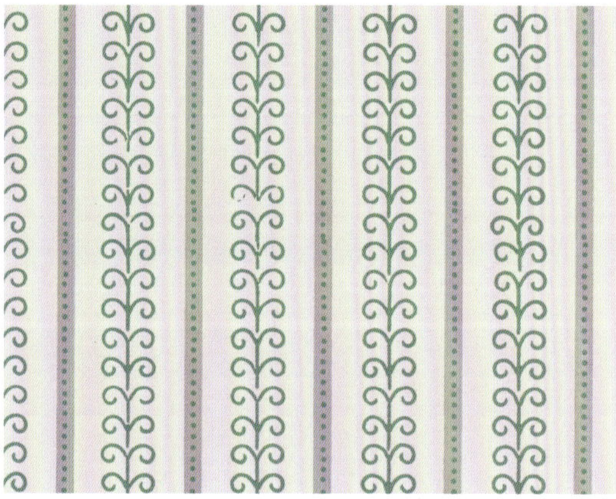

134. Philip Pompa, *Sprig*, c.1936–46, hand-printed from wood blocks with screened fillings, produced by John Line & Sons Ltd. V&A: E.2418–1980

135. Nat Hart, *Walbrook*, c.1936–46, trough and wood block print, produced by John Line & Sons Ltd. V&A: E.2435–1980

136. Henry Skeen, *Vanessa*, c.1936–46, colour print from wood blocks, produced by John Line & Sons Ltd. V&A: E.2430–1980

Overall, the exhibition favoured the products of only three companies – Cole & Son, John Line & Sons and Arthur Sanderson & Sons. Though key representatives of the wallpaper industry were involved in the early stages of planning for BCMI, some were unhappy about the selection of wallpapers. The Chairman of the Executive of the Wall Paper Manufacturers Limited (WPM)[14] reported that 'the recent selection séance gave great offence to two of our esteemed colleagues and the only result of their protest was to be told what was wanted in wallpaper was design of the character of Swedish glass'.[15]

By far and away the most successful wallpaper design seems to have been Sutherland's 'Abstract', first seen at the wallpaper exhibition in 1945 (when it was also reproduced as the cover of the *Architectural Review* for July 1945), then in BCMI, where it featured in 'A bedroom in a farm cottage'; it was still current in 1951 when it appeared yet again at the Festival of Britain. A note in the *Architectural Review* was very positive, describing 'Abstract' as 'a pattern which combines the strange animation of root or cartilage forms with the pleasant liveliness of traditional all-over designs. It is good to see that we can have busy, unostentatious, small-scale pattern without having to rely entirely on the chintzy flora of the past.'[16] The visitors to BCMI did not share the enthusiasm of the critics; there are few mentions of wallpaper in the Mass Observation (M-O) reports, beyond the comment, possibly prompted by the Sutherland design with its bright greenish-yellow ground, 'Bright wallpaper in the bedroom was said to cause nightmares'.[17] This had been a long-standing concern – evidenced in home decorating guides since at least the 1870s – and was apparently an issue that still resonated with the public. The artist Michael Ayrton, in a review of the 1945 wallpaper exhibition, expostulated at some length about the potential for wallpaper patterns to distress or disturb, concluding:

> Let it be a new and interesting paper, say by Graham Sutherland or Edward Bawden, or a delicate floral design by A.H. Williamson or Betty Tanner if I am fit and well, but let me not run mad counting blobs, leaves, or grass heads if I am in bed with 'flu.[18]

Nevertheless 'Abstract' was destined to become a minor design classic, appearing yet again as one of twelve patterns in *Wallpapers for the Small Home*, a brochure issued around 1950 by the CoID.[19]

The tastes of the general public were better gauged by the results of a Design Quiz [see **75**]. Exhibition visitors were offered a choice of three patterns – one traditional (A), one modern but unchallenging (B), and one more radical (C) – and asked to vote for their favourite.[20] Perhaps unsurprisingly they chose B, described as 'a beige ground with a regular pattern of off-white lines, with small off-white flowers of different shapes in the middle of each diamond'. The Design Quiz report concludes: 'The pattern preferred by the great majority would probably be the easiest to live with. It would tone with most furnishings but at the same time would be the least likely to become monotonous.'[21]

NOTES

1 T.A. Fennemore, 'Wallpapers', *Design '46* (London 1946), p. 85.

2 W. Shaw Sparrow, *The British Home of Today* (London 1904), cii. To provide a context for the pricing of wallpaper today, a basic pattern from a DIY store can cost as little as £9.99 per roll whilst a roll of hand-block printed designer wallpaper can cost in the region of £800–1000.

3 H.S. Goodhart-Rendel, 'Paperstaining', Review of 1945 Wallpaper Exhibition, *The Architect and Building News* (1 June 1945), p. 123.

4 Ibid.

5 Fennemore was himself a designer of wallpaper, as well as textiles and pottery. See V&A Archive, AAD/1983/3, T.A. Fennemore papers 1924–58.

6 *Catalogue to Exhibition of Historical & British Wallpapers*, Suffolk Galleries, London (May 1945), p. 7.

7 R.Y. Goodden, 'A talk at the "Salutation"', *Architectural Review* (July 1945), vol. 98, p. 21. Goodden was also involved directly with BCMI as the designer of the sports goods section.

8 Fennemore 1946 (cited note 1), p. 85.

9 V&A Archive, Exhibition File: AAD/1977/4/14A, Britain can make it: Council of Industrial Design 'Britain Can Make It' Exhibition. Information for Guides. Wallpapers.

10 P.J. Maguire and J.M. Woodham (eds.), *Design and Cultural Politics in Postwar Britain. The Britain Can Make It Exhibition of 1946* (London and Washington 1997), p. 149; ID312/14A Appendix to CM (45) 9 Scope of the Exhibition. Report by Sir Charles Tennyson.

11 From the time of Bawden's arrival in 1931 and Aldridge's in 1933 Great Bardfield attracted a number of artists and designers, and became for a while a notable artistic community. See G. Saunders and M. Yorke (eds.), *Bawden, Ravilious and the Artists of Great Bardfield* (London 2015).

12 Raymond Mortimer, 'Britain Can Make It!', *New Statesman and Nation* (September 1946), p. 220.

13 Pattern book in the V&A Museum no. E. 2296–2446–1980.

14 In 1899, 31 individual firms combined to form The Wall Paper Manufacturers Limited (WPM), which was organized as a number of branches, operating their own mills. Each had its own design department but certain functions such as purchasing, sales, advertising and promotion, and transport were managed centrally.

15 Directors' Weekly Reports, 1 March 1946; quoted in Lesley Hoskins, 'The Industry', *A Popular Art: British Wallpapers 1930–1960*, exh. cat. (Silver Studio Collection, Middlesex Polytechnic, 1990), p. 51.

16 *Architectural Review* (July 1945), vol. 98, p. 1.

17 Mass Observation Archive, University of Sussex, Britain Can Make It exhibition: SXMO A1/2126/1/A.

18 Michael Ayrton, 'Historical and British Wallpapers at the Suffolk Galleries', *The Spectator* (18 May 1945), p. 451.

19 V&A Museum no. E.1224–1974. Sutherland is the only named designer represented in the folio. The wallpapers featured were from Arthur Sanderson & Sons, Shand Kydd, the Wallpaper Manufacturers, John Line and Sons, and Cole & Son.

20 The patterns have not been identified but the description of B suggests a design very similar to that of Henry Skeen's 'Vanessa' for John Line & Sons, Ltd, V&A E.2430–1980.

21 V&A Archive, Exhibition File: AAD/1977/4/14A, Britain can make it: Design Quiz. Indications of public taste as shown by votes cast by visitors to BCMI between 23 September and 16 October, p. 3.

CATHERINE HOWELL

Things for Children

137. 33 children from the Hornchurch Chronicle Childrens' Corner visited BCMI on 7 December 1946. Here they are admiring the roundabout display in 'Shopwindow Street' designed by James Gardner, toy display arranged by D. Stuart Bell and Miss Franklin.

138. Patricia Leith designed the murals behind the display of children's toys in the 'Things for Children' section. She also designed the figures and decorative fascia over the canopy.

> ... the things are just new to them –
> they haven't seen anything like it before.[1]

This quote by a teacher bringing a group of schoolchildren to see the *Britain Can Make It* (BCMI) exhibition really sums up the view of most visitors regarding the 'Things for Children' display. This was Britain looking forward, seeing children as the future, but with an eye on the past.

Pre-war, British toy companies had done very well following the introduction of the Import Duties Act of 1932, which placed a 25% duty on toy imports, resulting in well over a thousand new toy companies being launched during the 1930s. This all changed with the outbreak of war.

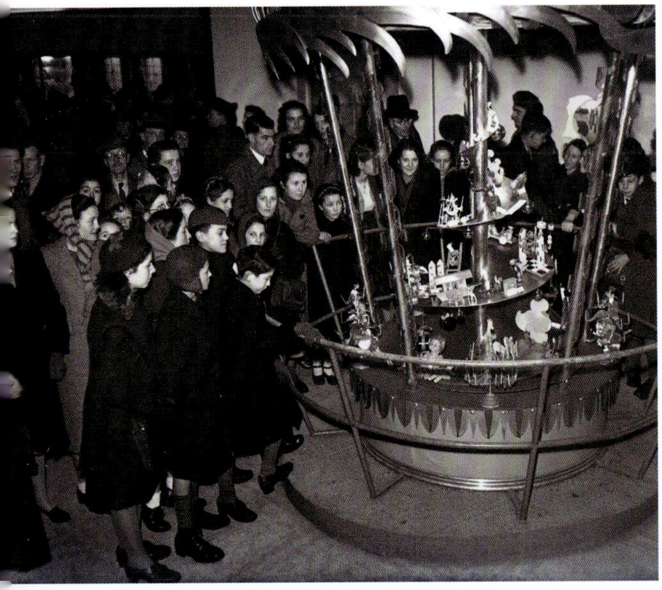

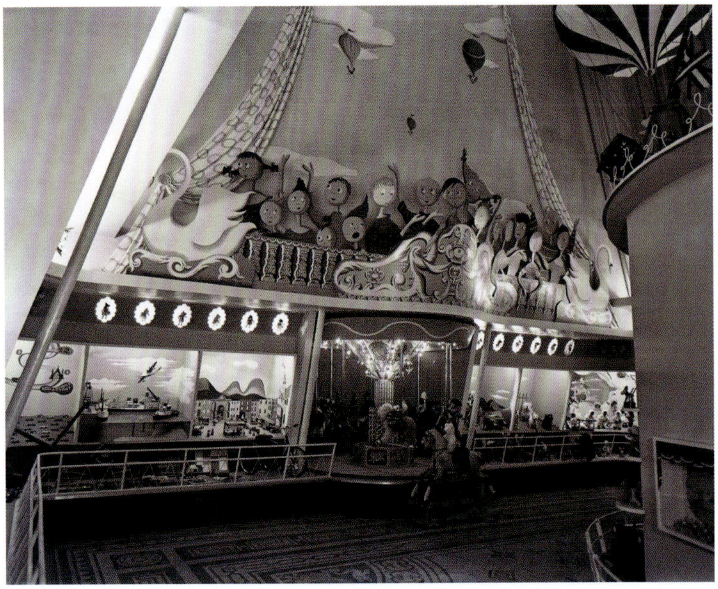

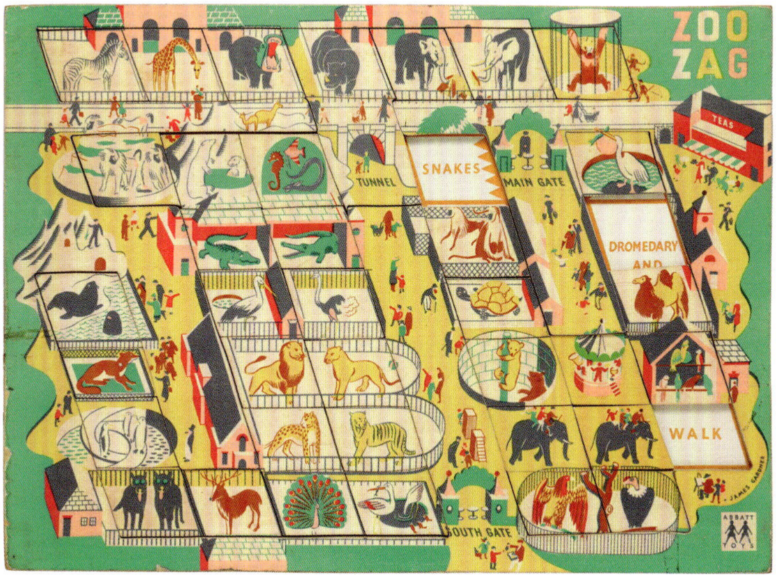

139. Designed by James Gardner, manufactured by Abbatt Toys, Zoo-Zag puzzle, c.1950, plywood with painted veneer, 45 x 32 cm. V&A: B.67–2009 (Given by Eve Harrison)

Children had had a lean time for playthings during World War II and manufacturers at BCMI felt that, as well as showing off their products, they needed to compensate children for their time of deprivation. During the war these companies, for the most part, had not been producing toys. Their factories had been contributing to the war effort, making weapons and parts for vehicles and fighter planes. In some cases toy designers had turned their attention to guns. Over 1,000,000 sten guns, redesigned by Walter Lines, were produced at the Lines Bros. Merton factory.[2] Meanwhile children had to make do with pre-war and home-made goods, toys in particular. The BCMI exhibition offered a glimpse into a bright and exciting future for its young visitors.

The children's section of the exhibition was extensive, with the main part devoted to clothing, furniture and toys, and other material featured in school and nursery room settings. The Children's Hall was designed by James Gardner and Basil Spence, and featured a large mural by Barbara Jones, depicting the nursery rhyme *Monday's Child is Fair of Face* [see **127**]. Gardner was himself a designer of toys, working with Paul and Marjorie Abbatt (of toy manufacturers Abbatt Toys), whose business ethos was underpinned by a deep interest in both toy design and the value of play [**139**]. The overall design of the 'Things for Children' display at BCMI was extremely imaginative. Toys were displayed in scene settings, each one relevant to the type of toy being highlighted. Teddy bears and soft toys were placed in a toy theatre, there was a carousel with wheeled animals and ride-on vehicles, and pirate figures dangled, cutlasses aloft, from a balloon overhead. A huge

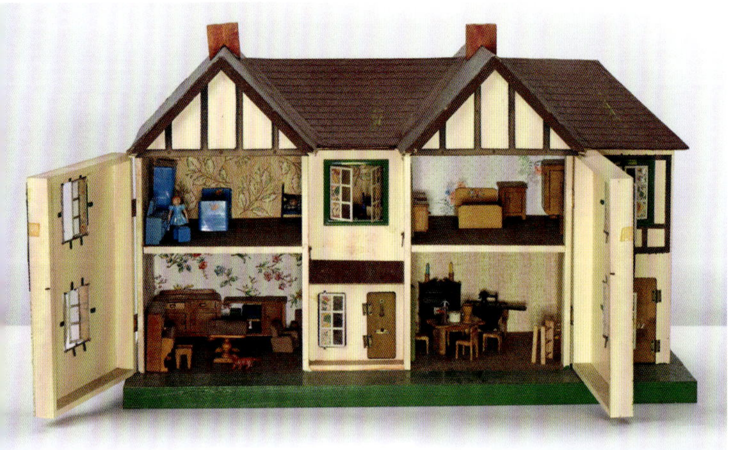

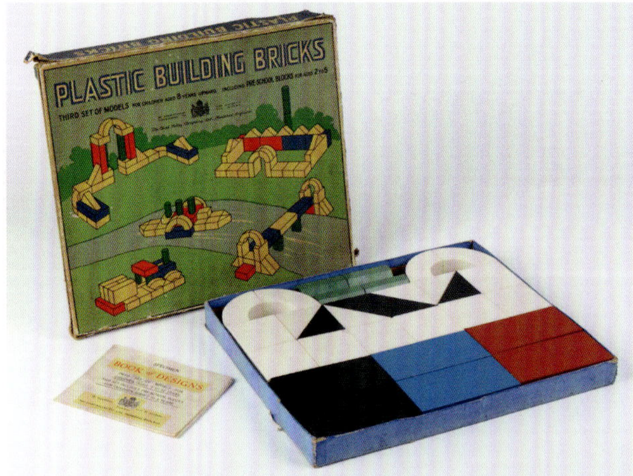

140. Manufactured by Lines Bros., dolls' house, c.1951–52, painted wood, 53.4 x 84 cm.
V&A: Misc.250–1987 (Given by Anne Susan King)

141. Manufactured by Chad Valley, building bricks, c.1946, plastic, 12.25 x 10.5 cm (box).
V&A: Misc.2–1989 (Given by Mr D. Gill)

birthday-cake paper sculpture by Tadeusz Lipski sat above the displays. Everything was designed to attract and delight.

Most of the major British toy companies from the past were represented at BCMI. The selection panel for the toys section comprised buyers from Harrods and Selfridge's; Lady Semphill, Beauty editor of *Vogue* and author of *English Pottery and China,* and the Vice President of the Design and Industries Association; and Barnett Freedman, who had been a war artist and had collaborated with James Gardner on the 1939 exhibition held at Shell-Mex House on the history of civil aviation, *How They Fly.* Possibly reflecting the personal choice of selector Freedman, International Model Aircraft Ltd, a subsidiary company of one of the best known toy manufacturers Lines Bros. (famous for their globally recognized Tri-ang brand) was well represented, with over four dozen model aircraft on display. Lines had already expanded its operations in the post-war period, opening new factories in Wales (1945) and Northern Ireland (1946) [**140**]. The company's tried and tested designs of dolls' houses, toy vehicles, pedal cars and doll's prams were augmented with several examples of a new toy – a push puppet called a Wakouwa.

As well as the major British toy companies that were represented at BCMI, there were also many smaller ones whose names are not remembered today. The Chad Valley Company was a pre-eminent producer of a wide variety of toys, from puzzles and games to teddy bears and soft toys [**141**]. For BCMI it chose to showcase its comparatively new use of plastic with examples of building blocks and toy vehicles. Although toy design and manufacture had been using plastics before the war, new formulas made it easier to mould and more durable. Significantly, it was to prove the material that globally revolutionized toy production.

Many companies only exhibited one or two items at BCMI, and it was obviously the intention of the organizers to show as many different products as possible. The long-established company of William Britain – famous since the late nineteenth century for its range of toy soldiers and figures made of lead – only displayed a few figures alongside a model of the state coach and a model Howitzer gun.

Barronia Metal Works was one of the companies singled out in the publicity for the exhibition, possibly because of its wartime record. It had made top-secret fuel valves for Lancaster bomber aircraft during the war and the precision-engineered toy mangles and cranes on display in the exhibition aptly demonstrated the ethos of one of the proposed titles for the exhibition, 'Swords into Ploughshares'.

Most of the favourite toy types among children were represented in the exhibition. There were construction toys by Lott's Bricks and Plimpton Engineering, who made the popular Bayko sets, and from Nicoltoys, who made Multibuilder. Teddy bears and soft toys by companies such as Pedigree and Deans were also included, as were toy vehicles, dolls, ride-on vehicles, toy figures and animals, dolls' houses and miniature furniture.

Kiddicraft and Abbatt Toys showcased their ranges of educational toys [142]. Hilary Page, founder of Kiddicraft, was an early pioneer in the use of plastics and had designed a range of 'Sensible' toys in the 1930s, and later versions of these classics were on show at BCMI. Paul and Marjorie Abbatt were at the forefront of producing well-designed toys. They had close connections with other creative people, including Ernö Goldfinger, who designed their London shop, and, as well as with James Gardner, worked with artists and designers such as Freda Skinner and Milan Morgenstern.

Although it was often the case that the director also acted as designer, some toy companies did employ designers. Toy train manufacturer W.J. Bassett Lowke was an active member of the Design and Industries Association and keenly employed designers, including Henry Greenly [143].

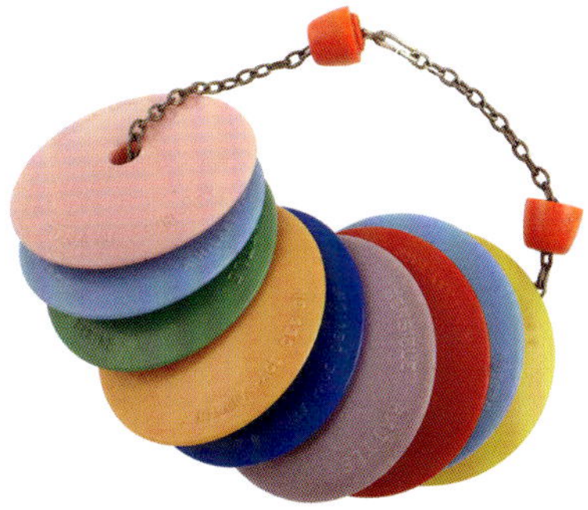

142. Designed by Hilary Page, manufactured by Kiddicraft, "Sensible" Rattle, c.1938, plastic and metal, diam. each disc 5 cm. V&A: B.72–2011 (Given by Tatiana House)

143. Manufactured by Bassett Lowke, Flying Scotsman locomotive and tender, c.1930s, painted metal, locomotive 33 x 11 x 6.5 cm. V&A: B.239–2012

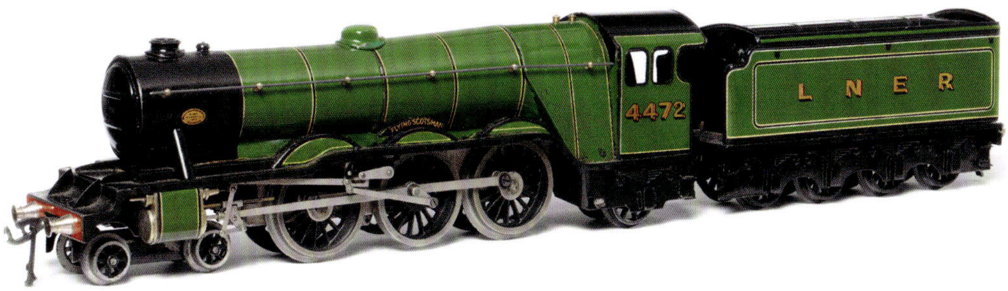

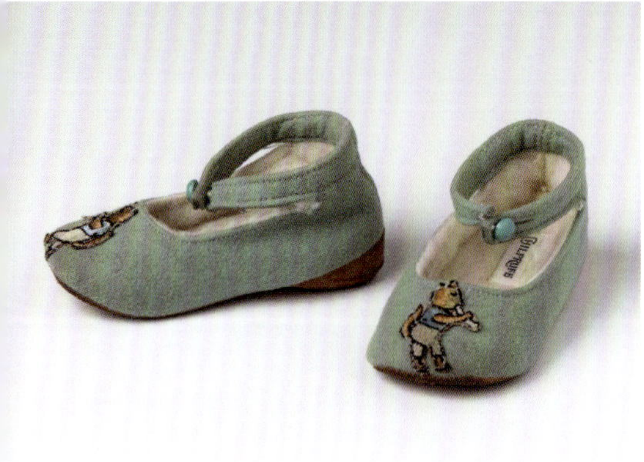

144. Manufactured by Chilprufe, pair of children's slippers, c.1940, felt, 12.1 cm. V&A: Misc.731–1992 (bequeathed by Mrs Edith Margaret Aston Hope)

Madeleine Biggs, who began making toys as therapy for soldiers following World War II, worked for the soft toy companies H.G. Stone (Chiltern Toys) and Pedigree, and Lott's Bricks were designed by the architect Arnold Mitchell. Heal & Son contributed several toys to the exhibition from the leading craft designers and toymakers of the day, including Yootha Rose and Mary Horder.

Compared to the very large area devoted to adult fashion, children's clothing, including Utility wear, formed only a small part of the children's section [**144**]. The selection panel for this area were buyers from Peter Robinson Ltd and Fortnum & Mason, and the editor of *Children's Outfitter* magazine. Most of the names featured are not remembered today, but well-known clothing companies such as Jaeger and Chilprufe exhibited, whilst footwear was provided by Norvic and C. & J. Clark. Most of the clothes were shown on mannequins (with amusing animal heads) made by costumier Hugh Skillen, and imaginatively displayed in a central garden area, in front of the Jones mural, as well as in the furnished room sets such as the nursery schoolroom. At the time, Mary Neale commented that the priority for children's wear was that it should appeal to children and be 'youthful, useful and pretty'; only then should design be considered.[3]

Visits by children to the exhibition – mostly as part of school groups – were well planned, and their reaction to the displays was documented by the Mass Observation (M-O) report commissioned by the CoID. Guides were provided for teachers planning school visits, and well-informed exhibition attendants helped explain the different displays.[4] A special guide leaflet for children with a colour sketch of the children's section by James Gardner was on sale for 1d, aiming to teach children what design was [**145**; see page 8]. The information for guides working in the area of the toy section highlighted the modernity and usefulness inherent in the designs. A toy sweeper 'works so well that an adult model on the same principles is now being made', whilst the use of plastic in model yachts is praised for its durability and cheapness. The claim for post-war construction sets is that they have been designed 'to teach boys not only to use their hands but also to use their heads'.[5]

At the royal opening of the exhibition, *British Pathe* reported that 'most attention went to the lavish display of toys'.[6] This was certainly true in the case of child visitors. Comments from the M-O report show that, whilst it was easy to collect children's comments from most areas of the exhibition, when it came to their own section the babble and hubbub was so loud that individual comments were often lost. There were some obviously gendered differences of opinion. Compare this, from a six-year-old boy – 'it's very nice. I liked the racing cars and the boats and the electric trains' – with this, from a 13-year-old girl, 'the fashion display

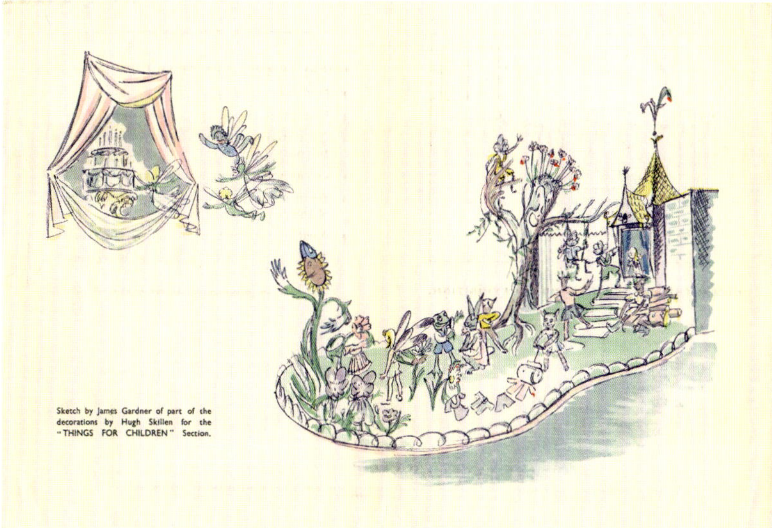

145. Inside page from the *Children's Guide to* BCMI, showing James Gardner's sketch showing part of the 'Things for Children' section

was marvellous ... it's all beautifully arranged'. However, both girls and boys appeared to like the typewriters and glassware.

This may be partly because some children parroted their teacher's reactions, but most were clearly the children's own opinions. A few comments made by older children would have pleased the CoID: 'I wanted to see some of the new things being designed', and 'we wanted to look ahead'. Products such as the new innovative designs of the streamline bicycle and modern kitchen equipment were definitely appreciated by this age group.

The classroom was the only area of the exhibition that came in for criticism by children. Girls from Adys Road Secondary School described the desks as 'horrible' with nowhere to put the books [see **110, 122**].[7] This comment, both practical and illuminating, would not have pleased the CoID, which was particularly focused on school design.[8] The information provided to guides in this area claimed that the school furniture was made from a combination of wood and metal and was light but strong, and intended to 'make young people familiar with objects of good design'. However, the children's opinion was that they did not look very strong or comfortable and one boy's comment was that 'we'd soon smash 'em up – they're like tishy [tissue] paper'.[9]

Writing in *Design '46*, Lady Semphill perceptively declared that 'British toys almost invariably have a lasting appeal for children, though it is not always easy for grown-ups to remember that this is the criterion by which all toys should be judged'.[10] Although British toy companies were keen to be represented at the exhibition, they did not view it as a trading opportunity. *Games and Toys*, the leading toy trade magazine of the day, reported on the toys in BCMI, but it is clear from their coverage that the exhibition was not considered as important as the major British toy fairs of the time. Consequently, only a handful of exhibitors used *Games and Toys* to advertise the fact that they were exhibiting at BCMI.[11] It is possible that those companies that chose to exhibit were keener to demonstrate the suitability of their products to the design remit of the exhibition than to use it as an opportunity to advertise.

However, *British Pathe* reported that this was where 'British toy makers will meet the world's future buyers at the nursery stage'. For these future buyers, the best show in the exhibition was the toys, 'better than those in the shops', and sadly for the children that was only too true. In common with a lot of other products on display, many of the toys were made for export only, as British manufacturers were seizing the opportunity to dominate international markets while countries such as Germany and Japan recovered.[12]

Five years later, half of Britain's domestic toy output was still being exported, which makes the fate of many of the toys displayed in the exhibition an interesting one. Instead of making their way into shops, these examples were distributed to the British Hospitals Association, and so some lucky children got to play with toys that most of their healthy counterparts would not get their hands on for some time to come.[13]

NOTES

1 Design Council Archive at the University of Brighton Design Archives (hereafter DCA)/903/A Note on the Reaction of Children to "Britain can Make It".
2 Museum of Childhood Archive, MOC/LINE/2/17.
3 Mary Neale, 'Children's Wear', *Design '46* (London 1946), p. 63.
4 V&A Archive, Exhibition File: AAD/1977/4/14, Britain can make it: Notes on "Britain Can Make It" for teachers.
5 V&A Archive, Exhibition File: AAD/1977/4/14A, Britain can make it, Council of Industrial Design 'Britain Can Make It' Exhibition. Information for Guides, Toy Section.
6 'Britain Has Made It – Shop Window', *British Pathe* (30 September 1946); Film ID 1410.22.
7 V&A Archive, Exhibition File: AAD/1911/4/14, Britain can make it: How a Child Saw the Britain Can Make It Exhibition, report compiled by Miss Dowler January 1947; Mass Observation Archive, University of Sussex: SXMOA1/2126/1/A.
8 See Jonathan Woodham, 'The Consumers of the Future: Council of Industrial Design Educational Strategies for Schools from 1944 to the late 1950s', *The Camberwell Collection: Object Lesson* (London 1996), pp. 16–25.
9 V&A Archive, Exhibition File: AAD/1977/4/14A, Britain can make it: Council of Industrial Design 'Britain Can Make It' Exhibition. Information for Guides, Educational Equipment; Mass Observation Archive, University of Sussex: SXMOA1/2126/1/A.
10 Lady Semphill, 'Toys', *Design '46* (London 1946), p. 66.
11 'Pre-View of New Toys', *Games and Toys* (September 1946), pp. 62–63; 'Britain Certainly Can Make Toys', *Games and Toys* (October 1946), pp. 58–60; *My Log* (editorial), *Games and Toys* (November 1946), pp. 52–53; *"Britain Can Make It"*, *Games and Toys* (December 1946), p. 84.
12 'Britain Has Made It – Shop Window', *British Pathe* (30 September 1946); Film ID 1410.22; DCA/903/A Note on the Reaction of Children to "Britain Can Make It". For further information on the post-war British toy industry, see Kenneth D. Brown, *The British Toy Business* (London and Rio Grande 1996); Kenneth D. Brown, 'Design in the British Toy Industry since 1945', *Journal of Design History*, vol. 11, no. 4, pp. 323–23; and Deborah Jaffe, *The History of Toys* (Stroud 2006).
13 V&A Archive, Exhibition File: AAD/1977/4/14, Britain can make it: British Hospitals Association ref. Dismantling.

PAUL JOBLING

Ruled by Fashion: Ashley Havinden and the Menswear Display[1]

The menswear display design

In a briefing paper presented to the Council of Industrial Design's (CoID) members concerning the proposed male dress section for *Britain Can Make It* (BCMI), Sir Charles Tennyson expressed the need to engage 'young men of imagination' in this section's evolution and construction.[2] The designer eventually appointed to this position was Ashley Havinden who, while 43 years old by the time the exhibition opened, was creative director at the advertising agency W.S. Crawford and highly regarded in his profession as 'one of the most brilliant exponents of modern-day advertising' [**146**].[3] A key reason for his involvement with the official poster design for BCMI (also discussed by Margaret Timmers in this book) was the use of clean lines and simplified treatment of form in the publicity he produced during the 1930s, which was regarded as both 'completely in accord with his time' and an unforced form of modernism that was 'natural to him'.[4] But it was probably his association with the retailers Simeon and Alec Simpson that secured Havinden's preferment as the designer of the exhibition's menswear section.

In 1932 Simpsons had pioneered Daks, a new style of flannel sports trousers that circumvented the need for braces, since they were held in place by rubber pads sewn into the waistband, and Crawford's won the account to publicize them shortly afterwards.[5] Havinden had evolved a national press campaign for the brand, incorporating cropped photographs of the product laid out with sans serif titles and asymmetric blocks of copy set in Grotesque light, which ran in a slightly modified stylistic vein until the 1950s and won the Layton Prize for the best British press advertisement in 1955 [**147**].[6] It was probabaly on account of his direction of such publicity for Daks that the 17-foot-high male figures that adorned the curved bays in the menswear section were tacitly attributed to Havinden [**148**]. However, as art director, he was not ever responsible for the creation of the photographs and illustrations used in Simpson's press publicity but for their overall appearance on the printed page.[7]

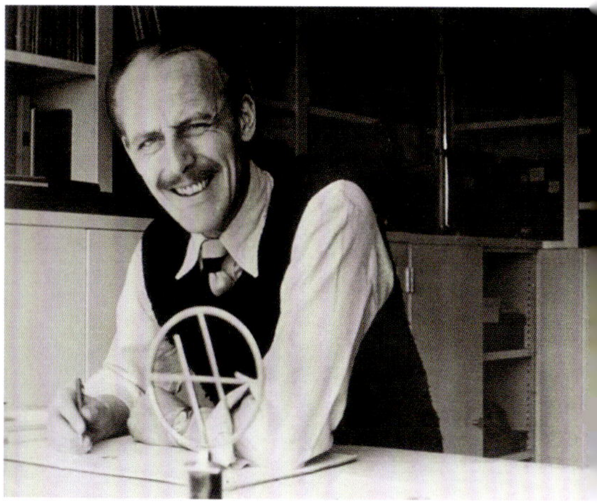

146. Portrait of Ashley Havinden, c.1947

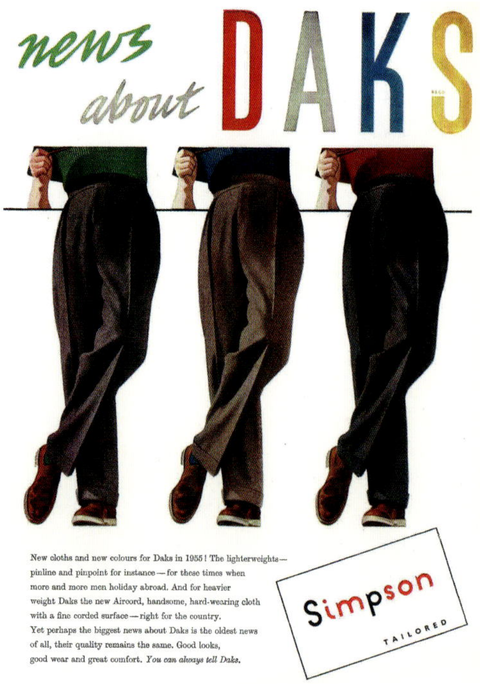

147. 'News about DAKS', advertisement in *Punch*, 6 April 1955; art-directed by Ashley Havinden for W.S. Crawford

The photographs of men wearing Daks trousers that Havinden cropped, for instance, were initially taken by Geoffrey Morris. And, although they were not explicitly credited to him in the BCMI official guidebook, the four male figures respectively wearing formal, town, country and sports attire were executed by Austrian-born artist Max Hoff (the alias of I.T. Pallas), who had moved to London in 1938 to escape Nazi persecution and worked on Simpson's advertising until the 1960s.[8] In treatment of form and style, therefore, they closely resemble the trope of the impeccably dressed and genial gentleman that he had already executed in interwar publicity for the company, while simultaneously providing visual compensation for the disembodied items of clothing and accessories on display.[9]

Hoff's monumental gentlemen were, of course, one of several ingenious design solutions that Havinden adopted to create a striking atmosphere for the awkward space that was devoted to the menswear section. It occupied approximately half of what is now the Islamic Middle East gallery at the V&A and entailed accommodating four of the structural columns that support the ceiling. So that the space could be experienced as a holistic entity, he created a series of sinuous screens to support the complementary organic display cases and curvilinear stands – the latter also incorporating vertical posts which echoed the four columns where the items of clothing were shown. By extension, the columns were clad in plastic and harnessed to the overall design, with small window panels cut into two of them at eye level through which accessories on red plush panels were to be viewed and three of them supported the display stands.[10] Indeed, between two of the columns Havinden inserted an orthogonal colonnade on whose posts items of knitwear and shirts were hung and which doubled up as a screen between the menswear display on one side of the room and the dress fabrics on the other [**149**]. Thus, as they entered the exhibit from the 'Things for Children' section, visitors encountered a sequence of seven clothing categories: town, and formal wear on the left; nightwear in the centre bay; country, and sportswear on the right; smoking jackets, dressing gowns and beachwear on a boat-shaped stand projected from the pillar on the same side [**150**]; and knitwear traversed left and right.

In carving out such a creative space, Havinden drew on the experience of exhibition design he had already gained with the books and printing stand for the British Pavilion at the 1937 Paris International Exhibition. But the organization of individual items in the menswear section was closely aligned with the inventive shop-floor and window displays at Simpson's Piccadilly department store that Havinden produced in collaboration with László Moholy-Nagy, who had relocated to London from Germany between 1935 and 1937. Simeon Simpson opened his flagship store, designed by the modernist architect Joseph Emberton, on 29 April 1936, and for the display of ready-to-wear and bespoke garments he employed Havinden as art director from the outset.[11] His striking conception for the menswear section, therefore, was more than just the consequence of his being a successful modernist in graphic design and advertising. It rested on his ability to cross over from one field of expertise to another, from the two-dimensional flat surface of the printed page or poster to the three-dimensional space of the gallery. The cross-fertilization between art and design in the treatment of form and space was regarded as an article of faith in European modernist practice of the interwar period and was something Havinden had inherited both from Crawford and through his membership of the Society of Industrial Artists (founded 1930). Accordingly, he argued that the designer 'should be familiar with all forms of modern expression in fields other than his own, because it is through the influence of modern architecture and the experiments of the purer arts of painting and sculpture that the new visual forms will be found'.[12] It was on this level that he was instrumental in shaping how the menswear would be displayed, and also received and understood by a post-war public in the institutional space of the museum. Thus he fulfilled the role of the 'author-function', as Michel Foucault termed it,[13] being not just a passive recipient of the patriotic discourse enlisted by the CoID but a proactive producer of such knowledge. He conjoined the modernist aesthetic of the publicity, carpet designs, shop floor and window displays that he had already elaborated for Simpson's to the spatial organization of the menswear gallery at BCMI in order to foreground postwar design as a nationally progressive and democratic force.

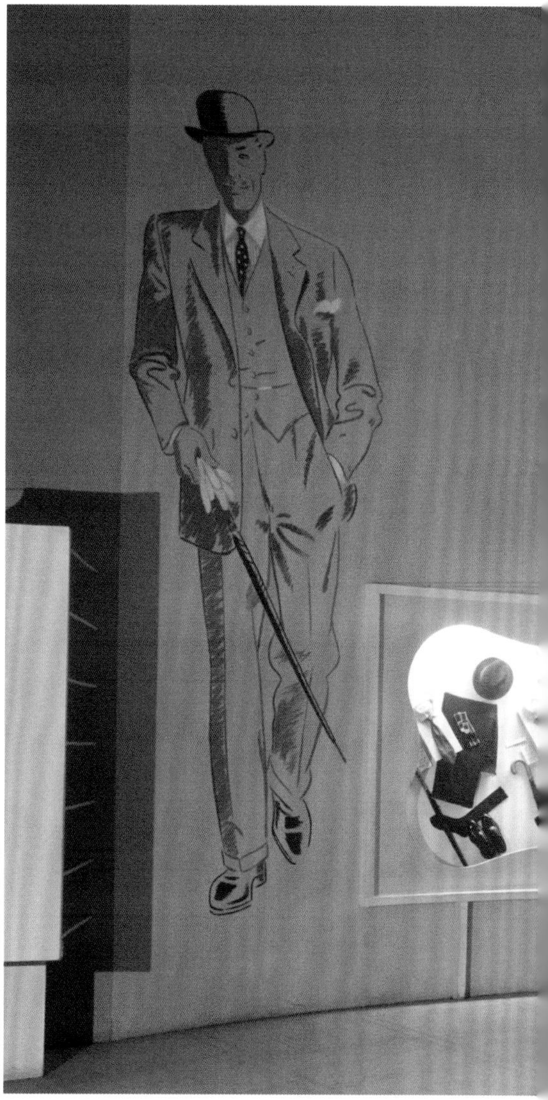

148. BCMI town and formal menswear display, design by Ashley Havinden with mural figures by Max Hoff

The merchandise on display

Havinden was as keenly interested in the merchandise itself as he was in designing a harmonious space in which to exhibit it, contributing a historical survey of male dress in Britain to *Design '46*, and echoing there an idea he had already expressed in 1940: 'Men are as much ruled by fashion as women'.[14] He was therefore an obvious choice for membership of the selection panel that decided which items would be put on display in the menswear section at BCMI, participating alongside two London-based tailors – G.F. Glenny and Charles C. Whitlock – and the journalist W.A. Swift. Together they chose a total of 231 bespoke and ready-to-wear garments and accessories, and 45 pairs of boots and shoes, which were dispersed across the seven main display categories.

The majority of menswear retailers and outfitters featured were based in Greater London. They included businesses from both the West and East Ends such as L. & R. Morley, Kingsway (who contributed pyjamas,

149. BCMI men's knitwear display, design by Ashley Havinden

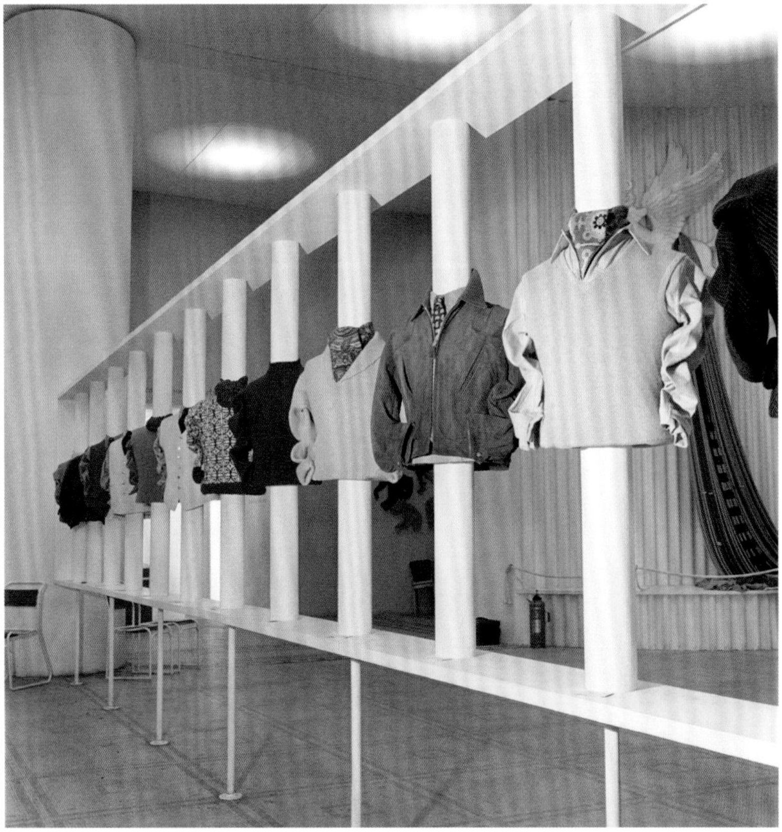

sports shirts, swimming trunks, gloves); Jaeger & Co. Ltd, Regent Street (formal and leisure shirts, ties, pyjamas, and dressing gowns); Curwen & Sons, Sackville Street (shirts); Holliday & Browne Ltd, Conduit Street (silk pocket squares); Welch Margetson & Co. Ltd, Long Acre (cashmere shirts and silk ties); Lillywhites, Regent Street (waistcoats); Sandbrook Witting & Co. Ltd, Bartholomew Close, EC1 (wool and silk ties, wool scarves); Allen Solly & Co. Ltd, Savile Row (woollen socks); and J. Lock & Co. Ltd, St James's Street (hats). Despite Havinden's association with Simpsons of Piccadilly and the popularity of the Daks brand, none of their merchandise was showcased. Moreover, items from only a handful of retailers and manufacturers from elsewhere in the UK were chosen for the display. Thus the panel's selection included cotton ties from Tootal, Manchester; various woollen garments from W. Anderson & Sons Ltd and Munro & Co. Ltd of Edinburgh; and boots and shoes by George Webb & Sons and Church & Co. of Northampton.[15]

For all its sense of invention in design, footfall to the BCMI menswear section was disappointing, with a Mass-Observation (M-O) report recording that a mere two per cent of the visitors they interviewed

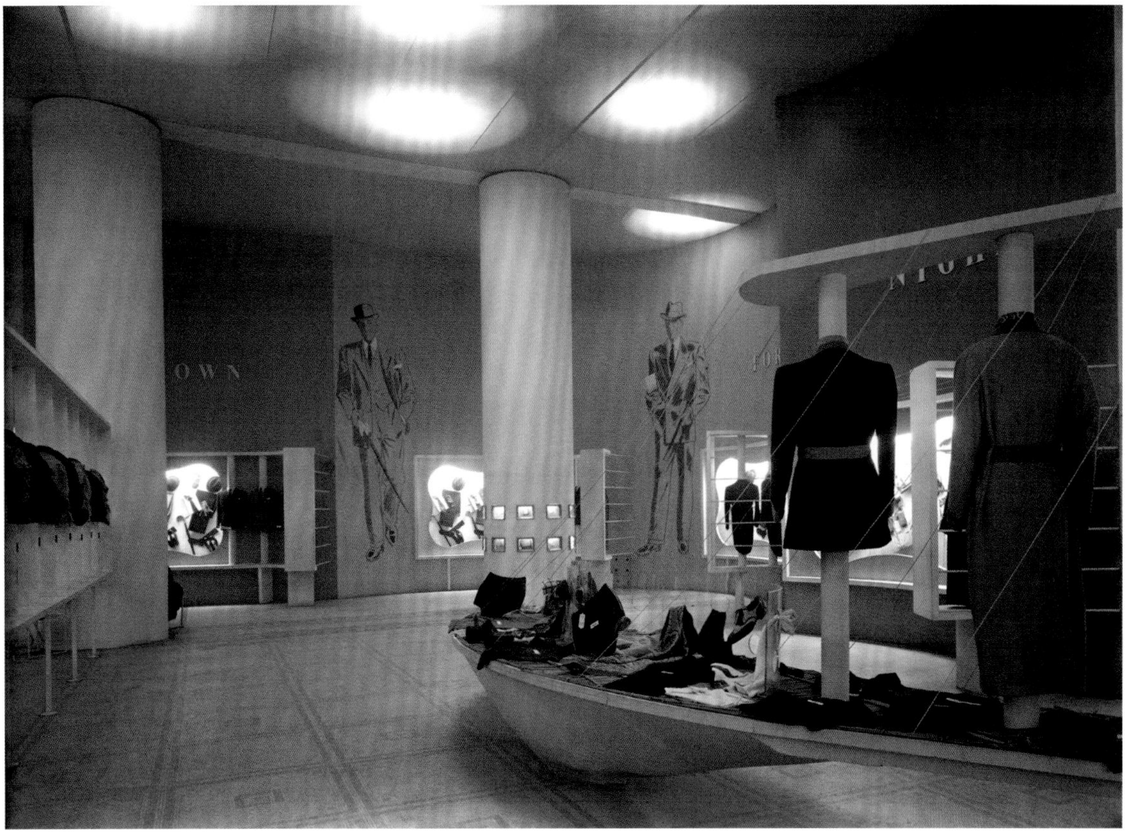

150. BCMI men's dressing gowns, smoking jackets and beachwear display, designed by Ashley Havinden

bothered to explore it. It attributed the paucity of visitors to the fact that the public was running out of steam by the time it reached the display, which was thirteenth of 22 in the exhibition.[16] Some critics also expressed harsh comments about the display of certain goods: trousers, sports shirts, shoes, hats and gloves, for instance, were singled out for adverse scrutiny in a review by Arthur Sadler.[17] As also for other parts of the show, one of the most common complaints to emerge was the unavailability for the postwar home market of the better-quality apparel exhibited, since approximately one third of textiles such as cotton and wool were earmarked by the government for the luxury goods export trade.[18] On top of this, the shortage of trained labour had curtailed manufacture, with the textiles and clothing industries witnessing an almost 25 per cent decline in their workforce, down from 1,803,000 employees in June 1939 to 1,405,000 by the close of 1946.[19] But the most significant hangover from the war effort to hamper the production and consumption of all clothing was the continuation of clothes rationing until May 1949 and of the Utility scheme until 1952, which latter had compensated to a certain extent for the shortage of raw materials and democratized to a larger extent the supply of good-quality clothing to all classes. In fact, during the war the British public was still intensely interested in buying clothes and, when asked which garment they would most likely have purchased had rationing ended in 1944, some 60 per cent of men from all classes expressed a preference for a new suit.[20] Even by July 1946, the majority of men on the panel of 100 families surveyed about the hardships of postwar living by the *Sunday Pictorial* said that they were 'especially put out by being unable to buy a new suit', and in a M-O directive one individual, a London civil servant, Anthony Heap, lamented that after trying to order a suit in grey tweed from his local Burton's branch at the Angel in November 1947 he would probably have to wait about nine months before he could have it.[21]

Following the cessation of hostilities, therefore, the display of male clothing at BCMI, like much postwar menswear publicity, ostensibly proffered nothing more than the bittersweet pleasure of delayed gratification to consumers.[22] And yet the very fact that it had showcased Utility garments alongside more bespoke items was also singled out for attention in several newspaper reports as a sign of progress.[23] A more generous use of cloth for double-breasted suits with wide lapels and longer coats was particularly commended, as were a more daring use of pattern for waistcoats, luxury fabrics in leisure garments, and a suit with jacket and trousers of different shades. *Punch*, for instance, remarked favourably on a 'velvet-faced cashmere smoking jacket in a dark wine shade with … a sash', while a correspondent in *The Times* averred: 'After all, why, save on the grounds of innate conservatism, should coat and trousers be of the same hue? We may be getting too set in our ways.'[24] It is by this measure, then, that Havinden – his protestation to eschew exhibiting any 'outré examples' of dress notwithstanding [25] – and the other members of the selection panel can be seen to have had a finger on fashion's pulse, portending what the 1950s 'new man' (as Geoffrey Gilbert was to call him) wanted to wear as well as the more flamboyant taste of the teenage male consumer with 'an eye on colour'.[26]

NOTES

1 Havinden had written in his essay 'Men's Wear', *Design '46* (London 1946), p. 73: 'The average man is in reality very concerned about his appearance, but in a different way from his great-grandfather. Men are as much ruled by fashion.'

2 Design Council at the University of Brighton Design Archives (hereafter DCA) 312/14A/ Appendix to CM (45) 9/Report by Sir Charles Tennyson.

3 'Creative Men of the Agencies No. 3: Ashley', *Art and Industry* (November 1938), p. 191.

4 'The Artist in Uniform', *Art and Industry* (1942), 32, pp. 90 and 93.

5 The trousers were called 'Daks' because it rhymed with the words 'slacks' and 'vacs', short for 'vacations': A. Havinden, 'The Importance of "Company Handwriting"', *Penrose Annual* (London 1955), vol. 59, p. 60.

6 'The Layton Annual Awards', *Advertiser's Weekly* (17 March 1955), p. 649. The modernism manifest in the campaign was closely aligned to the design ethos of *Der ring 'neue werbegestalter'* (The Circle of 'New Advertising Designers'), an international professional alliance of artists and designers, including Moholy-Nagy and Jan Tschichold, that was initiated in 1929 to promote the idea that graphic design would be 'part of the foundation on which the *new world* will be built'. In particular, The Circle argued that the style of both illustrations or photographs and sans serif typefaces, and their harmonious integration in layouts through the use of blank space and geometric or abstract forms, was paramount.

7 'Creative Men of the Agencies No. 3: Ashley' (cited note 3), p. 197.

8 P. Jobling, *Man Appeal: Advertising, modernism and menswear* (Oxford and New York 2005), pp. 102–03. When he first arrived in London his surname Hofbauer was abbreviated to Hof.

9 Sketches by Hoff also accompany Havinden's essay on menswear in the publication *Design '46*, but are not discussed by him. A letter from W.V. Cole to Mr Harker, dated 22 July 1946, states that 'Hoff's drawings are to be enlarged from 18ins to 17ft and painted in fascia with outline in one colour and modeling in another on top of the existing colour background of the fascia' (DCA/742).

10 Cited in 'Britain can make it', *Display, Design and Presentation* (November 1946), p. 17. Unfortunately, I was not able to trace any mention in the correspondence concerning the design of BCMI of the exact colours used to paint the fascia or Hoff's murals in the menswear display.

11 He also designed rugs that were installed in several of the store's departments and refined the distinctive company logotype that had been originated by architect Joseph Emberton for use on the façade of the Piccadilly store.

12 Ashley Havinden, 'Visual Expression', paper given at 63rd dinner of the Double Crown Club, 18 November 1937; reproduced in M. Bierut, J. Helfand, S. Heller and R. Poynor (eds.), *Looking Closer Three: Classic Writings on Graphic Design* (New York 1999), pp. 67–74.

13 Michel Foucault, 'What is an author?' (1969), *Screen* (Spring 1979), vol. 20, issue 1, pp. 26–27.

14 Ashley Havinden, 'Men's fashions in England and America', *Penrose Annual*, vol. 42 (London 1940), p. 63, and 'Men's Wear', *Design '46* (London 1946), p. 73.

15 Information distilled from *Catalogue Supplement* (London 1946), pp. 199–205.

16 DCA/742/Mass Observation, Report on Britain Can Make It, 1946/table 14 and p. 3.

17 Arthur Sadler, '2nd Viewpoint, The Publicity Manager's', *Display, Design and Presentation* (November 1946), pp. 10 and 19.

18 Three articles in *The Times*, 1 January 1947, dealt with this situation: 'Outlook for Cotton, new plant and new fabrics', p. 36; 'Wool Textiles, problems left by the War', p. 37; and 'Drapery and Dress Goods, efforts to meet an insistent demand,' p. 38.

19 George Schuster, 'Are We TOO Few for the Job?', *Picture Post* (19 April 1947).

20 Henryk Frankel and Peter Ady, 'The Wartime Clothing Budget', *Advertiser's Weekly* (21 June 1945), p. 27.

21.'Wanted: 100 Families to Speak for Britain', *Sunday Pictorial* (7 July 1946), pp. 4–5. Peter Hennessy, *Having It So Good: Britain in the Fifties* (London 2006), pp. 121–22 and 248.

22 For instance, the copy of 'Getting Back to it all', a quarter-page advertisement in *Punch* (6 February 1946) for Austin Reed, proclaimed: 'When man has been reconstructed and becomes a citizen at peace, we will once again be able to supply him with the good things he used to enjoy. In the meantime, a man will always find something at Austin Reed's to help fortify him against the weather. The necessities are there even if abundance is absent.'

23 See 'Two-colour suits, next year's styles in men's clothes', *The Times* (18 September 1946), p. 2, and 'More Variety in Men's Fashion', *The Daily Mail* (18 September 1946), front page.

24 John Basil Boothroyd, 'Britain Can Make it. A Late Look Round', *Punch* (30 October 1946), p. 374; 'Fine Shades', *The Times* (23 September 1946), p. 5.

25 Ashley Havinden, *Design '46* (London 1946), p. 74.

26 Geoffrey Gilbert, 'Meet the New Man', *Man and his clothes* (August 1953), p. 26; 'Today's Teenage Male', *Men's Wear* (7 April 1951).

JONATHAN M. WOODHAM

Great British Designers: Made in Britain

Prologue: Design '46[1]

Little has been written about the 'Great British Designers' (GBD) feature at the 1946 *Britain Can Make It* (BCMI) exhibition. On the surface, this is surprising, given that R. Dudley Ryder, the Council of Industrial Design (CoID)'s Chief Exhibition Officer and one of the key figures in bringing BCMI to fruition, stated in his introductory essay in the Council's *Design '46* publication that it was 'a small but important part of the Exhibition'.[2] Indeed, he claimed that its purpose

> ... was to remind us of the efforts and work of those who have gone before us, the pioneers of Industrial Design – such men as Josiah Wedgwood, Thomas Chippendale and Thomas Tompion, who have made our pottery, furniture and clocks famous throughout the world, and in more recent times, William Morris and C.F.A. Voysey.[3]

S.C. Leslie, the inaugural Director of the CoID, reinforced Ryder's outlook in his accompanying essay on 'Industrial Design'. He wrote of a diverse array of designers ranging from engineers in Birmingham designing an electric fire to a young woman colour-sketching a new print for curtains in a Manchester studio, together with others working on floral patterns for a carpet works in Kidderminster and a Sheffield master cutler designing a new kitchen knife. He saw them collectively as

> ... the people whose talent, skill flair, imagination are producing the forms, colours and patterns upon which depend the success of British firms and the pleasure of their customers at home and all over the world. These are the heirs and successors of Josiah Wedgwood the First, of Thomas Chippendale, of George Hepplewhite, or William Morris.[4]

This essay examines the ways in which design history and heritage *actually* played a role at BCMI, first through the GBD section itself and

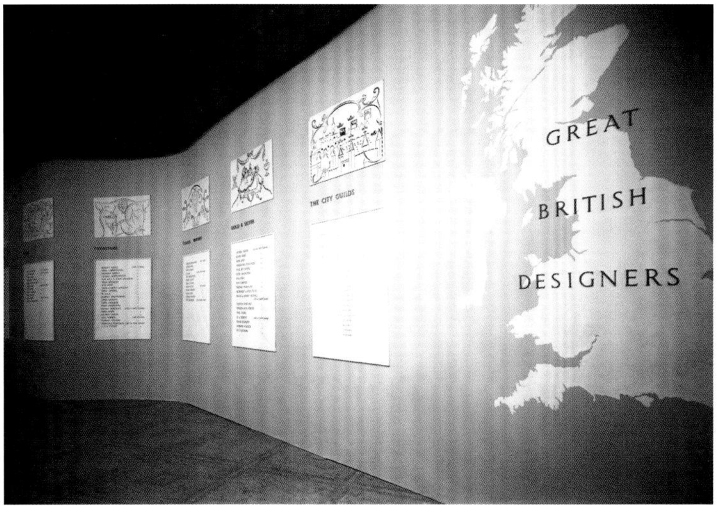

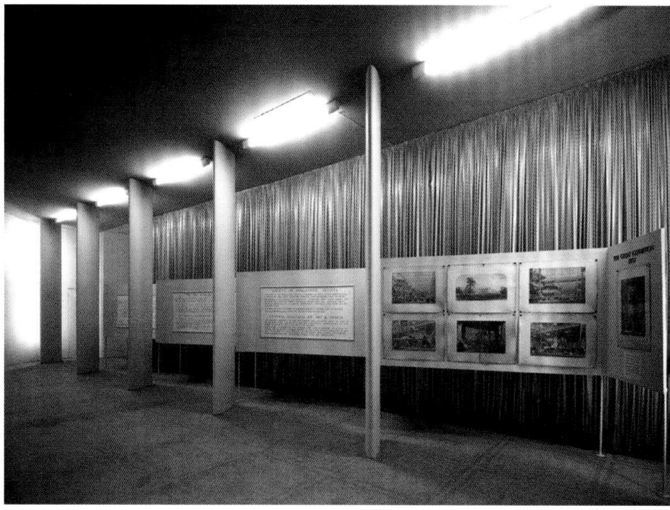

151. View of the Guilds and Pioneers section of the 'Great British Designers' display at BCMI

152. View of the design organizations section of the 'Great British Designers' display at BCMI

the ways in which visitors appear to have reacted to it; and, secondly, through an analysis of related CoID policy papers from 1945 and 1946, to show how design(er) history was 'lost' and 'found' at BCMI.

Great British Designers: the setting

Two surviving GBD photographs [151, 152] show distinctly unpolished presentations that fail to live up to the expectations conjured up by Leslie's and Ryder's words above. Illustration 151 shows one view of the display overseen by F.R. Yerbury and designed by R.S. Lavers.[5] The words 'Great British Designers' are emblazoned across an outline map of Great Britain, complemented by a series of wall-mounted poster-sized information sheets and, above them, related line illustrations. Somewhere a decision had been taken to step back even further than the eighteenth- and nineteenth-century design pioneers identified by Leslie and Ryder to London's City Guilds, with a first pairing devoted to a poster about the Guilds beneath a depiction of London in the Middle Ages. The poster proclaimed that 'Tribute is due to the Livery Companies of London who since the Middle Ages have done much to support education, good craftsmanship and design in industry'. Then followed a list of the 'Chief Livery Companies concerned with the industries represented in the Exhibition', ranging from the Broderers', Clothworkers' and Cordwainers' Guilds through to those of the Skinners, Turners and Weavers. Related posters included 'Gold & Silver' and 'Furniture', highlighting British design pioneers from the seventeenth to twentieth centuries.

Illustration 152 shows a different aspect of the GBD section, arguably even less engaging than the content seen in illustration 151. This commenced with a display of seven prints depicting the Great Exhibition of 1851, somewhat casually presented on one of a series of linked panel and pillar display boards. The other panels were devoted to information about leading British design societies and organizations of varying levels of importance and influence, including the Royal Society of Arts, the Faculty of Royal Designers for Industry, the Art Workers Guild, the Design & Industries Association, and the Central Institute of Art and Design.

Great British Designers: visitor reaction

The GBD display attracted little media attention – positive or negative. Some indication of visitors' experiences is recorded in the December 1946 Mass-Observation report on the exhibition, commissioned by the CoID.[6] Analysis was made of visitor reactions to individual sections of the exhibition in Section B. Table 13.[7] This indicated both the time that should have been spent in each section, assuming all items on display at BCMI were given the same emphasis, and the percentage of total time *actually* spent by visitors walking around the individual sections. In the case of Section 15: GBD, it was reported that although the time warranted by the exhibit would have accounted for 2% of the time spent in the exhibition as a whole, the time *actually* spent looking at it was too low to register. Such lack of public engagement was exceeded only by Section 17, devoted to the CoID, designed by Milner Gray of the Design Research Unit and Eric Fraser. It was estimated that the CoID exhibit's content would warrant only 0.5% of the total time spent at BCMI, although once again *in reality* visitors barely stopped to look.

Great British Designers: origins, enthusiasms and contradictions

One dynamic thread relating to the 'Great British Designers' section stemmed from the enduring design promotional enthusiasm of Gordon Russell, an inaugural appointee to the CoID.[8] Despite his work for the Board of Trade and the CoID, Russell maintained an ideological commitment to the cultural heritage of the arts and crafts, albeit moderated by the growing necessity of mass production and economic growth.

Soon after the July 1945 elections, Sir Stafford Cripps, President of the Board of Trade, called Sir Thomas Barlow and S.C. Leslie, Chairman and Director of the CoID respectively, to a meeting at the Board of Trade offices with a view to mounting a large-scale exhibition of British design in the following Spring.[9] Cripps was clear about his particular vision for the exhibition, as he emphasized in a handwritten note to Leslie, saying that there was 'one thing I forgot to say about the Exhibition. No "precious" stuff. All manufactured goods – not hand made'.[10] But many informed figures felt it premature to attempt an exhibition of high-quality British industrial design by Spring 1946, amongst them Gordon

Russell, Kenneth Clark,[11] Director of the National Gallery, London, Philip Whalley,[12] Director of Lewis's Ltd, and Alison Settle,[13] the fashion journalist.

That Russell was quick to see other opportunities, however, is shown by a set of handwritten notes sent from his home in Chipping Camden to Leslie in mid August 1945, enthused by a suggestion to hold an historical exhibition by Francis Meynell, a typographer, journalist, publisher and fellow member of the CoID. In what was almost a stream of consciousness Russell not only saw this idea as 'very sound' but also as highly exciting, as his frenetic handwriting reveals. Amongst the key aims that he expressed was his view that

> We should set out clearly the very splendid historical tradition of good design. Here we have something which could be used most effectively against American propaganda. We could show say 10th century writing as good as the finest in Europe [–] 12th century embroidery – the famous opus anglicanum[,] 13th century goldsmithing[,] 14th century perpendicular gothic building – an English product largely evolve[d] because of the need of light in northern climate, 17th & 18th century drinking glasses – the finest lead glass in Europe of great clarity of form .... Such an exhibition would be historical in the sense that the past, present & future are shown to be part of one whole. No sentimental worshipping of the past – no maudlin fear of the future. We have always been pioneers, sometimes without noticing it. Let's show that the age-old spirit is still lusty.[14]

A few days previously, on 10 August 1945, the CoID had drawn up a formal document for the 'Proposed Exhibition – Summer 1946' in which it was stated that

> ... reference to historical achievements should be limited. Logical though it is to argue 'Britain has always led – and she will still lead', in practice too much backward looking would confirm the critics and doubters of Britain's continued leadership.[15]

At this early stage of planning the ideas underpinning a display of GBD were not clearly articulated, although it was proposed that Section 4 of the exhibition would be entitled 'British Industrial Designers' with a mural displaying portraits of well-known British industrial designers of the past. Illustrations of their achievements would act as a prelude to the work of contemporary British industrial designers. Shortly afterwards Nikolaus Pevsner, the German-born art and architectural historian and author of *Pioneers of the Modern Movement: from William Morris to Walter Gropius* (1936), entered the fray. He wrote to Leslie on 4 October 1945 in response to the Council's official proposal for the 1946 exhibition, most particularly regarding the view that the contents of the exhibition should be 'unmistakably distinguished from the arts and crafts'.[17] Pevsner posed the question: 'would not the display of craft under the aspect of laboratory for industrial production be an asset?' Pevsner felt strongly that history had a part to play but also that the inclusion of contemporary designers in the then proposed Section 4 would be 'risky' as without the security of an independent jury it might become 'an affair of compromise'.[16]

An informal conference, by invitation only, was held by the CoID on 10 October 1945.[17] Several attendees supported a historical presence at the 1946 exhibition. For example, Leigh Ashton, Director of the V&A, where the exhibition was to be held, suggested that if it was decided not to include examples of 'English' design in the exhibition itself he could put on a series of small exhibitions leading up to the main display; Pevsner suggested the inclusion of a small show of seventeenth- and eighteenth-century designs 'of undoubted quality which still appeal today' and considered that 'the crafts cannot altogether be excluded'; T.A. Fennemore also felt that traditional designs should not be excluded but, opposing Leigh Ashton, within the exhibition itself rather than in the V&A.[18]

Allusions to the proposed GBD display itself occasionally surface in CoID notes and memoranda, often seemingly overshadowed by discussion of the fees offered to those working on it – F.D. Yerbury (75 guineas) and R.S. Lavers (50 guineas). The shift of GBD from Section 4 to Section 15 was accomplished almost by sleight of hand and detailed discussion about content was notable by its comparative absence. Furthermore, the official *Britain Can Make It* exhibition catalogue gave away few real clues about the section on 'Great British Designers', suggesting perhaps that not all was fully settled by the time the catalogue went to press. Even the later official *Catalogue Supplement* gave little away other than stating that the section was organized on behalf of the CoID by a committee representing the design organizations and institutions on display.

Conclusion

By the time BCMI opened, the ambitious aspirations of the previous autumn had been constrained by a series of compromises emanating from the considerable difficulties of presenting a successful exhibition of contemporary British industrial design in less than fifteen months. The 'Great British Designers' debate – whether to look forward or to celebrate the basis of the present and future of British design through a Pantheon of eminent British designers from the seventeenth to the twentieth centuries – in many ways reflected the contradictions in contemporary design debates and realities since the end of the First World War. What it underlines is that the interwar levels of confusion about the relative importance of craft, craftsmanship, heritage and tradition, industrial design, manufacturing processes, creativity and aesthetics that dogged many design organizations – professional or state-funded – were still in a considerable state of flux in the immediate aftermath of the Second World War.

NOTES

1 Council of Industrial Design, *Design '46* (London 1946).
2 R. Dudley Ryder, 'The "Britain Can Make It" Exhibition, *Design '46* (London 1946), p. 12.
3 Ibid. p. 11.
4 S.C. Leslie, 'Industrial Design', *Design '46* (London 1946), p. 9. New Zealand-born Leslie was educated at the Universities of Melbourne and Oxford, at the latter as a Rhodes Scholar and PhD candidate. After a brief foray into academic life he became involved with politics as assistant to the Australian Prime Minister in 1926, before becoming head of publicity at the Metropolitan Gas, Light and Coke Co. (1936–40), then a high-level civil servant in the UK during the War years. It is unclear to what extent he wrote this essay as an independent historically informed thinker.
5 R.S. (Ralph Stephenson) Lavers (d. 1969) was an ARIBA and held an AA (Architectural Association) Diploma. He designed the torch for the London 1948 Olympic Games. Lavers also designed a series of ironwork mural panels for the Science Museum in London. Since F.R. (Francis Rowland) Yerbury had been the Secretary at the AA, London (1911–37), it seems possible that he favoured AA graduates. Yerbury was also was one of the co-founders of the influential Building Centre, London, in 1931, as well as a staunch supporter of the Design and Industries Association (DIA), of which he became Vice-President.
6 Design Council Archive at the University of Brighton Design Archives (hereafter DCA): ID 903/Typescript report. 'A Report on "BRITAIN CAN MAKE IT" Exhibition: SECTION B Effects of the Exhibition: Taste, Design and Consumer Habits', prepared by Mass-Observation, 21 Bloomsbury Street, WC1, December 1946; http://arts.brighton.ac.uk/collections/design-archives/archives/design-council2.
7 Ibid., p. 2.
8 Russell set up his own furniture-manufacturing firm in 1926, the year in which he became a member of the Art Workers' Guild. During the Second World War he was a member of the Board of Trade's Advisory Committee on Furniture (1942) and Chairman of the Board of Trade Design Panel (1943–47).
9 Many details and documents relating to the setting up of BCMI 1946 can be found in Patrick J. Maguire and Jonathan M. Woodham (eds.), *Design and Cultural Politics in Postwar Britain: The Britain Can Make it Exhibition of 1946* (London and Washington 1997).
10 DCA/ID 312. Dated 10 August 1945, Summer Exhibition.
11 Ibid, letter to Leslie dated 18 August 1945.
12 Ibid, as recorded in the notes for the CoID's informal conference (the participants were by invitation from the CoID) held in connection with the Summer Exhibition [BCMI 1946], 10 October 1945.
13 Ibid. The Alison Settle Archive is held at the University of Brighton.
14 These notes, headed 'Cripps' Exhibition. Let's go on being pioneers', were not dated but had been passed on to Ryder by Leslie before 17 August 1945. DCA/ID 312 Summer Exhibition.
15 DCA/D312/45, 10 August 1945, THE COUNCIL OF INDUSTRIAL DESIGN, Proposed Exhibition Summer 1946.
16 Also located at DCA/ID 312.
17 See above, footnote 12.
18 Fennemore's papers are held in the V&A's Archive of Art and Design, AAD/1983/3. He was Registrar of the National Register of Industrial Art Designers (established in 1936) and founder of the Central Institute of Art and Design in 1940.

LESLEY WHITWORTH

Birth of an Eggcup: What Industrial Design Means

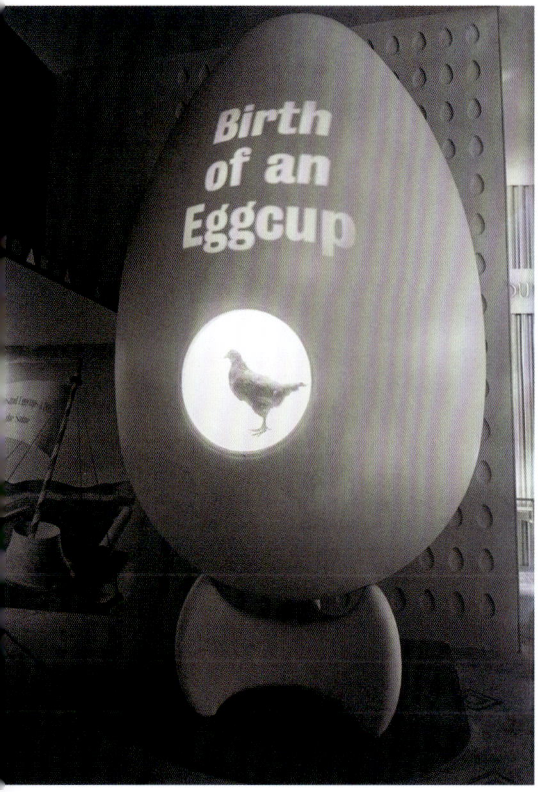

153. The egg at the entrance to 'What Industrial Design Means' (Section 16)

What can we say from the vantage point of over seventy years about Section 16 of the *Britain Can Make It* (BCMI) exhibition, 'What Industrial Design Means'? In particular, should we attempt a recovery of its historical reputation?[1] For those unfamiliar with its visual characteristics a description must first of all be attempted.

Section 16 came in the later part of the exhibition, and opened off the area in which the Council of Industrial Design discreetly showcased its own existence [see **7, 187**]. A giant illuminated egg sited on a plinth at its entrance attracted the attention of those who would otherwise have been drawn onward, along the exhibition's main drag. The appearance of a futuristic space-capsule piloted by a lone chicken must have achieved this with some success [**153**]. The Design Council Archive (Council of Industrial Design 1944–1972) contains a wealth of documentation concerning the exhibition's organization, but has not yielded the identity of the creator of this surprising feature, nor the detail of its composition.[2] For these we are indebted to *Display* magazine, which refers to a thirteen-foot high egg in fibrous plaster with a Perspex window.[3] Beyond the egg, extending along the right-hand margin of the room and drawing the visitor further in, was a whimsical depiction of the manufacturing process: a mural showing the productive capacity of a plastics press – three-thousand eggcups a day, all exactly the same. The artist's name is another absentee from the list in the *Catalogue Supplement* issued by the Council, but it is given in the *Display* piece as John Barker, MSIA, with the information that this feature was 24 feet in length [**154**].[4]

A remarkable suite of black and white photographs in the Design Council Archive makes it possible to regain a strong sense of this and other sections of the exhibition. The route around the 'Egg Section' as it was sometimes known was a long, attenuated U-shape. Once almost back at the beginning of the section, visitors' attention was absorbed by a 100-ton plastics press in full working order, on loan from T.H. & J. Daniels Ltd of Lightpill Iron Works, Stroud (Gloucestershire), for the purpose

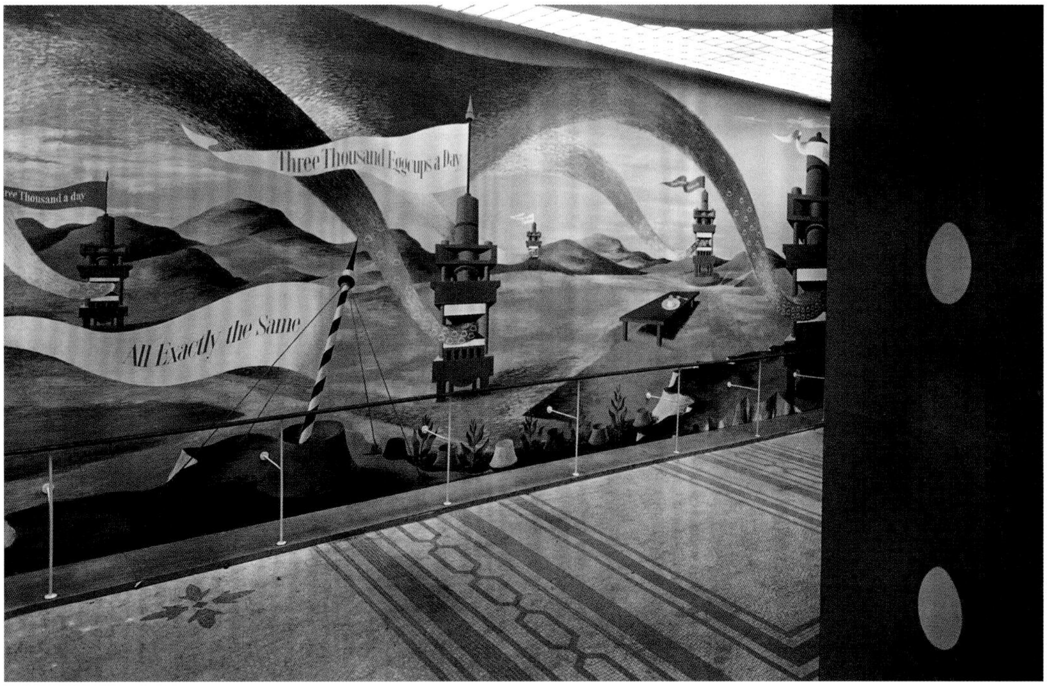

154. 'Three-thousand eggcups a day. All exactly the same' – section of John Barker's *Birth of an Eggcup* mural

of demonstrating that self-same relentless, but mindless, productive capacity. That the press in motion was genuinely captivating is attested to by evidence gathered in the report on the exhibition prepared by Mass-Observation (M-O).[5]

To a modern eye, the display strategies deployed between these two highpoints appear less satisfactory. A grid device and a thrillingly 1970s-style graphic frieze were evidently intended to unify many disparate elements, but discordant letterforms – for example – undid much of this good work. The conceptual scope of the display was vast and the results were unwieldy – a complaint that is comparatively ill-served by the images reproduced here [153–157]. Instead it is tempting to underpin these criticisms by recourse to commentaries on the content of a booklet that was proposed to accompany the physical display. Correlating closely with its content, and intended to do useful promotional work for the CoID long after the last elements of the exhibition had been dismantled, the booklet 'does not strike me at once as masterly or inspired', as one member of Council staff wrote. 'I think we're making an ant hill out of an eggcup. Can't such a theme be dealt with in 8pp & not drag in textiles [sic] furniture and jugglers?' was the damning indictment of another. Noel Carrington alighted on the most salient problem of all, that of *which* audience was being addressed?

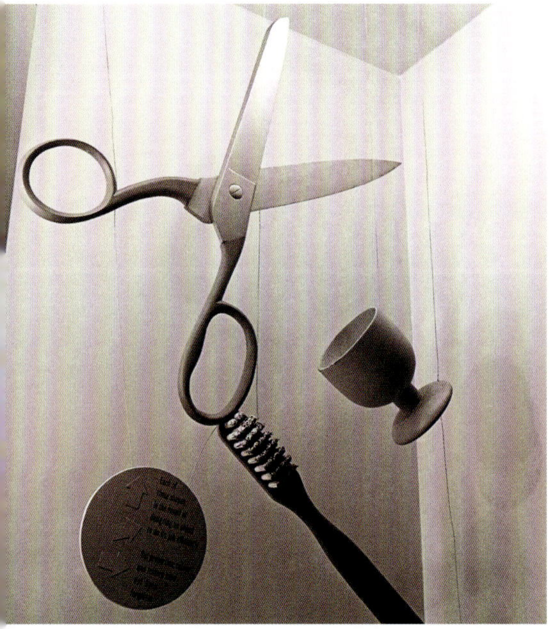

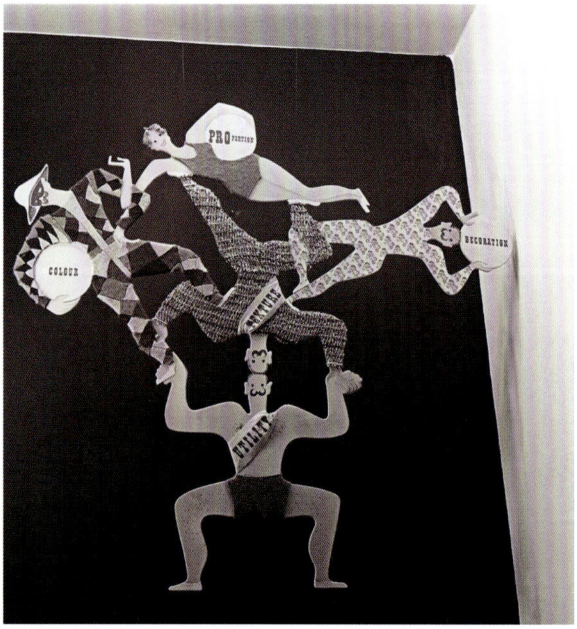

155, 156. Two ways of explaining function, colour, texture and proportion

> As a book for sale or education I see this falling between half a dozen stools. The images used are apparently for ten year olds; the language is generally the jargon of technical journalism. There are so many criticisms already of the inaccuracies or incomplete explanations that it is hardly worth saying more on that score.[6]

Credit for the design of the display is almost universally accorded to Misha Black OBE, FSIA, but the *Catalogue Supplement* to the exhibition notes the input of three associates, Bronek Katz, Dip ING ARCH; R (possibly Reginald) Vaughan, ARIBA; and Austin Frazer. Black, Katz and Frazer were all associated with the Design Research Unit (DRU); Vaughan remains more enigmatic.[7] Whilst revelatory of many glorious details of the BCMI production schedule, the *Catalogue Supplement* and archive File 742, Special Features (Constructional), are frustratingly mute on the subject of Section 16, and there are grounds for believing that DRU had a special status within the organizational bureaucracy of the exhibition. Black did not produce a script until the 23rd planning meeting on 8 July 1946, and a drawing was only approved at the 39th meeting on 6 August: this for an opening at the end of September. A memo of 4 September shows a grant from the Special Features' budget to Misha Black substantially in excess of others noted in the same place.[8] One possible explanation is that DRU was entrusted with its own commissioning process and operated in relative isolation from the main exhibition machinery.

Despite the claim of 'disarming simplicity' in one of the photographic captions, the 'Egg Section' was arguably anything but.[9] How could it be? As Norbert Dutton, FSIA, remarked in an interview at the time of the exhibition, the issue of what industrial design encompasses is 'rather a large question'.[10] Hence the design team made a bold attempt to deal with the following – although arguably not adroitly – industrial economics; the psychology of colour; the part played by emotion and sentiment in purchasing; manufacturing techniques appropriate to a wide range of different materials; working machinery; design as a team activity; and live hens. Its laudable ambition may not have been best assisted by the choice of focus: it is all too easy to side with the East End housewife described by M-O as 'looking at things throughout from a very practical standpoint', who commented that it was, 'Silly all this fuss about designing an eggcup'.[11]

Yet the effort had to be made. Intended to be British Industry's first great post-war outing and a 'gesture to the British public and to the world', design, conceived as a vital socio-economic imperative, was as yet unknown to vast swathes of the populace. 'Only designers talked about design,' as James Gardner, the exhibition's Chief Designer, observed of that period.[12] To address this vacuum a twin-pronged approach was inevitable. The exhibition therefore acted 'as an incentive to manufacturers, arousing their interest in the importance of design', and provided 'some exposition of the significance of design to the consumer'. That the main target for its propaganda message was to be 'home – the British manufacturer and public', appeared consistently in iterations of the exhibition's purpose on file.[13]

Section 16 bore almost the entirety of this educative burden: the 'attempt ... to explain what is meant by Industrial Design'. In fact, it is clear that in its final form 'Eggcups' represented a combination of two sections that were originally hypothesized by BCMI's planners – Fundamentals of Design and Process of Design. Despite the explanatory load implied by these terms Sir Charles Tennyson presumably envisaged a light-touch approach when he characterized them as a 'sideshow', and 'ancillary to the main displays, occupying very much smaller spaces, and not pretending to comparable interest'.[14] Any balanced response to Section 16 must consider the scale of the challenge faced by its designers, and the level of achievement commensurate to it.

Nevertheless a view that this part of the exhibition was not very successful has become widespread. Taking his cue from Mass-Observation (M-O), Jonathan Woodham notes the 'relative indifference of the public to a central focus of [the Council's] propaganda at the Exhibition: the "What Industrial Design Means" section'.[15] This, however, does not appear to do justice to the full spectrum of M-O's

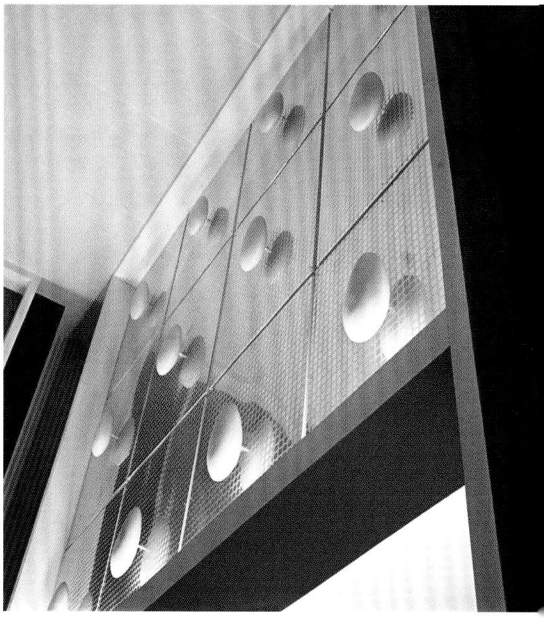

157. Sharply delineated grids in different materials, punctuated by ostrich-sized egg motifs, marked out smaller explanatory sub-divisions in individual bays

findings, any more than did the organization itself. For having stated that 'Very few were really impressed by this section', it makes a number of supplementary points that do not so much amplify this view as run counter to it. The admittedly difficult to interpret M-O Time Chart prepared in order to indicate visitors' response to different sections actually shows that 'Eggcups' held up well alongside acknowledged crowd-pullers like Women's Dress. The accompanying narrative affirms that 'The eggcup exhibit manages to hold interest', and furthermore that 'it is the only one in the second half of the Exhibition which did succeed in holding their interest' (emphasis in the original). This is noteworthy when one remembers the industrial design display was effectively a late diversion from an already demanding visitor route, and only accessible via the space in which CoID carried out its own austerely elegant piece of self-promotion.

The M-O report noted that 'The fatigue factor becomes particularly noticeable after Section 12' and, further, that 'Any or all of Sections 13–22 would attract greater interest if they were presented earlier in the itinerary, when people were less footsore and less dulled by the variety of what they had already seen'. A supplement to the *Board of Trade Journal* said that '16' was 'Perhaps the most important section ... from the point of view of the general public', and 'the key to understanding all the rest', so it could reasonably be argued that the whole order of the exhibition displays was unhelpful.[16] How much more intuitive it would have been if the early treatment of wartime innovations had led naturally into a discussion of new products and the role of the designer as an emergent professional in the field of innovation. How much more concrete and less abstract too. The decision to make it a later part of the sequence appears to have been taken very early on.

In conclusion, it is not fair to write off Section 16 as both a didactic and an actual *cul de sac*. Close attention to the photographic evidence available in the Archive indicates that Black and his colleagues asked a lot of their audience in terms of concentration. Apart from the theatricality of the entrance/exit, much in the belly of the space represented visual noise: it was horribly wordy and laboured, much of it seeming to fall foul of Black's own dictum expounded in his *Exhibition Design* of 1950: 'People come to exhibitions to see things, and being fobbed off with even the most tasteful typographic arrangements may irritate all but the already converted'.[17] But in 1946 *things* were in terribly short supply, and such things as there were appeared in quite distinct areas of the exhibition. As one, no doubt partisan, industrialist wrote to CoID's Director, 'the important thing is that for the first time in the history of this country you have made a lot of people conscious of design as an element in the product'.[18] An extended reading of Mass-Observation's report suggests that interest held up reasonably well considering the section's location and the intrinsic complexity of the issues involved. It is certainly not without relevance that BCMI took place early in the reconstruction period following the 1939–45 war, making another of Black's comments pertinent:

> Exceptional circumstances exist, however, when the subject matter is of such intrinsic interest that large numbers of the public have a passionate desire to obtain information about it, and, in those cases, exhibitions consisting largely of flat material can be very effective. This was true about some of the war-time exhibitions.[19]

NOTES

I would like to thank the editor for her unstinting enthusiasm for all things BCMI.

1 Among its critics are Jonathan Woodham and Lucy Bullivant. J. Woodham, 'The politics of persuasion: State, industry and good design at the *Britain Can Make It Exhibition*',; P.J. Maguire and J.M. Woodham (eds.), *Design and Cultural Politics in Postwar Britain: The Britain Can Make It Exhibition of 1946* (London 1997), p. 59; L. Bullivant, '"Design for better living" and the public response to *Britain Can Make It*', in P. Sparke (ed.), *Did Britain Make It? British Design in Context 1946–86* (London 1986), p. 147.

2 All archival material referred to in the main body of the text, including files, exhibition ephemera and photography, is located in the Design Council Archive at the University of Brighton Design Archives (hereafter DCA).

3 'Britain Can Make It. Birth of an Eggcup', *Display, Design & Presentation* (September 1946), pp. 1–12. A sub-contractor for fibrous plasterwork mentioned in the *Catalogue Supplement* is T.R. Rudd of John Tanner & Sons Ltd, 36 Victoria Street, London SW1. However, correspondence from R. Dudley Ryder of CoID to Haigh of the V&A, dated 28 March 1947 with following notes, suggests that it was contracted out by Beck & Pollitzer (Contracts) Ltd, of Universal House, Southwark Bridge, London SE1. V&A Archive MA/1/0981: Design Council (formerly Council of Industrial Design) 1945–1955.

4 Barker's name also appears in Council paperwork that resides with other BCMI material in the V&A Archive, Exhibition File: AAD/1977/4/14A, Britain can make it: Information for Guides.

5 The Council of Industrial Design commissioned this report, copies of which are contained in DCA File 903, Mass Observations.

6 DCA File 385/8, 'Birth of an Eggcup' Booklet by Misha Black (Fundamentals of Design Section. Britain Can Make It). The booklet was never put into production. A CoID questionnaire directed at teachers who had visited the exhibition with school parties asked especially about the impact of this section: sadly the children's responses have not survived. V&A Archive, Exhibition File: AAD1977/4/14, Britain can make it: School impressions November/ December 1946.

7 The most recent work on DRU, Michelle Cotton's *Design Research Unit, 1942–1972* (London 2010), says very little about *Britain Can Make It*. The article in *Display* also mentions work contributed by Bruce Angrave, MSIA, who had appeared in their August issue. On Angrave see also Susan Roberts, 'Advertising Art – the work of Bruce Angrave', *Art & Industry* (November 1946), pp. 136–38.

8 These appear in DCA File 58, Summer Exhibition Design and Policy Progress Meetings; and File 742, Special Features (Constructional).

9 The illustration accompanying the *Display* article is similarly captioned, 'An explanation of the function of industrial design in terms as simple as those of an Aesop Fable!'

10 'Design – Its place in modern industry. An interview with Mr Norbert Dutton, Fellow of the Society of Industrial Artists', *Future Books*, vol. IV, 1946, pp. 22–24.

11 Mass-Observation (cited note 5).

12 'Exhibition design', Giles Verlade interviews James Gardner, in Sparke (ed.) 1986 (cited note 1), p. 10.

13 See especially DCA File 312, Summer Exhibition 1946.

14 DCA File 312, Summer Exhibition 1946, document numbered 31, 'Proposed Exhibition – Summer 1946' ; and DCA File 312, (Tennyson) document numbered 100, Appendix to CM(45)9 Minutes of Council meeting, 12 October 1945.

15 Maguire and Woodham 1997 (cited note 1), p. 59. See also Bullivant 1986 (cited note 1), p. 147.

16 'Design and the Designers', *Supplement to the Board of Trade Journal* for 28 September 1946, p. 14. Interestingly, a *Punch* journalist recounted how hard it was to make the journey back-to-front, should anyone make the attempt. John Basil Boothroyd, 'Britain Can Make It: A Late Look Round', *Punch* (30 October 1946), pp. 374–75.

17 Misha Black, 'Exhibition Design', M. Black (ed.), *Exhibition Design* (London 1950), p. 12.

18 DCA File 911, Letters containing criticism, complaints and suggestions. Letter from Hugh Weeks of Cadbury Brothers Ltd., to S.C. Leslie of CoID, dated 1 November 1946.

19 Black (cited note 17).

ALEXANDRA GODDARD

Utility and Other Furniture

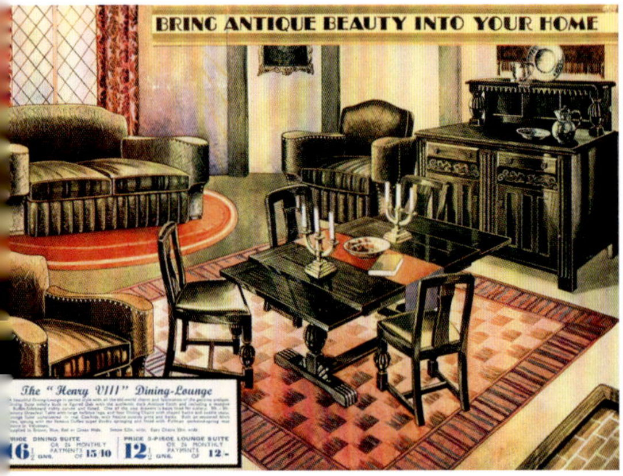

158. Bolsoms: *Quality Furniture* catalogue, c.1930–40. Geffrye Museum of the Home (inv.no. 91/1978-4)

> I believe that our own people, after the long years of strain and denial, will draw encouragement from the sight of the things here which, from now on, will become available to them in increasing numbers.
> *—King George VI in his opening speech, 24 September 1946*

Britain's furniture trade had long been a source of economic power via domestic, Empire and international markets, and it had potential to help the drive for post-war recovery that *Britain Can Make It* (BCMI) was aiming to boost.

Fuelled by London's skilled immigrant workforce, the East End was a powerhouse of manufacture that had 'furnished the world' in the late nineteenth and early twentieth centuries.[1] The firm Harris Lebus, founded by Jewish refugees, outgrew its Shoreditch premises and in 1900 moved to a site in Tottenham, north London, which by 1950 was the 'biggest furniture factory in the world' at more than one million square feet.[2] Meanwhile, High Wycombe, Buckinghamshire, with its plentiful supplies of local raw materials, developed from a centre of vernacular chair-making to a hub of modern cabinet-making that thrived in the inter-war and post-war periods.

Furniture could be found throughout the exhibition. It played an important role within the 'Furnished Rooms' section; other carefully selected items enhanced the Leisure and Dress Fabrics sections, as well as equipping the rest spaces and Tea Lounge, where visitors could actually use, rather than merely view things. In addition, two sections were given over specifically to the display of furniture, Section 6 – 'Furniture and Furnishing Fabrics' – and Section 16 – 'Utility and Other Furniture'.

BCMI took place at an unprecedented time for the furniture industry, when the Utility furniture scheme was still in full force. The scheme was a form of rationing, similar to that of food, put in place by the government in reaction to a singular set of circumstances created during the war.

Initially it was a response to a shortage of raw materials brought about by timber blockades, which meant that wood for furniture-making was in short supply. The Blitz (September 1940–May 1941) then destroyed or damaged more than a million homes across the country and a new category of furniture customer was created, the 'bombee'. The Government introduced 'Standard Emergency Furniture' in February 1941, a simple basic set of designs which served as quick replacements for what had been lost. This was the foundation on which the Utility scheme was based.

As the months wore on the situation became acute, and it was clear that a more thoroughgoing scheme was necessary. Profiteering was rife, with some poor-quality furniture selling for more than 300% of pre-war prices. The Board of Trade set up the Utility Furniture Advisory Committee, whose members included Elizabeth Denby, designer of one of BCMI's furnished rooms, Herman Lebus, Senior Partner of Harris Lebus, and furniture designer Gordon Russell, later chair of the committee and Director of the Council of Industrial Design (CoID) from 1947. A group of furniture designers were approached to submit designs for a set of 22 pieces of furniture, which then went into production. All other furniture manufacturing then ceased, by law. Only 150 manufacturers were initially licensed to make the designs and each item was stamped for authenticity. As well as bearing the manufacturer's license number, each item was marked with the CC41 mark. CC stood for 'civilian clothing' – the first type of goods to be subject to this type of rationing – and 41 is an abbreviation of 1941, the year in which it was introduced. Prices were set by the government, and were exempt from the 33⅓ % sales tax to protect consumers. Items could only be purchased using tokens allocated according to need, with priority given to 'bombees', newly-weds and growing families.

This hiatus to normal production caused by the war also presented an opportunity to influence public taste – a chance to wipe the slate clean and democratize 'good design'. The Utility scheme was a social leveller and everyone had to have the same. Most of the early Utility furniture was designed by Herbert Clinch and Edwin Culter, both High Wycombe-based. It was necessarily simple to construct as most skilled workers had either been conscripted or were making munitions, and the less-important work of furniture-making had to be done by unskilled women and older men. It was also imperative that the construction was light on materials. This provided an opportunity to move away from the heavy, material-hungry historicizing designs popular in mainstream taste in the 1930s [158]. Utility's lighter-weight and more modern design was now becoming the norm, albeit independent of consumer choice.

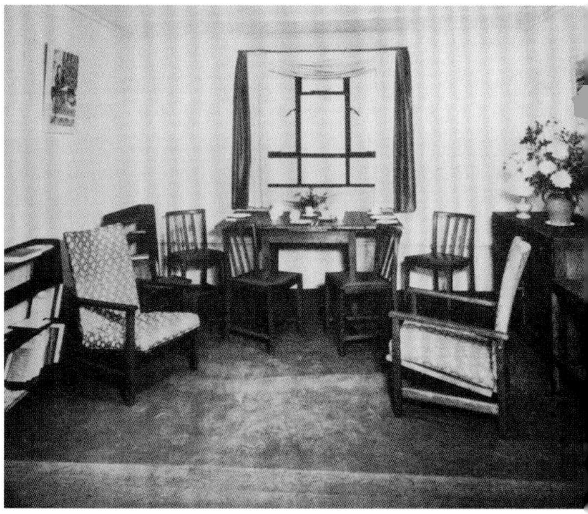

159. Utility living/dining room, photograph taken at the first exhibition of utility furniture at the Building Centre October 1942. From the *Utility Furniture and Fashion 1941–51* exhibition catalogue, originally published by the Inner London Education Authority, 1974, 2nd edition published by the Geffrye Museum, 1995

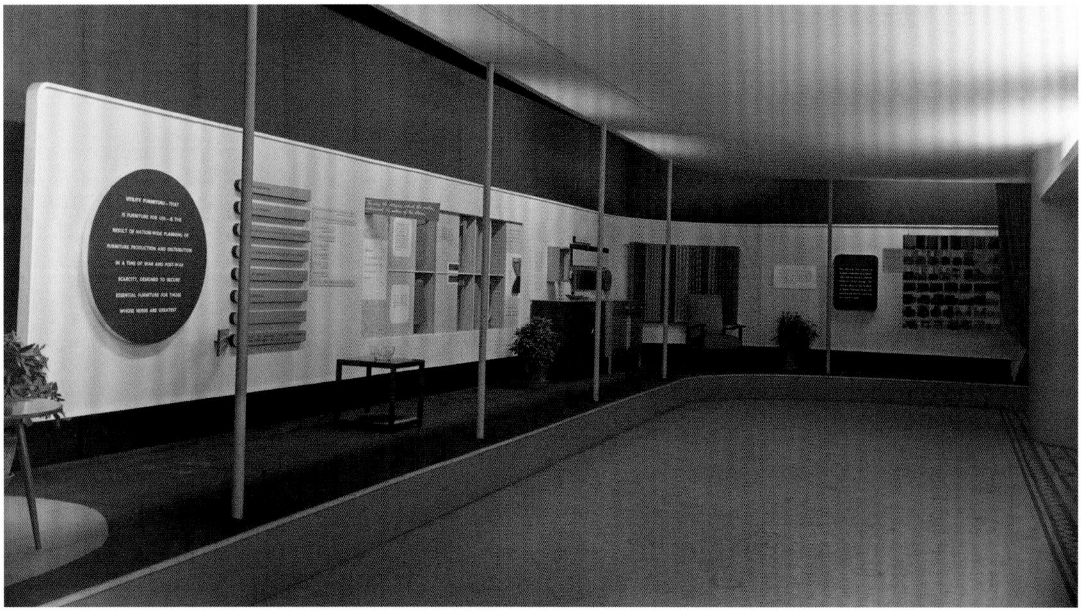

160. View of the 'Utility and Other Furniture' section

Gordon Russell is generally credited with bringing this agenda of design education to the masses. His background was furniture designing in the Cotswolds, where he was steeped in the legacy of the Arts & Crafts Movement. Although his firm belief in the civilizing effect of good design owed much to those roots, his fundamental hope was that it could be brought to the mass market using the power of the machine, rather than remaining the preserve of the craftsman. In his autobiography he states 'I felt that to raise the whole standard of furniture for the mass of the people was not a bad war job'.[3]

In 1946, public choice was still very limited. Although additional Utility designs had been introduced since the initial 22, all consumption was still being controlled by the Board of Trade and choice and availability were minimal. It was not until November 1948 that 'Freedom of Design' was introduced, which allowed manufacturers to make furniture of their own design, provided it conformed to prescribed usage of materials and was approved and labelled or stamped. It was not until the end of 1952 that the scheme was dissolved entirely, enabling a return to the pre-war open market.

In the exhibition catalogue all items were coded according to their availability for home and export markets. For home '1' denoted something was available now, '2' soon and '3' later. Likewise export used 'A' for now, 'B' for soon and 'C' for later [see **2**]. A large proportion of furniture products on show were labelled 3A, meaning they were not currently available at home but could be manufactured for export. This

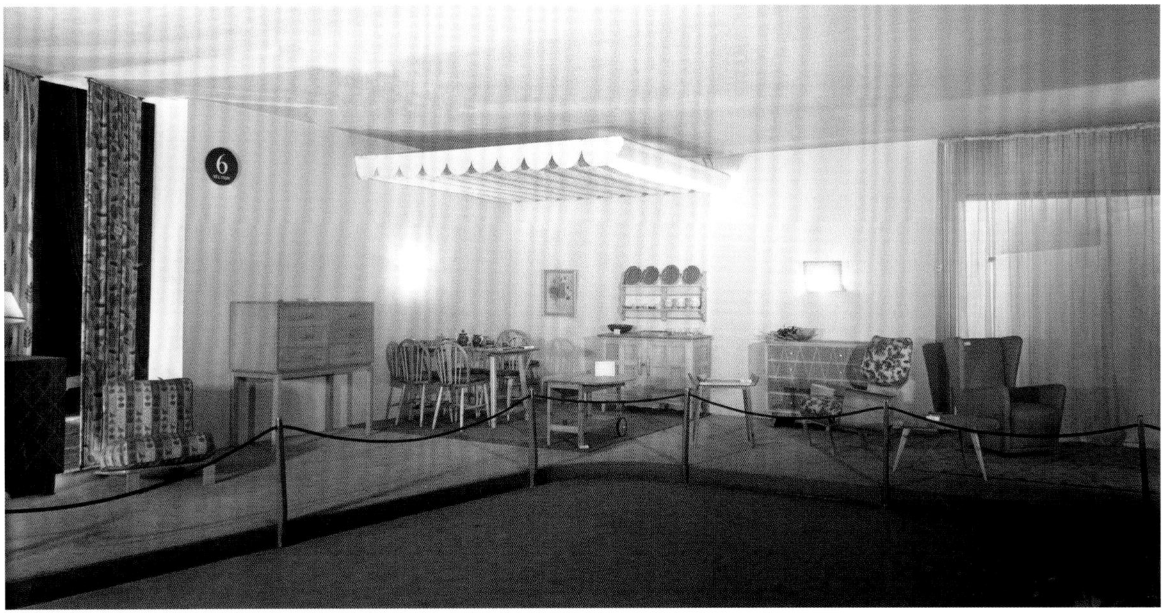

161. View of 'Furniture and Furnishing Fabrics' section, designed by Jacques Groag, showing a desk designed by Christopher Heal for Heal's Wholesale and Export Ltd on the left

was likely to have been a source of frustration for the visiting public and may have been a contributing factor to the exhibition's popular-press pseudonym 'Britain Can't Have It'. Organizers were however, very clear about the availability of what was on show. 'Practically the whole of the wooden furniture exhibits are designs which cannot at the present moment be made for the home market, but are indications of the way in which our manufacturers are thinking for future production when restrictions are lifted.'[4]

The display in the 'Utility and Other Furniture' section (put together by Jacques Groag of the Board of Trade's Utility Furniture Design Panel) was relatively sparse, and almost appropriately austere [**160**]. The public would be quite familiar by now with the Utility designs, even though initially supply had been so limited they could not be put on display in shop windows. Instead they had been shown at an exhibition held at London's Building Centre, in the West End, in October 1942, which then toured to other major cities, and could also be seen in the exhibition catalogue of January 1943 [**159**]. At BCMI, Utility was explained and certain key examples were on show, including the recently introduced occasional table from the 'Cotswold' range. However, devoting display space to it would not have been a particular priority given that it was a reminder of wartime privations and carried negative connotations for many. According to Mass Observation 'People were just not interested in the Utility Furniture', and associated it with flimsiness and poor quality. If people liked it at all it was for its simplicity and plainness.[5] The decision

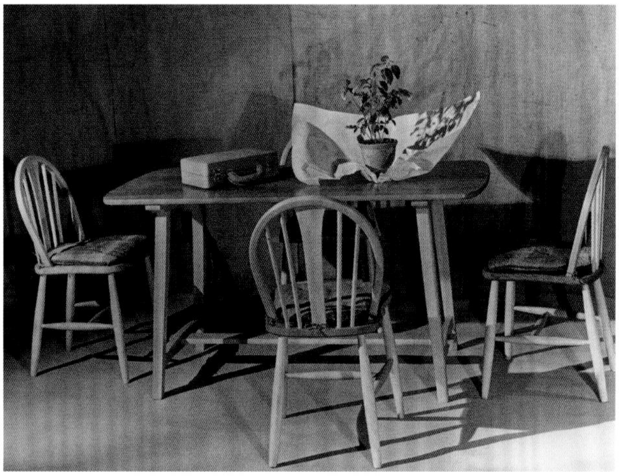

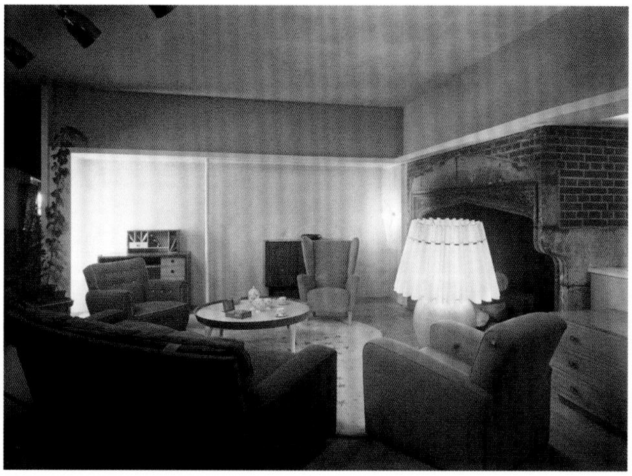

162. Model 4A 'wheel back' Windsor chairs and 121 table designed by Lucian Ercolani for Furniture Industries Ltd

163. Desk designed by Christopher Heal for Heal's Wholesale and Export Ltd on the left of this view of the 'Furniture and Furnishing Fabrics' section

to display Utility alongside 'other furniture' suggests it was taking its place in the design history narrative, rather than being prioritized.

BCMI's main showcase for contemporary furniture was therefore Section 6, 'Furniture and Furnishing Fabrics', located in 'pride of place in the exhibition' directly before visitors entered the 'Furnished Rooms' section [**161**]. The designs were considered to be 'most indicative of a freshness of outlook on the part of the manufacturers' by the selection committee and their two technical assessors, Lucian Ercolani and Anthony Heal.[6]

Ercolani's design work, as part of Furniture Industries Ltd., was well represented in the exhibition. The Board of Trade had asked him to take the Model 4A traditional Windsor chair from the Utility range and make it suitable for mass production. Doing so was a complex feat, marrying the traditional form with efficient modern production methods. It was ready in time for BCMI and featured in several of the Furnished Rooms, perhaps a realistic reflection of its ubiquity given the fact that the Board of Trade commissioned 100,000 of them. A variant, with a flattened central back splat (described in the catalogue as 'wheel back') was shown in the 'Furniture and Furnishing Fabrics' section alongside two new designs by Ercolani, the Model 121 dining table [**162**; see **119**] and Model 120 sideboard. Variations on the Windsor went on to become the stock in trade of Ercol (as the company became known) and are still in production today.[7]

Heal's contributed the largest number of furniture exhibits to BCMI, with several designs by Christopher Heal featuring in this section, including a writing desk with removable shelves [**163**]. Although timber continued to be in short supply for furniture-making directly after the war, Heal's and others made innovative use of materials which were now plentiful, such as plywood and aluminium. New materials and technologies developed for munitions manufacturing during the war were repurposed for domestic furniture-making in a process which has become known as technology transfer. Clive Latimer designed a range of furniture for Heal's using a new material called Redux, a veneered aluminium sheet which 'combines the strength of the metal with the attractive appearance of the wood', using a fraction of the amount of wood used in traditional manufacturing [see **79**].[8] Ernest Race developed a range of aluminium-framed furniture made from scrap metal left over from wartime aircraft, combined with a laminated plastic sheet material called Holoplast, which had been developed for marine construction. The BA 3 chairs were used for public seating in the Tea Lounge [see **97**, **102**; see also **101**].

Many of the companies featured in the exhibition went on to flourish in the post-war years. Heal's continued to be a well-respected retail furniture brand throughout the rest of the twentieth century, and is now reviving its reputation as an innovator in contemporary furniture design and production. Ercol has also maintained production of the successors of the Windsor chair, benefiting from the twenty-first-century revival of 'mid-century modern' design. Race Furniture built on the success of Ernest Race's BA 3 chair, producing furniture for the Festival of Britain, a range of whose designs have been recently relaunched.

Furnishing fabrics were highlighted within the furniture displays. Under Utility regulations the range of pattern and colour available to the public was very limited and the post-war years presented an opportunity for innovation. Although those on display at BCMI did not draw much public attention or approbation, designers like Alastair Morton of Edinburgh Weavers (Morton Sundour) and Jacqueline Groag went on to become enduringly popular and influential.

Given the importance attached by the organizers to the furniture in the exhibition, its reception among the general public was generally unfavourable. Despite having been a motivating factor in encouraging people to visit, 'the furniture display that they saw on the whole disappointed them'.[9] They did not dwell long in the display of 'Furniture and Furnishing Fabrics', with only half of the visitors even stopping to look. By the time they reached the Utility section near the end they were even less interested. The most popular item was a desk by Heal's; the least popular an occasional chair by the Co-operative Wholesale Society. People's comments mostly centred on comfort and practicality, and their judgements of particular pieces were focused on that. 'People are tired of Utility, and are looking for furniture that is well-made from good materials, as well as being well designed'.[10] This seems like a fair aspiration from a public who had put up with the privations of war, but it was to be few years yet before they could make their dreams a reality.

NOTES

1 'London Furnishes the World' was the motto of the London Cabinet Trades Federation in the early twentieth century; cited in P. Kirkham, R. Mace and J. Porter, *Furnishing the World, The East London Furniture Trade 1830–1980*, (London 1987), p. vii.

2 'Biggest in the world – The Harris Lebus Factory at Tottenham', *The Cabinet Maker and Complete House Furnisher* (1 July 1950), p. 60.

3 Gordon Russell, *Designer's trade: Autobiography of Gordon Russell* (London 1968), p. 200.

4 V&A Archive, Exhibition File: AAD/1977/4/14, Britain can make it: Council of Industrial Design Information for Guides: Furniture.

5 Design Council Archive at the University of Brighton Design Archives (hereafter DCA)/903/Mass Observation report Section C, pp. 31–33. For a discussion of the furniture at BCMI see Jonathan Woodham, 'Design and everyday life at the Britain can make it exhibition, 1946: "stripes, spots, white wood and homespun versus chintzy armchairs and iron bedsteads with brass knobs",' *The Journal of Architecture* (2004), 9:4, pp. 463–76.

6 V&A Archive, Exhibition File: AAD/1977/4/14, Britain can make it: Council of Industrial Design Information for Guides: Furniture.

7 Lesley Jackson's thorough and detailed book *Ercol: Furniture in the Making* (Shepton Beauchamp 2013) gives details of Ercolani's contribution to the development of the Windsor chair.

8. *Redux Furniture at the BRITAIN CAN MAKE IT Exhibition*, Aero Research Technical Notes Bulletin no. 49 (January 1947).

9 DCA/903/Mass Observation report Section C, pp. 31–33.

10 Ibid.

MARTIN ANDREWS

Books and Printing

> … the health and vigour of typography in any country has to be judged, not merely by the appearance of a few elegant limited editions of books, but rather by the extent to which one can expect the local job-printer to produce decent and attractive bill-heads for the local butcher and baker.[1]

Having braved the queue to get into the exhibition and joined the throng trailing around the maze of displays, the average visitor must have been exhausted by the time they reached the 'Books and Printing' section – the second from last display before the exit. Relieved to see signs for the restaurant and 'rest lounge', visitors passed through an area

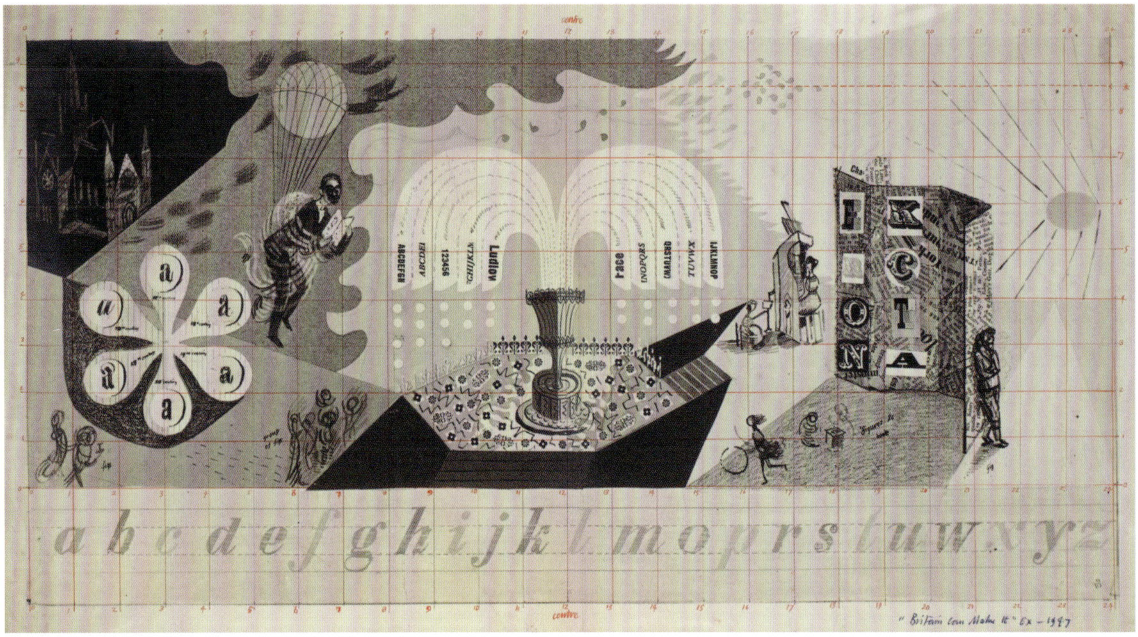

164. Preliminary design by Laurence Scarfe for the mural in the 'Books and Printing' section

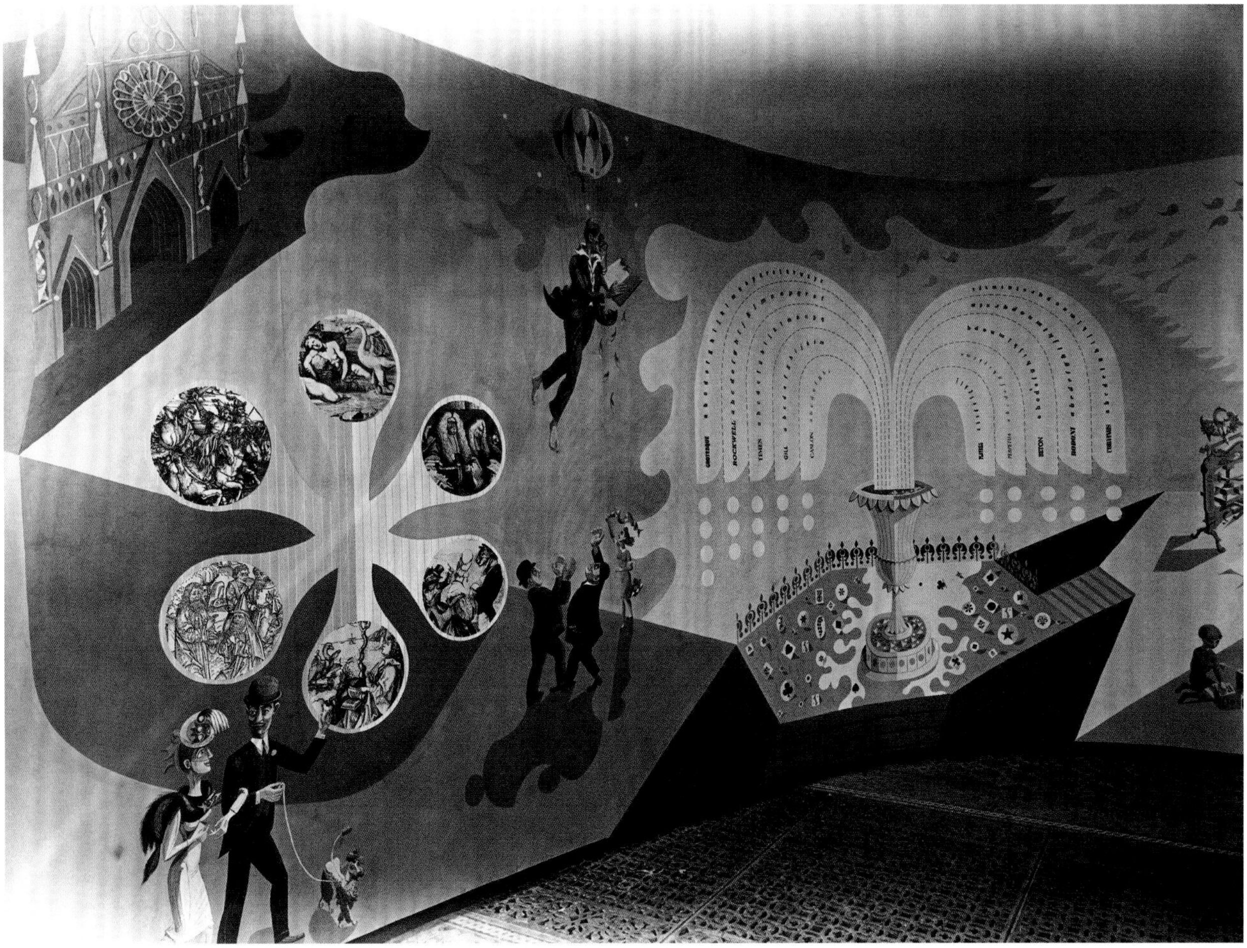

dedicated to leisure, with displays of sports and camping equipment, and a dramatically lit raked central island display of gardening tools.[2] The visitor's eye would then have been caught by a glimpse of a colourful mural by Laurence Scarfe depicting a dreamlike world of books, letterforms and printing, representing an allegory of enlightenment through the printed word, painted on a curved wall that formed a short passageway [**164, 165**]. On the left of the mural Scarfe depicted a Renaissance church, together with details of famous book illustrations over the centuries, including work by Dürer. To the right, a scholarly gentleman was illustrated reading a book, rising like an angel into the sky, suspended from a hot air balloon, scattering knowledge on members of the public below. The central motif of the mural depicted a fountain cascading alphabets of classical and contemporary typefaces from the Monotype range into a pool of floating printer's flowers and symbols. To the right of the fountain was a Columbian hand press, the workhorse in many jobbing

165. Laurence Scarfe's mural in the 'Books & Printing' section

166. The display of Magazines and Journals in the 'Books and Printing' section

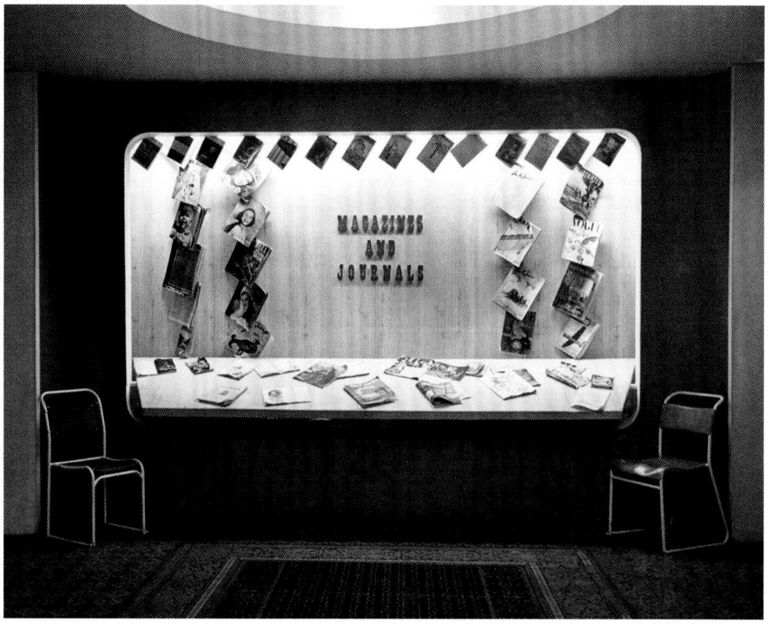

printing establishments in the nineteenth century. A screen covered in a collage of newspaper pages and three glorious examples of display typefaces was further to the right. The fashion for reviving nineteenth-century ornamented types later became a strong motif in the graphic design of the Festival of Britain. In the shadow of the screen children (representing the future) are learning their letters, and two youths playing in the street run towards the light of the sun.

This led to an eye-catching display of familiar popular magazines and periodicals – the beginning of the section dedicated to books and printing. The 'Books and Printing' section was rather out on a limb from the general displays: to the left was a narrow gallery lined with cases full of books and printed material, providing a survey of the development of printing and typographic design over the previous 25 years. To the right was a gallery displaying contemporary publications, that led to a 'reading lounge' and 'book store'. After the visual excitement of glamorous fashion displays and room reconstructions, full of imaginative and theatrical designs, the presentation of the section on books and printing must have appeared rather tame and subdued, and a challenge to the designers, James Gardner and his assistant Edward Barr.[3] But at least all the material on show was published and available, unlike many of the products in other displays, which were subject to the jibe 'Britain Can't Have It' [**166**].

Books and examples of everyday printing are difficult to display in a creative way, en masse, and in a limited space. They are relatively small objects that need to be viewed closely to appreciate the subtlety of

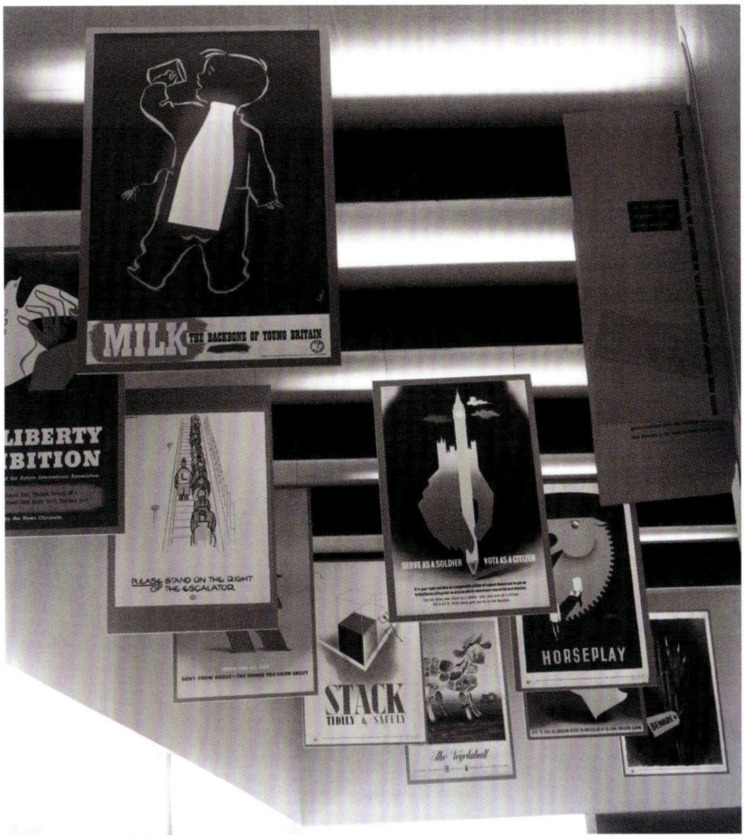

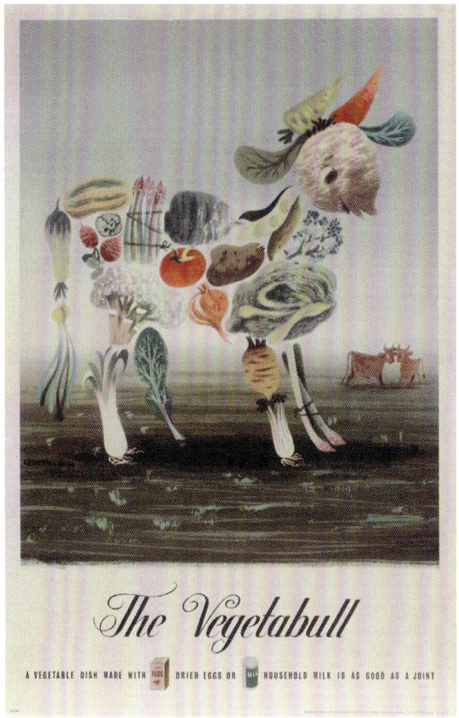

167. Posters displayed in the 'Books and Printing' section

168. Designed by Jan Le Witt and George Him, *The Vegetabull*, 1941. The poster gives advice about diet and rationing in the context of dwindling supplies of meat and other produce during the war. Colour lithograph, 75 x 49.5 cm. V&A: E.1083–2004 (Gift of the American Friends of the V&A; Gift to the American Friends by Leslie, Judith and Gabri Schreyer and Alice Schreyer Batko)

design. Behind glass, the tactile, three-dimensional qualities of the books are lost – they need to be handled. An extremely innovative aspect of the display at BCMI was the idea of providing a 'book store' and 'reading lounge' to address this problem. In an information sheet issued to the official guides to the exhibition, staff were encouraged to engage the public by offering them the opportunity to look at duplicates of the books on show: 'it is hoped that those visitors who are sufficiently interested will ask the attendant if they can examine these, as it is only possible to judge a book's worth by handling it'.[4]

The construction of the display followed the same style as the rest of the exhibition, with pastel colours, curved and angled shapes, and decorative motifs, although the internal layout of the cases was more constrained. To add flair, posters were mounted on the walls and suspended from the ceiling, hanging like flags. The designers of them included many of the great names in poster art of the era – amongst them Abram Games, F.H.K. Henrion, Edward Bawden, Ashley Havinden, Fougasse, Tom Eckersley, Jan le Witt, George Him (known as Lewitt-Him) and James Fitton [**167, 168**].

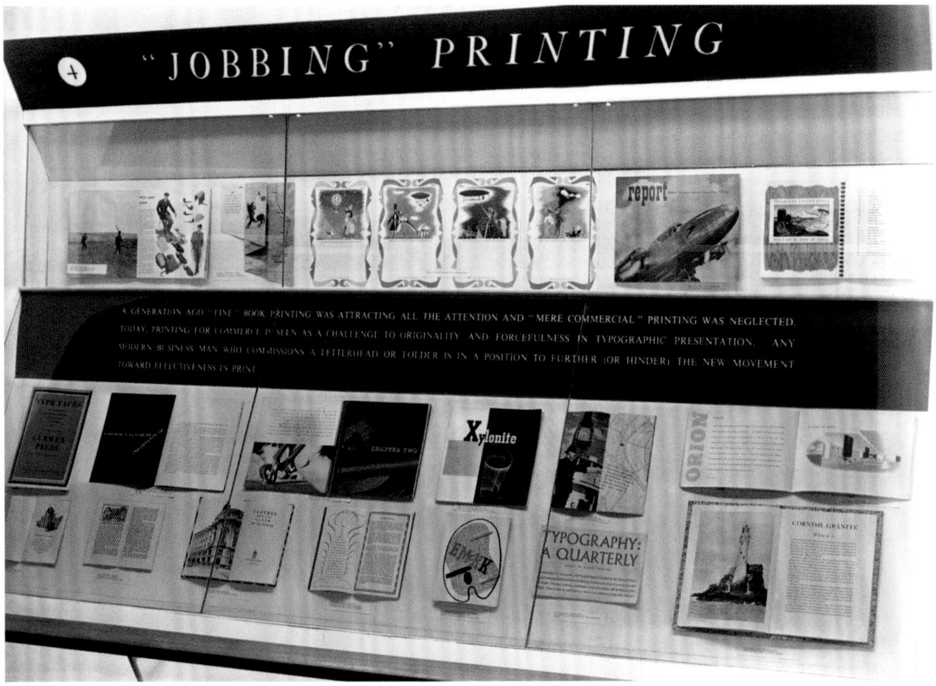

169. The display devoted to Jobbing Printing

Following the aims and objectives of the exhibition in general, the selection of material set out to promote the printing and publishing industries, and to demonstrate how Britain was at the forefront of good typographic design. The selection panel was made up of distinguished practitioners: Walter Lewis, typographer and printer, who in 1945 had retired as Master Printer of Cambridge University Press after 22 years in the post; J. Cameron Smail, a prominent educationalist specializing in technical education and the training of printers, and Principal of Heriot-Watt College in Edinburgh; and the recently knighted Sir Francis Meynell, a poet and influential typographer and printer who founded the Nonesuch Press and wrote widely on the subject.[5] A preliminary selection of material was made by John Tarr, in his role as 'Technical Assessor', assisted by Alan Jarvis as 'Industrial Liaison Officer'. John Tarr was the Head of the Type Drawing Office at the Monotype Corporation, and a prominent member of the printing community.[6]

The overall display was divided into five different categories: 'Britain can make it known' included examples of advertising selected by the Advertising Creative Circle, an organization representing the leading London advertising agencies; 'General printing' illustrated a cross section of 'jobbing' printing, that is printing used in everyday life that is not a book, or to quote from the guide's information sheet, 'anything from a dance ticket to a booklet' [169]. The range of 'General printing' was wide and included Ordnance Survey maps, income tax booklets and housing manuals alongside more glamorous and visually appealing items. This section also included a display of posters. 'Magazines and Journals' showcased current trends in layout and design, whilst the section on 'Books' included a range of between fifty to sixty books of all descriptions and for all age groups, and also featured a display of book jackets. Most of the major publishers were represented with a range of books to suit all pockets and tastes, from refined illustrated editions from the private presses and university presses to cheap paperbacks and colourful children's books [170].

The final section was called 'The Typographic Renaissance', and the storyline was written by Beatrice Warde, another distinguished member of the typographic establishment.[7] All the personalities involved knew each other well and were part of an elite group. A significant factor that united them was their association with the Monotype Corporation, the market leader in Britain in the manufacture of typesetting and type casting machinery.[8] This was one of only two displays in the exhibition that gave an historical perspective rather than concentrating on the future – the other section was a general historical overview entitled 'Great British Designers' – which included the work of early typographers, such as William Caslon, John Baskerville and John Bell, who were represented alongside more contemporary figures, such as Eric Gill.[9] The information sheet for guides attempted to summarize (in a rather simplistic way) the thinking behind the concept for the display:

> This section has been called the Typographic Renaissance. It tells its own story, which is the story of British typography during the last 25 years. Many of the examples displayed are not necessarily of good design, but are there for the purpose of illustrating the story. This story is, roughly, that most letter faces 25 years ago were of a very inferior nature, but that largely due to the Monotype Corporation and a handful of printing houses new letter faces were cut and new standards in typography set up. Britain has very definitely led the way in this restyling, an achievment [sic] of which I trust the guides will feel justly proud.[10]

This is perhaps a rather exaggerated claim: European modernism had profoundly effected typographic design, and there was little evidence of this in BCMI. Nevertheless, in 1944 Meynell had commented in his introduction to Tarr's book *Printing Today* that England 'can hold its head up in any international company'.[11]

Beatrice Warde expanded and developed a subtler argument for the basis for the theme of the display in an essay entitled 'Books and Printing' in *Design '46* (the book accompanying the exhibition). Here

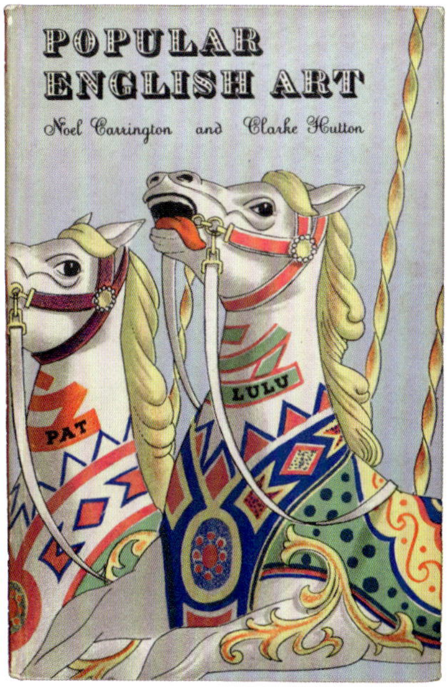

170. *Popular English Art* by Noel Carrington and Clarke Hutton (1945), one of the books included in the 'Books and Printing' section

Warde put forward the idea that in industrial manufacturing good design has 'some connection with profits', and that over the previous 25 years the realisation by manufacturers that design could make a difference had resulted in better and more successful products. Countering the views of earlier critics, including William Morris, who attributed the decline in printing during the nineteenth century to mechanization, Warde challenged the cynical to 'admit that progress in the machine age is not all down-hill'. She continued:

> In Britain today the moral can be pointed in terms of shoes or ships, bungalows or fountain-pens; but never more impressively than with pieces of printing, in any of a hundred categories from books to time-tables, that illustrate the impact of the widespread movement for typographic re-styling. That movement originated in England and is now spreading throughout the civilized world: fine British type-designs, and model examples of British typographic reformation, have had a profound effect upon international style. The booklover thinks first of the striking improvement in the appearance of the normal 'trade' book that has taken place since 1924, and the way in which the gospel of good design has been spreading into long-neglected corners of the book world, to the great benefit of schoolbooks, cheap reprints, cheap Bibles and paper-back 'sixpennies'. But the 're-styling movement' has now swept farther afield, and has called the best typographical brains of the country to the rescue of every sort of industrial, institutional and transport printing – with time-table reform as a specially absorbing problem in functional design.[12]

The design of the Post Office Telegraph Form and *The Times* newspaper were highlighted by Warde as examples of dramatic 'reform'. These were items that 'could be picked up free or for a few pence. Design-for-print is necessarily based, as democracy is based, on the assumption that the Many matter more than the Few.'[13] Unusually for the time, but in the spirit of the BCMI, there was an emphasis on showing printed items that were part of the everyday lives of ordinary people – display cases featured timetables, school books, bibles, newspapers and 'jobbing' printing.

Warde attributed the typographic renaissance to the creation of a new profession, that of the typographic designer, as 'agent or middleman between printer and customer'. In the past she argued, design was a matter of evolution – the result of practical trial and error. When printers and type compositors arrived at a suitable solution or 'the right way', clients tended to ask for more of the same, and so traditional formats were established. However, Warde claimed that the 'intervening century has been one in which printers have had all they could do to adjust their minds to a succession of new mechanical processes, all of which tended to undermine the craftsman's sense of responsibility for the final appearance of the job'. Responding to a demand for something 'different and distinctive', 'creative printing offices' established layout and art departments and introduced new courses in typographic design for apprentices. In 1946, printing was still one of the country's major industries. Warde estimated that over 8,000 printing offices were in operation, and she hoped that good design would filter down to the 'little jobbing-printer around the corner', who would recognize the importance of 'fitness for purpose' in typography.[14] Essential to this aim was the availability of well-designed typefaces, and in her essay Warde praised the contribution of the Monotype Corporation and Stanley Morison's programme reviving classic typefaces and introducing contemporary designs.[15]

These were the themes that underpinned the display. Examples of 'before and after' were included in Warde's display to demonstrate the effectiveness of good design. Her rather patriotic trumpet blowing and democratic fervour almost certainly reflected the mood of the time. During 1940 paper rationing was introduced, restricting supplies to 60% of a company's pre-war usage. Two years later the War Economy Standard was introduced to printing and publishing, a voluntary scheme that demanded smaller type, the use of less white space on the page, inferior paper and binding, and regulations governing unnecessary space for prelims. Although demand for books increased during the war, the public had got used to inferior quality and the dullness of utility.[16] It was against this background that optimists looked forward to an increased quality of design and more general enjoyment of illustration and colour and the publishing of books cheap enough to be available to all. There was also a belief that good design of everyday things such as timetables, forms, catalogues and information literature could help to make everyone's lives easier in the future. What the exhibition could not foresee was the dramatic technological changes that would revolutionize design and printing processes over the next fifty years of commercial photocomposition – and then the coming of the digital age.

NOTES

1 Beatrice Warde, 'Books and Printing', *Design '46* (London 1946), p. 121.

2 A number of eminent figures associated with typography and publishing served on other selection panels, such as Jack Beddington, who as publicity director at Shell had commissioned many of the leading artists of the period to design posters and instigated, with John Betjeman, the famous series of *Shell Guides*. Both he and Noel Carrington, editor, publisher, book designer and originator of Puffin Books, rather surprisingly sat on the panel to select garden tools, hand tools and lawn mowers for the Leisure Section of the exhibition.

3 Edward Barr was one of James Gardner's assistants, who joined the team on 1 May 1946. He had two main jobs – exhibition typographer and working with Gardner on the design of the Books and Printing Section. On 19 February 1947 Barr wrote to W.H. Newman at the CoID responding to a request for feedback on the project and in this letter Barr outlined his programme of work. When he first took up the post details of headings and text had not been finalized so he helped with colour schemes and design details. By June/July 1946 he was working on cut-out display letters and getting them made. August was spent concentrating on the 'Book and Printing' section and in the last three weeks before the opening in September he was busy with preparing work for typesetting, mounting cut-out lettering and organizing a team of professional signwriters. His critical comments were that some directional signage, section titles and sections selling publications were not finalized in time as no one seemed to have responsibility for organizing this. He felt that inadequate attention was given to the 'ticketing' of items [labelling] – the small individual labels for each item proved unsatisfactory and had to be replaced with more legible labels during the course of the exhibition – Barr thought that a firm of caption writers should have been employed under the direction of the typographer and display team. Finally Barr observed that the typeface chosen for all the signs (Times Roman) was a mistake: '... it is an excellent and very carefully designed letter shape, but its excellence is too easily lost by signwriters who work quickly and too much by tradition for exact imitation of such a precise face': V&A Archive, Exhibition File, AAD/1977/4/14A, Britain can make it.

4 V&A Archive, Exhibition file: AAD/1977/4/14A, Britain can make it: Information for Guides, Books and Printing section.

5 In 1925 Stanley Morison was appointed typographic designer to Cambridge University Press, and under his influence Walter Lewis stocked most of the types Morison had commissioned for Monotype. Lewis showed that, when mechanical type composition was used intelligently and with sensitivity, the result could be both handsome and functional, and Cambridge books produced under him were recognized for their clear typographic structure and legibility. Walter printed books for the Nonesuch Press for Sir Francis Meynell. The Nonesuch Press was known for high standards of design and production, producing books with the qualities of 'fine printing' but using commercial processes. Meynell was a close friend of Stanley Morison and all three men, Lewis, Meynell and Morison, showed that mechanical type-setting could be used for the finest bookwork and, rather than lowering standards, provided an opening for a whole new field of endeavour – reflected in the work on show in the exhibition. For a full account of the Monotype Corporation, see A. Boag and C. Burke, *History of the Monotype Corporation* (London 2014).

6 In 1944, John Tarr published a book entitled *Printing Today* (Oxford 1944) with an introduction by Francis Meynell, in which Meynell acknowledged the necessary change to typographic design, 'Unless English manufacturers abandon their complacent attitude towards design – which in a machine age has superseded 'craftsmanship' as a fact ... they, and we with them, are going to have a sorry time when the world trade flows again'. In a chapter on 'Design in Printing', Tarr recognizes the contribution of the European New Typography, exemplified in the work of the Russian typographers, the Bauhaus and typographers such as Tschichold, 'The storm of the more experimental "modernism" having passed, the atmosphere is now cleared of many meaningless traditions and craft preoccupations in printing design ... whatever styles may eventually be exploited, they will need to be capable of mechanical execution by modern machinery, methods, and materials': pp. 166–67.

7 For thirty years, the American Beatrice Warde was the Publicity Manager for Monotype and an eminent practitioner, writer and commentator on typography.

8 The Monotype Corporation manufactured typesetting and type-casting machines in their factory in Salfords, near Redhill in Surrey, which they built in 1899. The company became one of the leaders in the field and put money and effort into developing typefaces and raising standards in typographic design.

9 Supplement to the Board of Trade Journal for 28 September1946. Britain can make it. Design & the Designers. Pioneers of Design, pp. 14–15. Copy held in the V&A Archive, AAD/1977/4/17. For the display on 'Great British Designers', see Jonathan Woodham's essay.

10 V&A Archive, Exhibition file: AAD AAD/1977/4/14A. Britain can make it: Information for Guides, Books and Printing section.

11 John Tarr, *Printing Today* (Oxford 1944), p. 6.

12 See Beatrice Warde, 'Books and Printing', *Design '46* (London 1946), p. 121–25.

13 Ibid., p. 121.

14 Ibid., p. 122.

15 Stanley Morison was a dominating figure in the typographic world at the time: amongst many other achievements he was Typographic Adviser to the Monotype Corporation and largely responsible for their series of type revivals and new designs. In 1932 he redesigned the typographic style of *The Times* newspaper and designed the Times New Roman type.

16 Paper rationing continued until 1949. In a letter to *The Times* in 1947, Francis Meynell replied to a letter from a Mr Lockwood who had made a plea for more paper, saying that 'attractive packaging and well-designed catalogues and stationery can greatly help export'. Meynell commented that more paper would not guarantee good design and observed that 'The exhibits of commercial printing were very sparse at "Britain Can Make It" for the reason that little could be found of a very high standard of design' (*The Times*, 24 September, 1947, p. 5). To improve standards in government departments, Meynell took on the role of Honorary Typographic Adviser to Her Majesty's Stationery Office, in the same year as BCMI. A separate display on paper, showing the various uses of paper and card, was included in the 'What the goods are made of' section of the exhibition and examples of design and printing were to be seen in a display on packaging, and in books and magazines that were used to dress the room interiors.

DIANE BILBEY

Designers Look Ahead: The Hall of the Future

> You are now entering a hall of the future. While this is the last section of the Exhibition it is really only the beginning, because here British designers show things YOU may be using in the next five to twenty years.
> *Pause ... Repeat.*[1]

Hall of the Future

These aspirational words are the transcript for the recorded narration played at the entrance to the 'Hall of the Future', from one of the special sound features placed around the exhibition. It was spoken by broadcaster Frank Phillips, the homely, familiar voice of the BBC Home Service during World War II, in a repeated recording played from a speaker in the belly of one of the 'Mr Designer' mannequin figures [see **10, 85**]. This was a dramatic introduction to the final and most ambitious section of *Britain Can Make it* (BCMI), the 'Designers Look Ahead', or 'Hall of the Future' display. Basil Spence designed the Hall, with the all-seeing eye of the designer (a theme used throughout the exhibition) looking down on the expectant public and exhibits below. This was a futuristic and theatrical scheme, including a ceiling made from an elaborate cross-hatching of flood-lit heavy nylon cord. It was to show visitors the shape of things to come, and leave them with something to think about as they left the exhibition [**171**].

In its review of this section, the journal *Display* wrote: 'The somewhat Wellsian peeps into the future seem an anticlimax after the concrete realities of design as it is to-day, but display in this section by Basil Spence, the exhibition architect, is effective, with lofty draped roof, and the designer's eye surveying all from above'.[2]

Exhibits in the 'Hall of the Future' focused on two key areas – travel and labour-saving domestic products – both suitably aspirational themes for a Utopian future. Designers were invited to submit designs and encouraged to 'Look Ahead' and 'anticipate some of the things that people will probably be buying or using five, ten or twenty years hence'.[3]

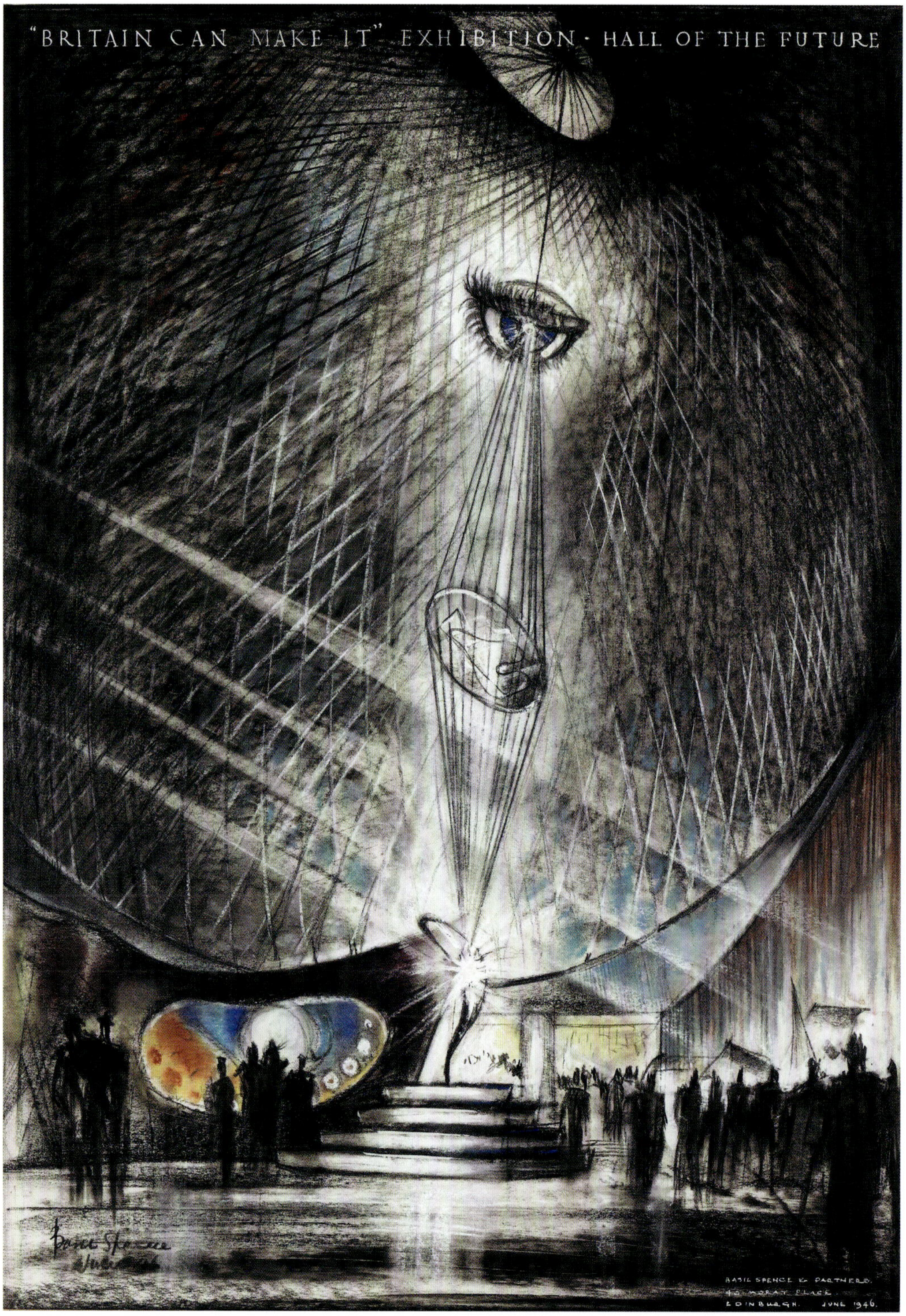

171. Perspective sketch of the 'Hall of the Future' or the 'Designers Look Ahead' section by Basil Spence

The Council's call for designers prompted a good response, but of the many preliminary proposals submitted, only twelve designs were selected and worked up into prototypes – an air-conditioned bed [174], a wall radio [175], a television projector [176] a sewing machine [180], an electric cooker [181], a suitcase, a bicycle [182], a taxi [183], a railway car [173], a caravan trailer, a catamaran [184] and a space-ship [185, 186]. These were made using many of the materials for which research and development was accelerated during the war, including Perspex, plastic and aluminium.

A mural by the New Zealand artist John Hutton, featured on one side of the Hall, provided an 'abstract background to design of the future' [172, 184].[4] It also provided the perfect backdrop for Warnett Kennedy's Perspex 'space-ship' model.

The novelty and engineering innovation in the design of the 'Hall of the Future' was featured in many contemporary publications. The *Architect and Building News* commented:

> A great hyperboloid of crossed strings hangs suspended between two ovoid metal loops and forms the roof to this hall. A great eye – symbolic of vision in design – is suspended in the valerium space. From it there shines a light, the beam from which pierces a perspex disc and strikes a special feature on a raised dais. Along one wall are exhibits of the future: on the other Mr. John Hutton has painted a mural in free flowing lines to represent the freedom and potentialities of the future of design. This hall is the climax to the Exhibition; its purpose is to give the departing visitor food for thought.[5]

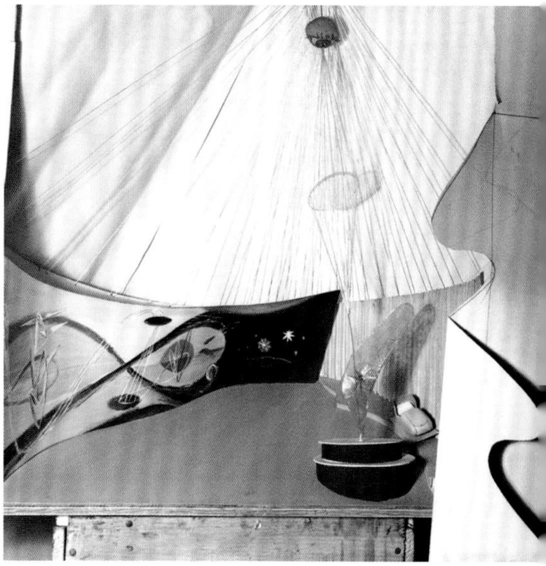

172. Model for the 'Hall of the Future' or 'Designers Look Ahead' section showing a mock-up of John Hutton's mural. The giant designer's eye motif hangs from the ceiling on nylon cord. Beneath is the air-conditioned bed.

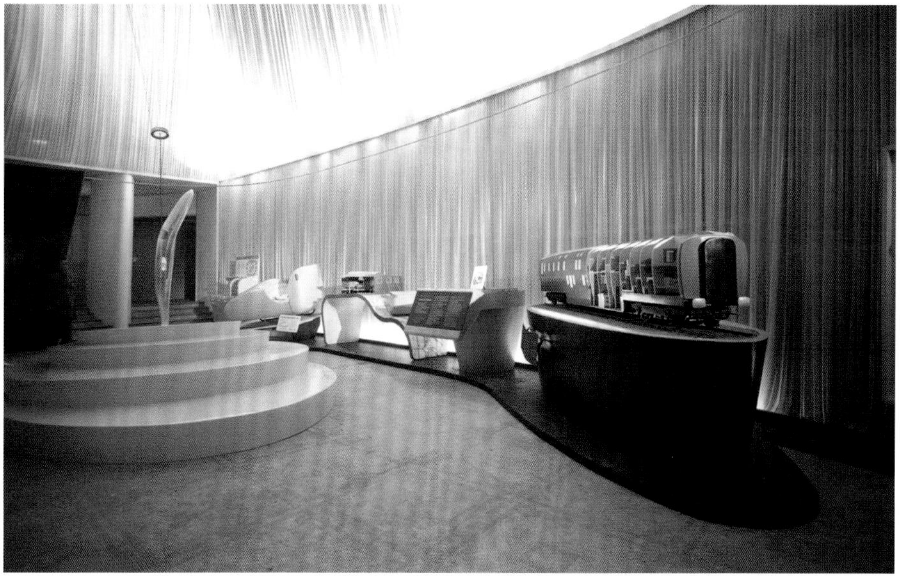

173. A view of the 'Hall of the Future' or 'Designers Look Ahead' section showing some of the exhibits. The model of the third-class sleeping car designed by Misha Black and Reginald (William) Vaughan (Design Research Unit) is in the foreground.

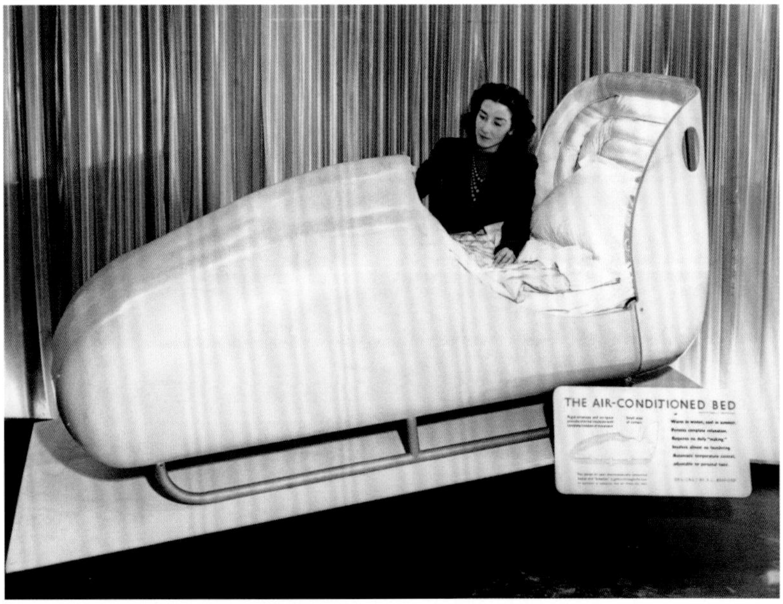

174. Air-conditioned bed. Designer F.C. Ashford. Design and research costs £26 5s; model £125.

'The average man spends one-third of his life in bed – that is, about a quarter of a century. Mr Ashford's design will ensure that those twenty-odd years be spent in comfort and health' (BCMI *Exhibition Catalogue Supplement*, p. 213). The supposed benefits of the air-conditioned bed were that it was cool in the summer, yet warm in the winter. It was also labour-saving as it required no daily making. The temperature was controlled electrically by a dashboard style panel. The sleigh-like bed also included a lid, which when not required as a headrest could be shut to make it appear even more streamlined. According to an article in the *Essex Newsman*, the bed's design was the result of a former Naval Lieutenant-Commander (i.e. F.C. Ashford) having cold feet in bed one night: 'The Commander, who designed it, could not get used to a civilian bed after his comfortable hammock on the lower deck, and wardroom bunk when he was elevated to the upper deck. "The blankets," said the Commander, "just wouldn't stay put, and my feet were cold." So he set to work to invent a bed on which the blankets would stay put – and succeeded in inventing one which did not need blankets at all. The air-conditioned bed, he claims, needs only a pillow, a mattress cover, and a top sheet. All the rest is done by a thermostatically controlled engine in the foot of the bed, so that the temperature can be adjusted according to the weather!'. The bed also featured in the *Daily Mail* (23 September 1946, p. 3), 'Sleep "with the lid off" in the electric bed'.

Early plans

Less than a year before the opening, the CoID organized a meeting to discuss the content of the exhibition as a whole. Attended by leading industrial designers of the day, it set out provisional plans for what was considered to be a special section of the exhibition – 'Designs of the Future'. Minutes record that 'The phrase "designs of the future" considered as a title for the section was criticized as too hackneyed and as having some unfortunate past associations. The title "Here is Tomorrow" was provisionally accepted.'[6] 'Here is Tomorrow' was still being used as the working title for this part of the exhibition in May 1946, when members of the Design Committee were asked to attend a meeting to decide which submissions to accept from those put forward by designers. The CoID clearly changed their minds on the title, replacing 'Here is Tomorrow' with 'Designers Look Ahead', a more accurate description of what the section aimed to illustrate. The motivations for the final section of the exhibition were grand – that it 'should serve to stimulate the imagination of the visitor, and send him away from the Exhibition feeling that "Something has happened to England"'.[7]

The selection

Decisions as to what to include were made by members of the Design Committee of the CoID. Members, including Sir Kenneth Clark, Gordon Russell, Sir Francis Meynell and the Hon. Josiah Wedgwood, were invited to a meeting just four months before the opening of the exhibition, on 27 May 1946, to make the final selection.[8] The importance of this final section of the exhibition was made clear in notes provided for the Design Committee: 'In particular, the Council wishes to give emphasis to a section, which may be the concluding feature of the Exhibition, on designs of the future'.

The grounds on which exhibits were to be selected were twofold – the realistic and the futuristic. The realistic category was to include items that were considered 'not too far ahead of current manufacturing possibilities and techniques ... it should contain designs planned as within the range of practical possibilities by (say) 1951'. The more futuristic ideas were to consist of 'designs of a more imaginative and "daring" kind, not however consisting of mere fantasies, but based on reasonable assumptions about the developments likely to take place in the next ten or twenty years in new substances, processes and techniques'.[9]

The 'Note to Industry' included in the *Catalogue Supplement* for the exhibition made it clear that this section of BCMI was potentially an exciting spring-board for British industry:

> This section of the Exhibition is not a fanciful vision of the future. It anticipates in a practical way the current trends both of popular requirements and technical possibilities ... The Council invites manufacturers to interest themselves in the ideas displayed here, and will readily put them in touch with the designers who retain the copyrights. One idea springs from another. A contact so established may lead to further developments.[10]

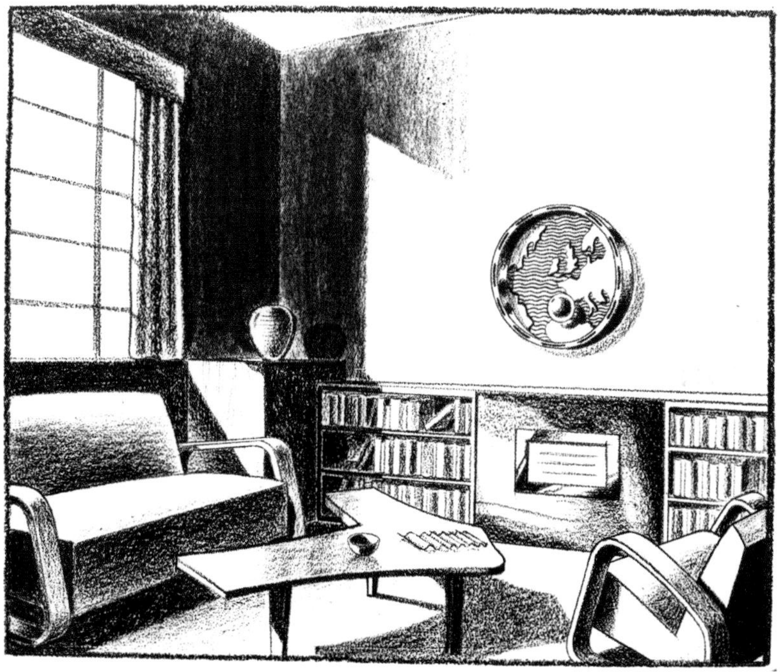

175. Wall radio. Designer Warnett Kennedy. Design and research costs £157 10s; model £100.

'Mr Warnett Kennedy aims to show that the design of a radio set may be treated as wall decoration; the working parts, too, may have their own fascination if made visible. The entire effect may be that of an abstract design' (BCMI *Exhibition Catalogue Supplement*, p. 214). The appliance worked by using a removable 'plug', that was inserted into the relevant slot to pick out the radio station. Loudspeakers were formed by the area on the 'map' showing the oceans. Warnett Kennedy was shown demonstrating his radio as part of a British Pathé news feature on BCMI filmed on 30 September 1946, entitled 'Britain Has Made It – Shop Window'. The prototype radio, sponsored and developed by Seymour Products Ltd and A.H. Cornwall & Sons Ltd, constructed by Armstrong & Co. Ltd, was included in an advertisement they placed in *Display, Design & Presentation* (November 1946, p. 39), highlighting the collaborative effort between designers and technicians. The advertisement explains: 'This Built-in-Map Radio has a movable, Perspex globe that houses the all-mains receiving unit. Slip it into the station desired, and indicated on the map – and lo! ... the music comes.'

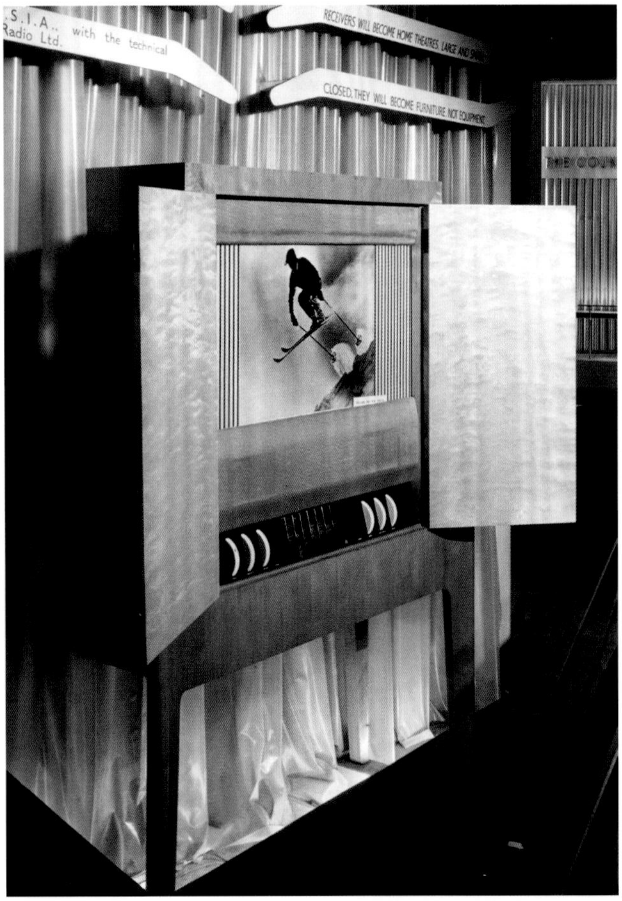

176. Flat-screen television projector. Designer Eden Minns; technical assistance Murphy Radio Ltd. Design and research costs £105; model £112 10s.

'In a few years television will be as common a form of domestic entertainment as sightless radio is now' (BCMI *Exhibition Catalogue Supplement*, p. 213). This insightful statement was quite a leap of faith for 1946. It seems hard to believe that in 1945 there simply was no broadcast television. Pre-war television had only been available as a luxury item to around 20,000 homes, and the BBC shut down broadcasts on 1 September 1939 in preparation for the onset of war. It reinstated broadcasts on 7 June 1946, and a combined radio and television licence came into effect in 1947. Eden Minns had worked for Gordon Russell, designing cabinet-style radios for Murphy Radio Ltd, and the link between early television cabinet design and that of radio is clear to see. The television displayed in the 'Hall of the Future' was only a mock-up 'television projector'. It was slim-line and flat-screen, and had doors that could be closed when the set was not in use. Radio manufacturers had the most to lose from the popularity of television, so it was forward-thinking of Murphy Radio to collaborate on this prototype, and to diversify into television manufacture. Although over 100 radios were displayed at BCMI, only six other television sets or projectors were shown.

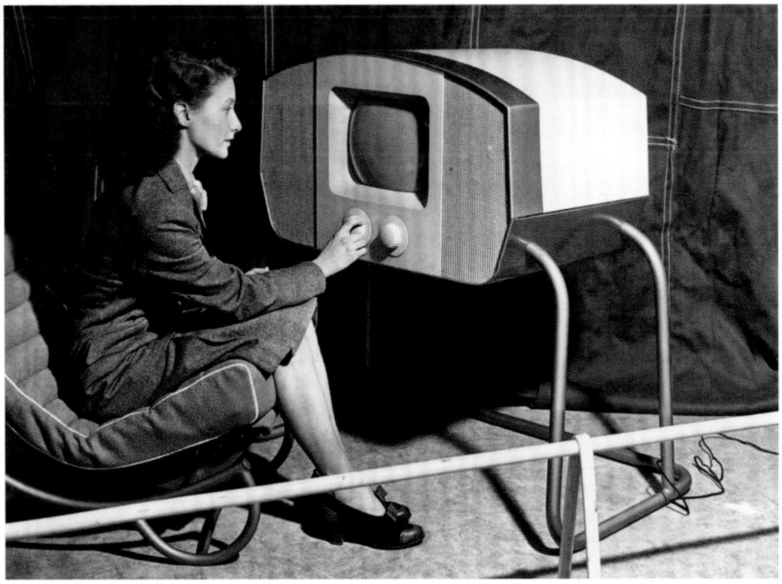

177. The console television manufactured by Ferranti Ltd was one of the small number of television sets displayed at BCMI, not in the 'Hall of the Future', but in the Television Research Engineers office in the 'Furnished Rooms' section, designed by Christopher Nicholson [179]. Nella Last wrote diaries for Mass-Observation and was known as Housewife 57. She visited BCMI on 15 November 1946 and wrote a special report. It included a description of the green tubular framed occasional chair manufactured by Christie-Tyler Ltd in front of the television: 'I was looking at some chair that looked as if made for people with permanently curved spines & my son said "aw come on now—where's your sense of humour!" A woman standing by smiled & said "you would need a rather strong sense of humour to buy & use those I think"' (Mass Observation Archive, University of Sussex, SxMOA 1/2/26/1/C/3). For the Television Research Engineers office design see Christopher Nicholson, RIBA Drawings Monographs no. 4 (London 1996), pp. 102, pl. 79, p. 75, l. 52.

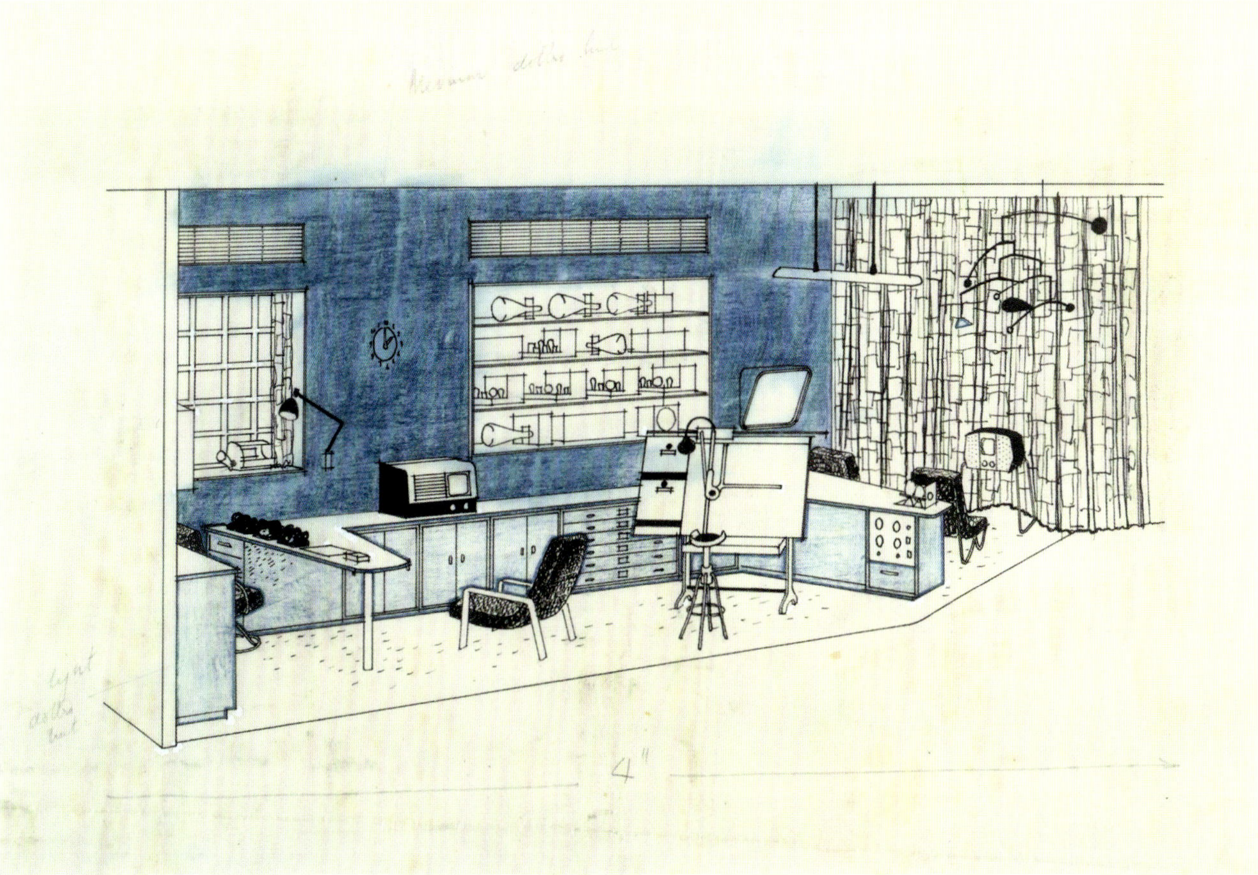

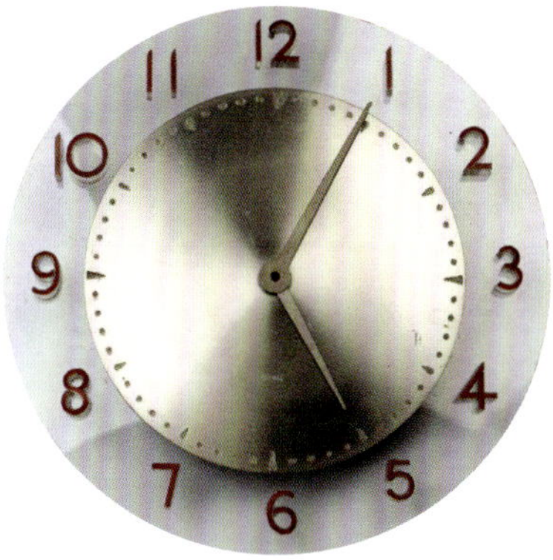

178. Christopher Nicholson, made by Ferranti Ltd, clock, 1946, painted acrylic with metal dial, diam. 33.5 cm. V&A: W.4–1998

179. Christopher Nicholson's design for a technical office was the most forward-looking room set in the 'Furnished Rooms' section. Here, the inclusion of television sets reflect the futuristic and professional, not yet for the general public. It was described in the *Architect and Building News* (27 September 1946): 'The back wall is painted dark green and the end wall light green .... Bench and desk tops are covered in white leather cloth. Rust-coloured fabric with white piping is used for chair covers. The floor is finished with blue linoleum.' Three versions of the television are shown in this room, all produced by Ferranti Ltd: a table model (much like a radio set), the console model (with the kinetic sculpture mobile overhead), and an experimental television projection set-up. Ferranti's acrylic and lacquered metal electric clock (also designed by Christopher Nicholson) is displayed above the table console television [178].

180. Sewing machine. Designer F.H.K. Henrion and W. Joseph Woods. Design and research costs £262 10s; model £300.

'The sewing machine designed by Mr. Henrion and Mr Woods is less than 2ft long, less than 1ft high and 4 in. wide. It weighs less than a normal six months old baby' (BCMI *Exhibition Catalogue Supplement*, p. 214). The bright, light, portable electric sewing machine, with its built-in carrying case, was far removed from the heavy machines found in most homes. In a recent monograph on its designer, F.H.K. Henrion, Adrian Shaughnessy comments, 'Had he [Henrion] developed it in any other context than in post-war Britain, he might have owned a world-beating product to rival Olivetti typewriters or Braun electric razors. But his invention – with its brushed aluminium pre-echo of Apple laptops – was only ever to exist as a prototype.' M-O investigators picked up some comments on the sewing machine. Its investigators recorded that an attendant in the 'Hall of the Future' had remarked that 'usually the bike's the first thing they ask for when they get into the exhibition – they generally ask where the streamline bicycle is', and went on to tell that the 'Poor old sewing machine gets completely ignored yet it's so much more useful than the bike'. The M-O investigator went on to note: 'Indeed few people bothered to look at the sewing machine'. Mass Observation Archive, University of Sussex, Draft transcript report on Britain Can Make it, SxMOA1/2/26/1/A. The prototype is now part of the collection held by the University of Brighton Design Archives.

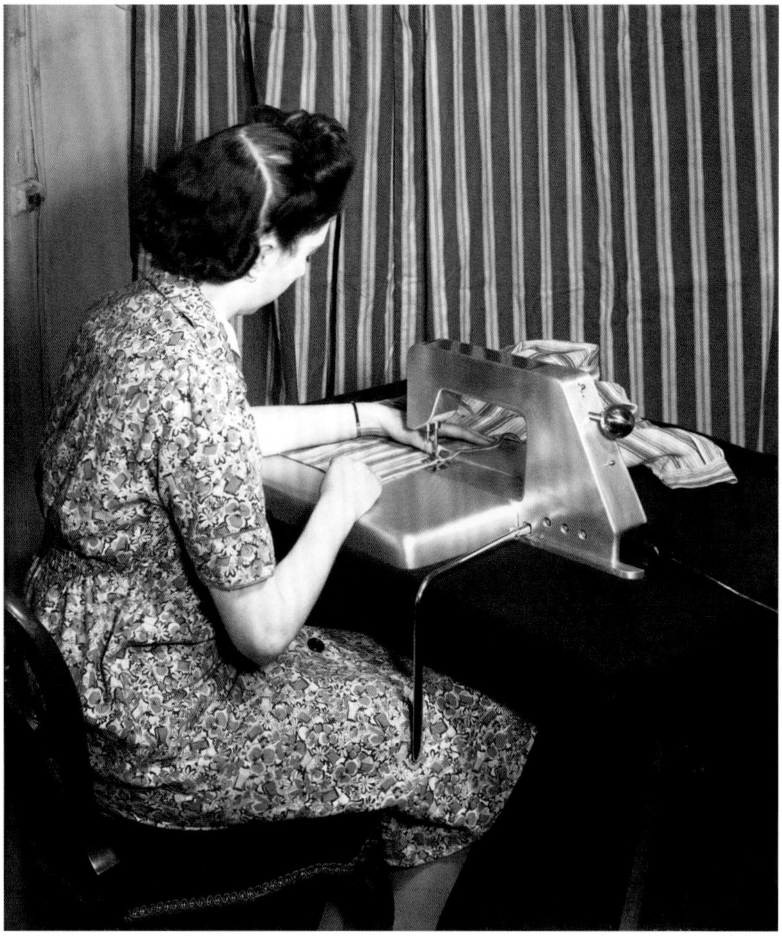

The BBC

The CoID had a strong relationship with the BBC, often suggesting themes the broadcaster might like to consider as subject-matter for their programming. Clearly the Council felt that the 'Designers Look Ahead' section was a particularly important one in terms of public interest, and this resulted in the BBC producing a short programme devoted entirely to it. Filmed before the opening of the exhibition on 17 September 1946 at Alexandra Palace, it took the form of a series of interviews with designers standing next to the models or drawings of their designs. Basil Spence was also on hand, filmed next to a model showing the layout of the display.[11]

The outcome

A report produced in January 1948 on the outcome of the 'Designers Look Ahead' section highlighted the issues and difficulties it had presented for the organizers. The contractual arrangement between designers and the CoID was particularly problematic. Although

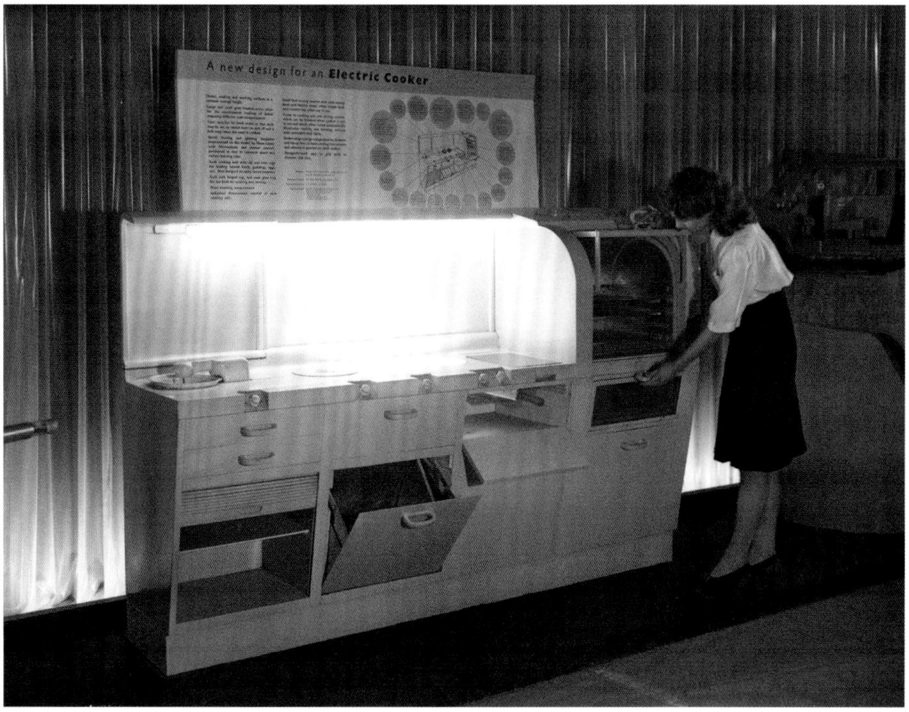

ownership of copyright remained with the designer of a particular design, the actual models became the property of the Council, and any royalties or other payments received in respect of the design were to be used to repay the not inconsiderable costs incurred by the Council for research, design and studio costs.[12]

Although 'Designers Look Ahead' proved to be a popular section of BCMI, particularly for children, who flocked to see Benjamin Bowden's streamlined bicycle [182], others were disappointed. William Gaunt wrote in the *Evening Standard*:

> It was in the most would-be sensational aspect that I thought 'Britain Can Make It' least successful. I am not speaking of the setting, brilliantly carried out by James Gardner and Basil Spence in an imaginative way which gave verve to the show. I mean its Hall of the Future, its designs for to-morrow. A Freudian or surrealist bicycle did not impress me by its appearance. An air-conditioned bed, like a speedboat on sleds, left me cold. As to the space ship. I am happy as a writer on art and design in having no competence to speak.[13]

A tongue-in-cheek review in *Newnes Practical Mechanics* highlighted the fact that, frustratingly, most of the products were merely mock-ups or prototypes and did not actually work. It placed the items on display into three categories. In its first category, described as 'things that actually

181. Electric cooker. Designer Milner Gray and Reginald (William) Vaughan; associate designer Michael Percival; technical consultant W.M. Ackery (Design Research Unit). Design and research costs £210; model £275.

'The horizontal oven saves space and makes stooping unnecessary. Special cooking vessels, shaped for maximum heat retention, are built into the structure. Electric power being at hand, it is used to operate a food-mixer. Heat is thermostatically controlled. The absent housewife will be heartened by the inclusion of automatic time-switches' (BCMI *Exhibition Catalogue Supplement*, p. 215). This was one of two designs for 'Designers Look Ahead' section, contributed by Milner Gray and Reginald (William) Vaughan of the Design Research Unit (DRU), the other being a taxi-cab. The cooker was a radical design, described in a caption to a *Picture Post* image at the time of the exhibition as an 'Atomic cooker .... This electric cooker impresses everyone, though some find it a little frightening.' It included many built-in features reminiscent of today – a timer, double oven, splashback, concealed lighting and glass doors. An article aimed at women in the *Yorkshire Post and Leeds Intelligencer* (4 October 1946, p. 6) reported on the new electrical gadgets at BCMI; it highlighted the cooker: 'An electric clock with time switches enables the cook to pre-set the times at which the oven-heat is to be switched on and off, and an alarm bell rings when the meal is ready.'

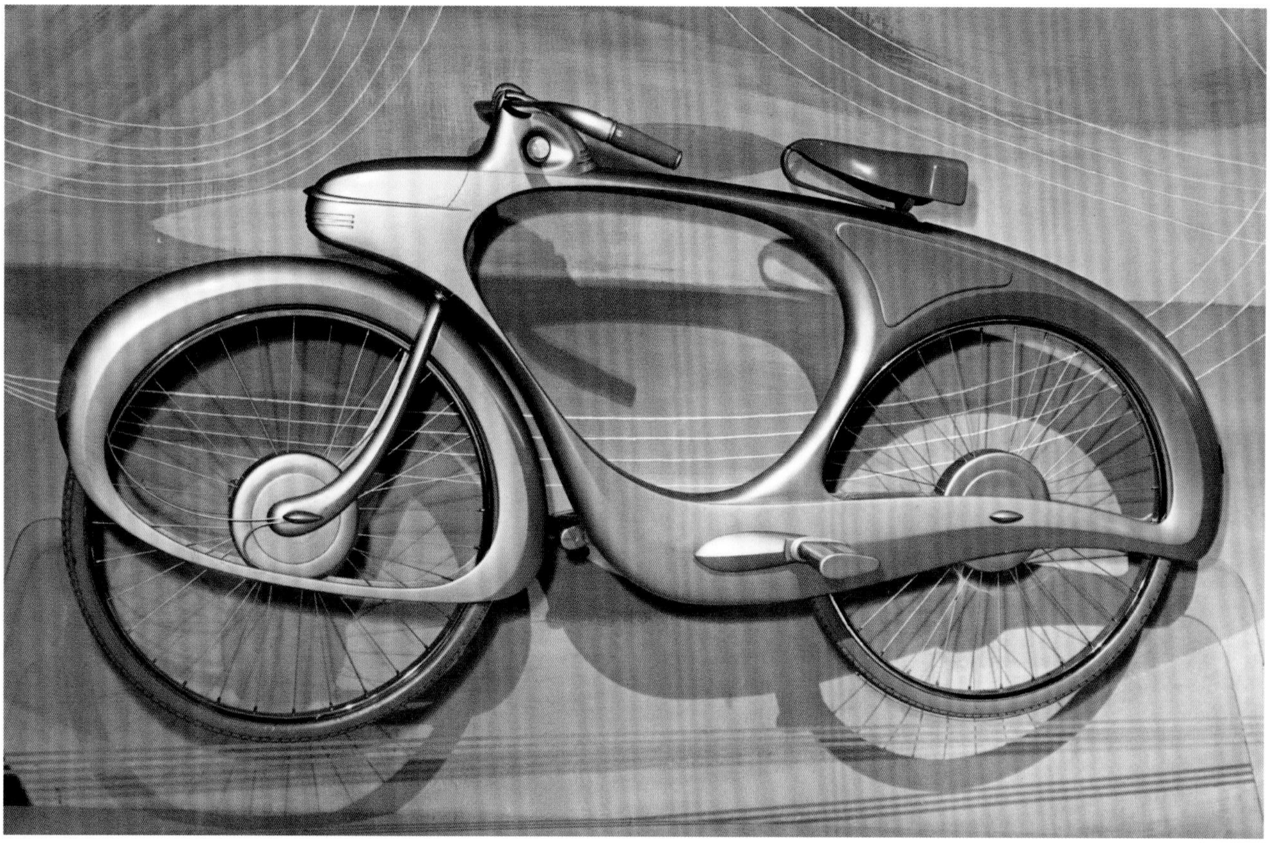

182. Aluminium bicycle. Designer Benjamin (Ben) Bowden; Allen Bowden Ltd. Design and research costs £100; model cost £300.

'The framework is light, strong, compact and well sprung. Built-in brakes are operated by twist-grip controls. A variable throw mechanism evens out the pedalling effort. The whole construction is easy to keep clean. Mr B.G. Bowden calculates that by the time his cycle can go into production there is no reason why the costs should be higher than prices paid to-day' (BCMI *Exhibition Catalogue Supplement*, p. 213). Part of the caption to the reverse of a photo of the bicycle held at the University of Brighton Design Archives: Design Council Archive notes: 'Here you see the bicycle of to-morrow – bicycle of the 'make-life-easier-and- better era'. Bowden's bicycle made a great impact on visitors to BCMI. Its innovative streamlined design and lightweight pressed alloy material made it by far the most popular item in the 'Designers Look Ahead' section. Bowden had helped design armoured cars during the war, and this experience informed his design and choice of materials. A newspaper report in the *Nottingham Evening Post*, 'Made in Britain. Striking Goods On Show To-morrow' (23 September 1946, p. 5), suggested that the potential revolutionary effects of the Bowden bicycle 'might make the present models as obsolete as the old "penny farthings"'. Sadly for Bowden, this was not to happen in Britain. Mass Observation (M-O) investigators recorded comments made by a group of boys, aged between 12–14 years, as they entered the 'Designers Look Ahead' section. "They're rushing around, first to one side and then the other. One points to the bicycle "Caw – look at that Sir." And the master in a very casual voice answers: "All right. Come on. You'll never see it in this country." This statement proved to be correct. Disillusioned by the lack of interest for his revolutionary bicycle from British manufacturers, Bowden initially left for South Africa, later emigrating to the United States, where his design did go into production, but not until 1960. Manufactured in fibreglass rather than the alloy design at BCMI, and in a range of pale colours, it was marketed as the 'Spacelander'. Surviving examples are now rare and highly sought after. See Paul Clark, 'Ben Bowden's "Bicycle of the Future" 1946', *Journal of Design History*, vol. 5, no. 3 (1992), pp. 227–35.

183. Taxi. Designer Milner Gray and Reginald (William) Vaughan; associate designer George E. Williams (Design Research Unit). Design and research costs £210; model £200.

'By placing the engine to one side and moving the driver forward they have made room for a second fixed seat inside. "For Hire" signs light upon both sides at night as soon as the meter flag goes up' (BCMI *Exhibition Catalogue Supplement*, p. 215). This space-saving design, including the repositioning of the engine, provided added comfort for the passenger and more space for luggage. It also provided the driver with a better seating position, giving him an uninterrupted view of the road ahead. A sliding door provided ease of access to and from the taxi. The most significant feature, the roof light 'For Hire' indicator, was quickly taken up by all London taxicabs and is still in use today.

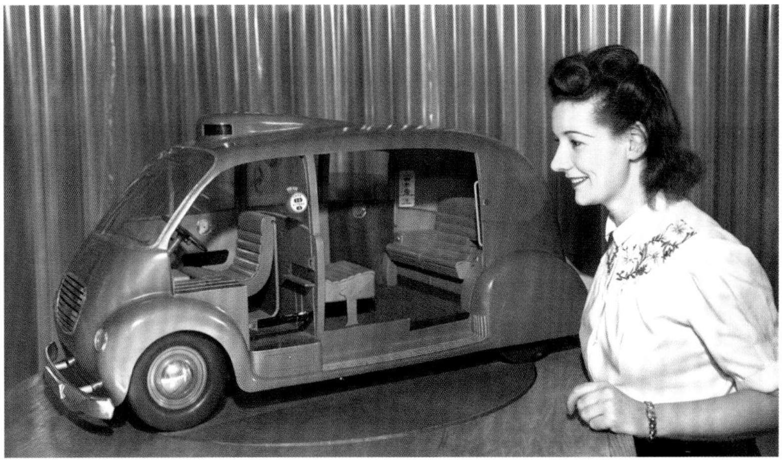

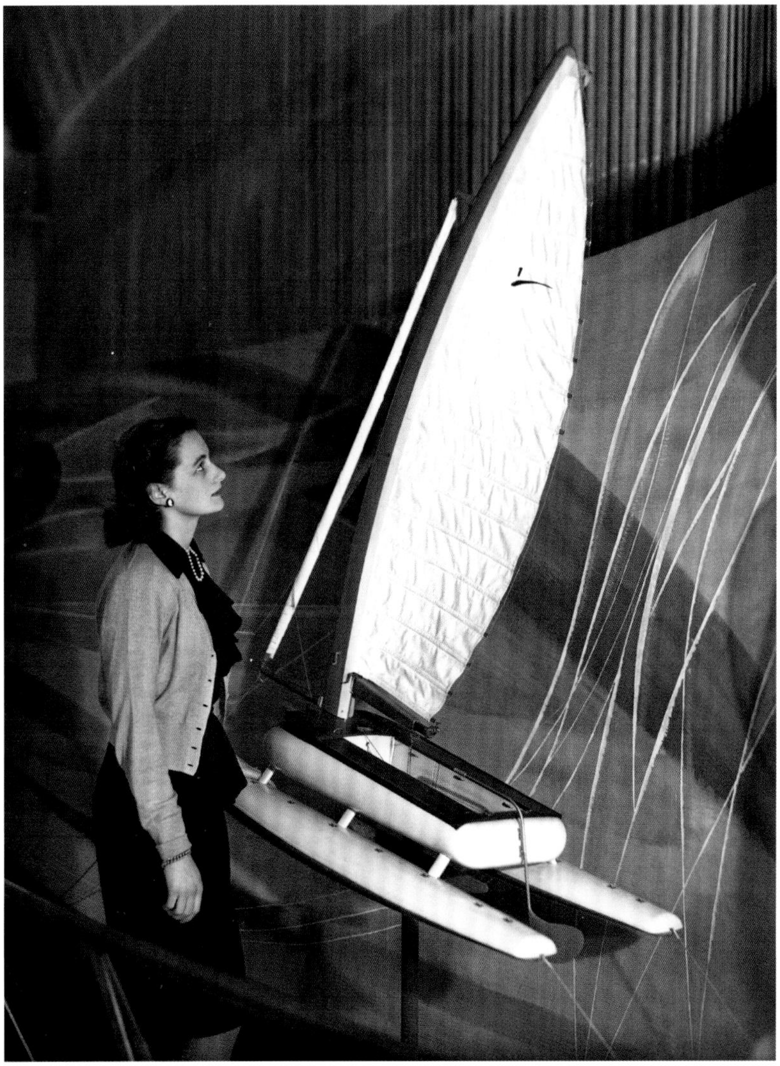

184. Wingsale catamaran. Designer Wells Coates; with Leslie Appleton, consulting engineer, and Reginald Magna. No design and research costs charged; model £315.

'This design … comprises two projects: a new form of mast and sail, and a modern version of an old type of sailing vessel: the twin-float "Catamaran"' (*BMCI Exhibition Catalogue Supplement*, p. 216). Wells Coates was a successful architect and industrial designer. His architectural projects included the Lawn Road flats, known as the Isokon building, widely regarded as a key example of British modernist architecture. Coates put forward three design ideas to the CoID for inclusion in the 'Designers Look Ahead' section, but only the catamaran was accepted. He was passionate about sailing – the 'wingsail' was Coates's own term for a sail that was shaped like the wing of a bird. In a letter to his daughter, he described his involvement in the exhibition: 'My exhibit – a quarter full size model of the 'Wingsail Catamaran' – was finished just in time, and we were televised, filmed, photographed, interviewed, and generally pushed around in a flood of publicity'. At the royal opening, Coates's was presented to the King and Queen. According to his daughter, Laura, he was delighted. He wrote in a letter to her that 'His Majesty, being a yachtsman, made the most intelligent comments and asked just the right questions, and we chatted for some minutes about it. Finally he said – after I had explained the high performance qualities of this aerodynamic sail, and that it would be introducing, in the catamaran or twin float version, a new sport 'high speed sailing' – that it would "revolutionize sailing and require new classes or categories in racing" because – he said – with a knowing look – "you know, yachtsmen are very conservative …."' Laura Cohn, *The Door to a Secret Room. A Portrait of Wells Coates* (Aldershot 1999), pp. 173–74; see also pp. 177–79. The model generated interest at BCMI, and encouraged Coates's to build a 16ft version that was sailed on the Norfolk Broads. The interest was, however, not sufficient for the catamaran to be manufactured commercially. Today, aerodynamic designs and catamaran hulls are commonplace. The model is shown here in front of John Hutton's mural.

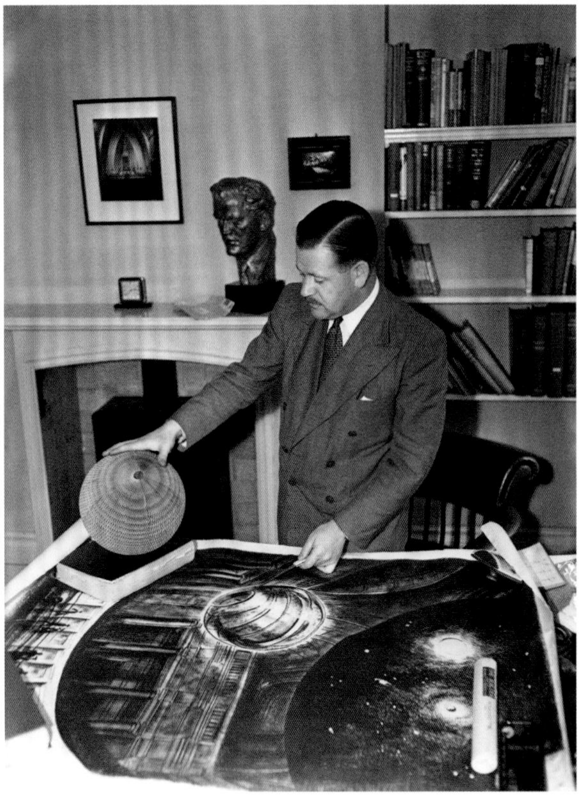

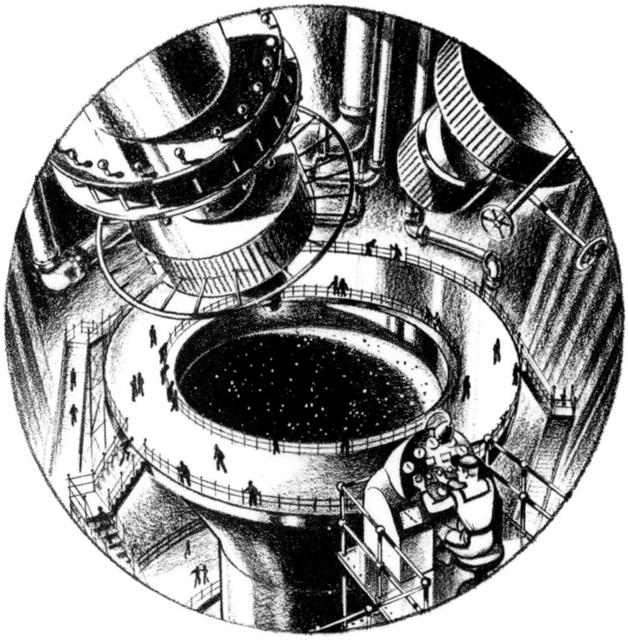

185, 186. 'Space-ship – 'The Man-made planet'. Designer Warnett Kennedy. Design and research costs £525; model £475.

'The possibility of travel through inter-planetary space must now be considered seriously. The control of atomic energy brings it within the compass of reality' (BCMI *Exhibition Catalogue Supplement*, p. 213). Designs by the Scottish architect Warnett Kennedy featured prominently in the 'Designers Look Ahead' section. His contributions were varied and eclectic, and covered practical lightweight luggage, a wall radio, and a space-ship. The spherical geodesic model for a space-ship, or 'Man-made planet' as Kennedy described it, with living quarters (like today's space stations), foresaw the transition into the space age. Kennedy's 44-inch glass, Perspex and Dufaylite model was accompanied by a series of photographs, illustrating significant technical developments of the rocket, from its earliest inception to the atomic bomb. It was displayed 'hovering' near to Hutton's mural, and visitors could view it by peering through a hole in this 'moon wall'. *Newnes Practical Mechanics* devoted a whole article to Kennedy's atomic-powered 'Spaceship', 'The Man-made Planet' (25 September 1946, pp. 134–35), and placed it in context: 'Until the later stages of the war public opinion was sceptical of plans for inter-planetary travel. A space-rocket was regarded as a glorified firework. Derision ceased, however, when the V2 rocket bomb turned the firework into a dangerous weapon; and rocket fuels themselves were outmoded as a source of power for space flight by the atom bombs at Hiroshima, Nagasaki and Bikini.' The article went on to explain that the unusual spherical shape of Kennedy's design suggested that the concept of the streamlined space-ship was outdated: 'It became legitimate to assume that a space-ship need not begin its journey at violent speed .... It became possible also to devise an alternative to existing designs; for a space-ship having a nose and tail would have to revolve in space at a point where the gravitational pull of another planet began. Otherwise the ship would land on its tail or the passengers on their heads .... Mr Warnett Kennedy felt the most sensible shape for a space-ship, not limited by considerations of streamlining, was the sphere. The earth, moon and other bodies are spherical. Why not a space-ship designed as a "man-made planet?" "I do not claim that the space-ship is immediately practicable, but I do regard it as an intelligent anticipation of the trend of future design."' However, Kennedy's model became a figment of fun in contemporary newspaper reports. The Barkley family's visit to BCMI was the subject of a jokey article by William Barkley in the *Daily Express* (25 September 1946, p. 5), in which he wrote, 'The last exhibit is the Space Ship which can travel to the moon in about two days. Mrs Barkley studied it and said: – "Better get some of these goods into the shops before they discover that the moon is inhabited and send everything there".' At the royal opening of the exhibition, Queen Elizabeth is reported in an article in the *Western Morning News* (25 September 1946, p. 3) to have said, 'What fun it must have been designing that. It's good to look sometimes into the future.' A further article by Sylvia Pollak in *Architectural Design & Construction* (October 1946), p. 273, undermined the whole idea: 'Perhaps we shall rocket without qualms to the moon. "But," said a down-to-earth critic of this conception, "Have we not got enough troubles to settle here? Do we need to go to the moon?"'

worked' – it tellingly placed nothing; in its second category – 'things that could work' – it placed the railway car, the catamaran, wall radio, taxi cab, portable sewing machine, television projector and caravan. Its final category – 'things that might work someday' – it included 'that bicycle' and the space-ship.[14]

Rather than quietly closing the door on BCMI, the 'Hall of The Future' aimed to help the exhibition go out with an inspirational bang. Most of the designs clearly did not become commercially viable products, but some of the ideas, or at least adaptations of them, did find their way into production. Of course, most of the designs displayed were unrealistic, impractical or just over-ambitious. Whatever its contemporary detractors may have said or written, no other section of the exhibition better evokes its ambitious and creative spirit.

NOTES

1 Design Council Archive at the University of Brighton Design Archives, hereafter DCA: 742/ 'Special Features' Sound 23 July 1946.
2 '"Britain Can Display It" Illustrated review of some of the leading display features at "Britain Can Make It" which opened on September 24', *Display, Design & Presentation* (October 1946), p. 10.
3 *Britain Can Make It Exhibition Catalogue Supplement* (London 1946), Group XX. Designers Look Ahead, p. 212.
4 V&A Archive, Exhibition File: AAD/1977/4/17, Britain can make it: Information for Guides, Designers Look Ahead.
5 'Britain Can Make It Exhibition at the Victoria & Albert Museum', *The Architect and Building News* (20 September 1946), p. 177.
6 DCA/361/The Council of Industrial Design. Policy Committee. 1946 Exhibition. Special Section on Designs of the Future; The Council of Industrial Design. Proposed Exhibition – Summer 1946. Draft Plan (2). Swords into Ploughshares: British Goods for the New Age.
7 DCA/361/The Council of Industrial Design. Proposed Exhibition – Summer 1946. Draft Plan (2).
8 Other members of the Design Committee included Sir Thomas Barlow, Mrs Mary Harris, Allan Walton, and Mrs Tomrley. *Britain Can Make It Exhibition Catalogue Supplement* (cited note 3), p. 227. Tate Archive. Kenneth Clark papers, letter to Sir Kenneth from C.A. Mesling 16 May 1946 ID.606.
9 Tate Archive. Kenneth Clark papers PAX/P/6 The Council of Industrial Design. "Britain Can Make It" Exhibition 1946 Special Section on designs of the future. Copy also held at V&A Archive, Exhibition File: AAD/1977/4/13 and DCA/361/Council of Industrial Design Policy Committee 1946 Exhibition. Special Section on designs of the future.
10 *Britain Can Make It Exhibition Catalogue Supplement* (cited note 3), p. 217.
11 BBC Written Archives Centre: TVART 1 Council of Industrial Design 1946–1954, File 1.
12 V&A Archive, Exhibition File: AAD/1977/4/15, Britain can make it: Letter 6 July 1946 from H.G. Crabtree to Mr Banham, COID, report dated 28.1.48 BCMI Designers Look Ahead section. V&A Archive: AAD/1977/4/14, Britain can make it: Disposal of Council Property.
13 William Gaunt, 'Design makes good. Britain Can Make It ... A survey by William Gaunt', *Evening Standard* (25 September 1946), p. 9.
14 Marquis of Donegall, 'Britain Can Make It Exhibition. Novelties, Inventions and Gadgets', *Newnes Practical Mechanics* (January 1947), p. 132.

ANDY GROARKE

Surrealist Strategies in Exhibition Design

Almost fifty years after its staging, James Gardner, the designer of the Victoria and Albert Museum's most popular exhibition *Britain Can Make It* (BCMI), recalled how daunting the responsibility for orchestrating the project had been:

> Cripps, deciding he must give the public something to look forward to, laid on an exhibition of 'things to come'. This was to be set up in the galleries of the vast Victoria & Albert Museum – then empty; its treasures having been evacuated (with the children) to the country .... Cripps pointed out that, as no one had a clue as to what would turn up, I must introduce 'décor' that would give people a lift, even if the exhibits didn't.[1]

The fixed-route exhibition was arranged thematically and included six thousand items divided into twenty-two sections, arranged in a fixed circulation route of about one-third of a mile [**187**; see **7**]. Visitors were taken on a sensational and sometimes bizarre journey that began with the aftermath of war and culminated with a glimpse of exciting and life-enriching future possibilities.[2]

Gardner confessed that he was preoccupied with the dilemma of how to design a show that would look complete in every detail, even if the audience did not recognize the exhibits which were being produced for the show.

> This inspired a new kind of layout plan; instead of presenting goods to the eye, as one would in an open market – and that is how exhibitions have evolved – I tucked them round corners, behind screens and in little enclaves, so at first the visitor would see lots of décor but no goods – wouldn't even notice if there were no goods at all.[3]

This strategy of displaying objects in a stylized and obscure manner was very different from the prevailing ideas about exhibition design

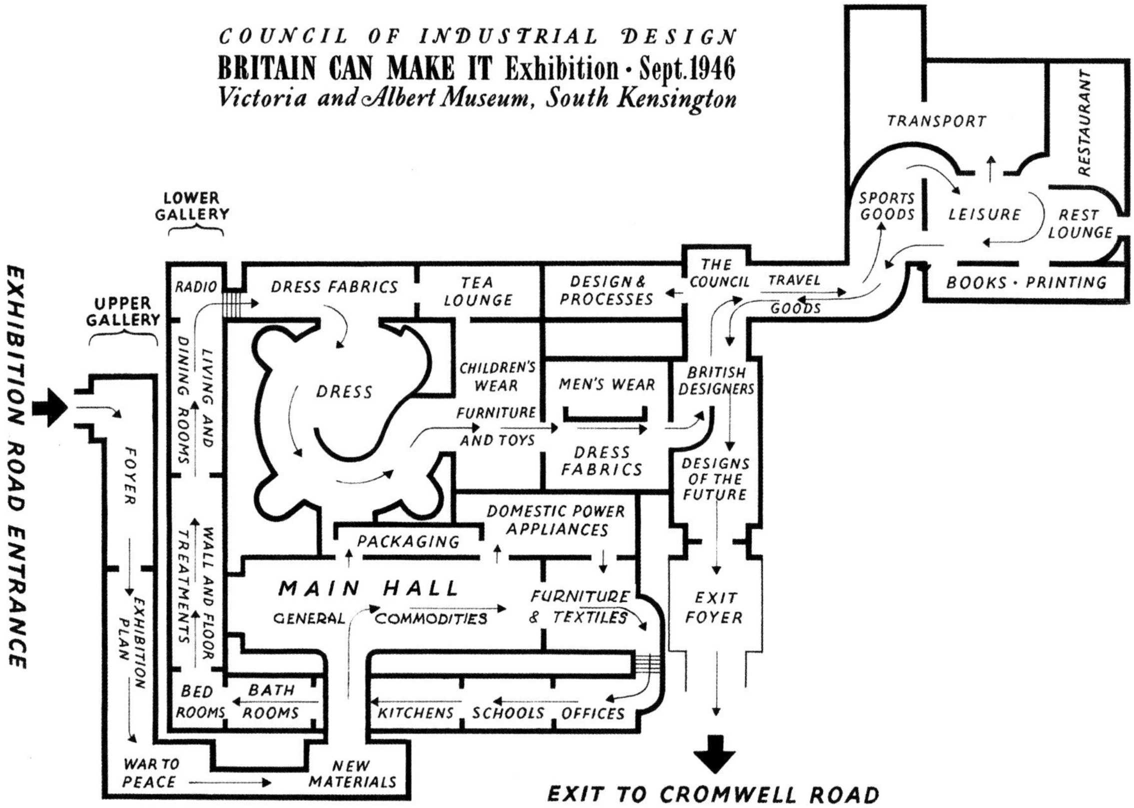

187. Diagram showing the plan and different sections of BCMI

during the period.[4] In London, design had rarely been presented in a museum prior to BCMI and local press compared the style of the show to the art of merchandise and window display. Pilkington Jackson wrote a review of the exhibition in 1948 for *The Studio* magazine. In the article, entitled 'Surrealism Comes Alive in the New Art of Display', he explains that the first signs of this 'new art' appeared in shop windows, particularly in Paris, London and Edinburgh in the 1920s: '"Britain Can Make It" demonstrated that Display had arrived as an art in itself'.[5]

Exhibition design meets Surrealism

The proposition underlying Surrealism, a movement that embraced politics, literature, painting and sculpture, was to reveal the unconscious and reconcile it with rational life.[6] Many of the visual and decorative elements of Surrealism had been widely appropriated by the advertising, graphic, fashion, and display industries since the 1930s and remained popular throughout the 1940s. Surrealist exhibition design became a creative practice that captured the varying dimensions of architecture in a performative mode of construction, interruption and dismantling.[7]

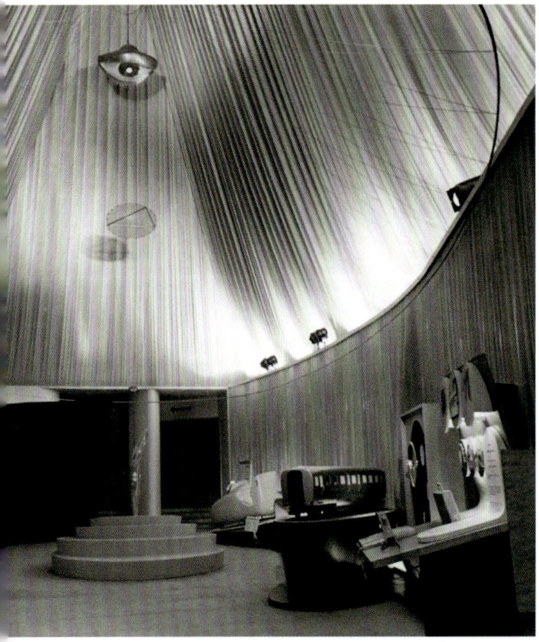

188. The 'Designers Look Ahead' section of BCMI featured the all-seeing eye looking down on the designs of the future. This picture includes a spaceship, an air-conditioned bed, a model sleeping car, a wall radio and a portable sewing machine.

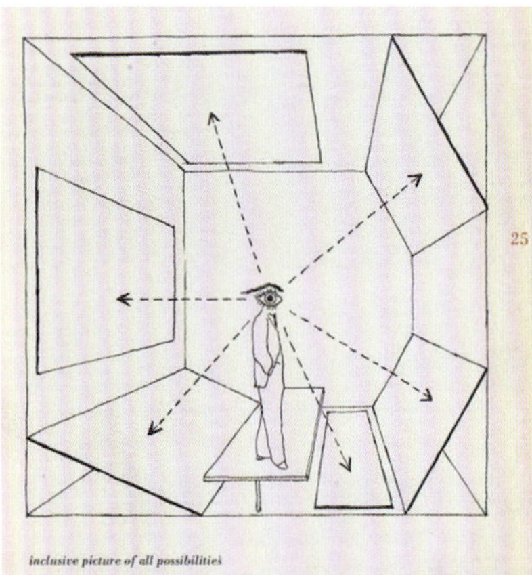

189. Herbert Bayer, 'Fundamentals of Exhibition Design', 1938. Bayer wrote that the exhibition space is available to the individual eye and should obtain its forms from the qualities of the eye itself. Here he has diagrammed all possibilities for engaging the viewer's field of vision.

By the late 1930s, Surrealist exhibition installations (themselves ambiguously domestic spaces where adhoc furniture lurked among incongruous recreations of natural or urban environments) offered their public a chance to see for themselves just how effective the Surrealist transformation of objects and space might be. Salvador Dalì was the promoter of this spectacular and highly influential 'fantasy Surrealist' style, upon which much of the discussion of the encounters between Surrealism and architecture has been based.[8]

Spurred on by Dalì's preferences, the desire to arouse curiosity and pleasure through association and eccentric juxtaposition informed the design of the 1936 *Exposition surréaliste d'objets*, at the Galerie Charles Ratton in Paris. Nodding to the *grands magasins* with their luxurious interiors lined with display cases to seduce the bourgeoisie, the show staged fashion, objects and furniture in a peculiarly Surrealist shopping bazaar.[9]

Two years later, at the *Exposition Internationale du Surréalisme,* in Paris, the emphasis was not so much on paintings as on objects and the creation of a whole environment. Marcel Duchamp was the curator and conjured up a holistic and startling presentation of Surrealist art work which was staged in three sections, which were full of paintings and objects as well as unusually decorated rooms and mannequins **[190]**.

The *forte* of Surrealism with its ambiguities and unexpected correlations worked to the BCMI designer's advantage in regards to the arrangement of content. In the plan of the exhibition, one can see highly improbable thematic juxtapositions almost like a string of experiments: 'Children's Furniture and Wear' following the 'Tea Lounge' and 'Women's Dress' is at once awkward and unorthodox yet also provokes unexpected associations between the categories and their objects. Furthermore, spaces were specifically designed in fantastical or exaggerated shapes and sizes which governed and revealed the objects they enclosed **[54–56]**.

BCMI was a show born out of post-war austerity and its designers were forced to be resourceful and imaginative; reusable framing (tubular scaffolding sheathed with prefabricated sheets of plywood fabric) and plastic films (packing materials theatrically lit) were employed to provide structures in the numerous spaces. Multi-sensory visitor experiences were constructed, which highlighted the importance of touch and allowed visitors to appreciate the materiality of products directly. Perfume was misted into the air in the women's fashion section and music was played within the galleries.[10] In the context of 1946 these practices were considered highly innovative, shifting the focus from the exhibits to the creation of an immersive and memorable experience.

The 'Timber' display by Gardner (illustrated on the front cover) was lauded in a review, 'Round the Exhibition' by Sylvia Pollak, who made the direct comparison between Gardner's design techniques and Surrealist art:

> 'Timber' ... has the quality of a Surrealist picture, built up of architectural forms stretching back into infinite perspective, made of various types of wood, veneers and laminated shapes, with admirable effect. Surrealism as expressed by the painter many years ago, has been reduced by these clever technicians to a display medium, cunningly using lights and shadows to create the perspective effects and introduce the slightly humorous notes which make the designers' point [see **42**, **44**].[11]

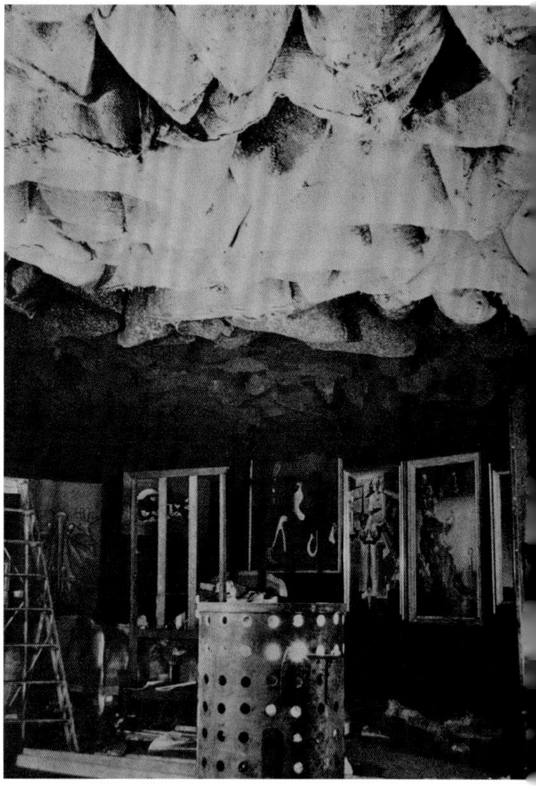

190. *Exposition Internationale du Surréalisme*, 1938. Marcel Duchamp curated the exhibition which was presented at the Galerie Beaux-Arts in Paris. The exhibition was staged in three sections which featured design objects, fine art, decorated rooms and mannequins. In this room, Duchamp hung 1200 bags of coal from the ceiling, creating a striking and immersive environment.

The legacy

The legacy and influence of BCMI and its use of Surrealist 'spatial strategies' have been to reinforce the breakaway from the formally classical treatment of display space, which featured a prescribed sequence of hall, focal point and main axis. Gardner provided a platform for exhibition design to embrace disparate types of display to encourage collaborative design across fine and applied art disciplines – and to be an experience in itself.

Within the spatial practice of architecture, the design of exhibition environments introduces a very specific relationship between visitor experience and a 'sense of place' or a relationship to one's physical context. An exhibition's physical relationship to the real world is often detached and ephemeral, leaving a long-term legacy of only photographs and stories. Frequently, the intention of creating a memorable exhibition experience is one of deliberate disorientation, disguise and disconnect from context and from the everyday. An exhibition space has the potential to place the visitor 'in the frame' by creating a unique sense of place, set in the context of ideas and reinforced by the items on display.

The role of an exhibition is arguably to provide a vivid framework for a visitor's imagination to make connections between things which previously have had no rational points of contact. Although using fewer theatrical or scenographical techniques than BCMI, the exhibition design for *Surreal House* at the Barbican Art Gallery in 2010 created an immersive installation of four hundred pieces of fine art, architecture, film, photography and design within the abstract scenography of a maze of houses. Using a subliminal composition of vaguely archetypal house forms and spaces, the design played on familiar associations of the domestic. The configuration deliberately attempted to disorientate and detach the visitor from the immediate context of the Barbican Art Gallery, allowing a person to momentarily escape into a psychological labyrinth

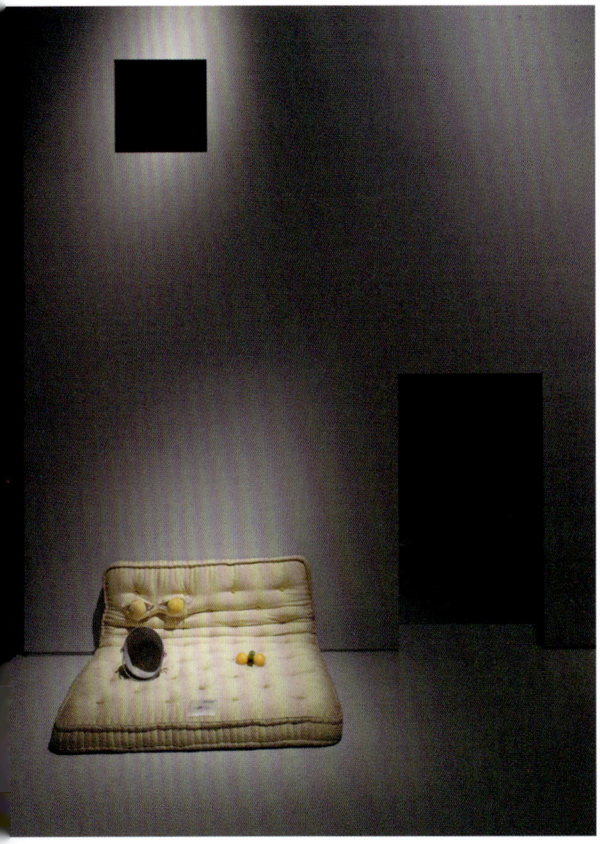

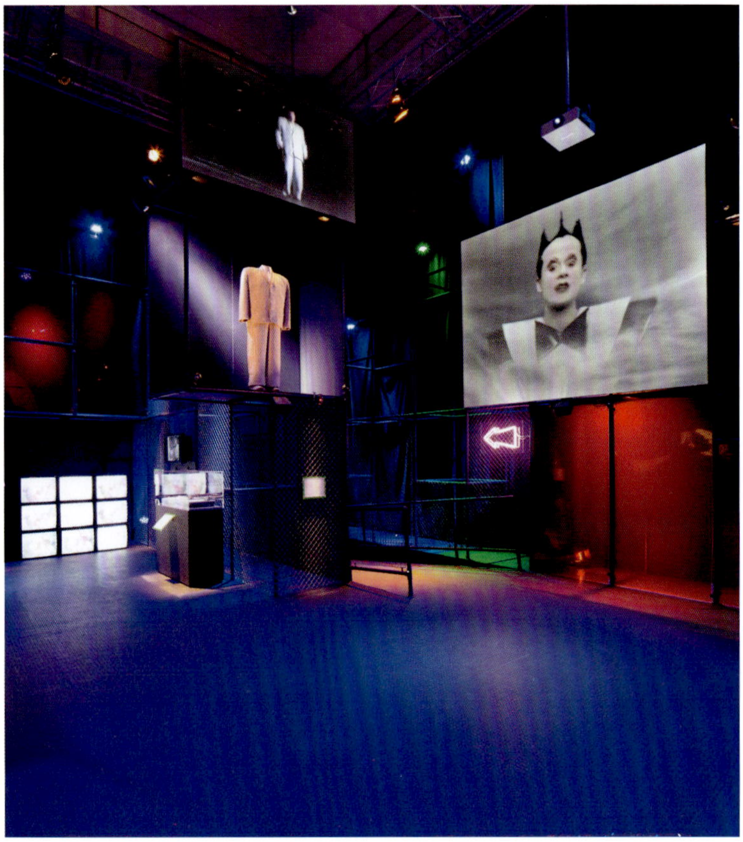

191. *Surreal House,* 2010. A series of archetypal 'house' forms were clustered together in the Barbican Art Gallery in London to create a bespoke backdrop for the presentation of Surrealist art, sculpture, film and architecture. The overall installation punctuated by framed views blurred the subject/object relationship of the Surrealist art, the gallery and the visitors.

192. *Postmodernism: Style and Subversion, 1970–1990,* 2011. The exhibition at the Victoria and Albert Museum provided a series of very dramatic and immersive spaces designed to display over 250 exhibits of art, architecture, design, fashion, graphics, film, performance and video.

of the house. Within the foreground of this fabricated context, lighting, materials and colour were used to visually and physically connect pieces of this diverse, surreal collection, allowing the visitor to make their own interpretive links between ideas and objects [191].[12]

BCMI's accomplishments confirmed that the appreciation of an entire exhibition is greater than the sum of its parts on the one hand, yet the individual works and exhibits can also be interpreted as a series of episodes. *Postmodernism: Style and Subversion*, presented at the V&A in 2011, was designed as a sequence of experiences to present an extremely eclectic collection of art, design, architecture, fashion and film in a way that constructed stories through objects [192]. 'Postmodernism' also conveyed the zeitgeist of the movement and the more ethereal concepts of 'identity', 'commodity', 'quotation', 'bricolage' and 'consumption' – notions that resonate through the display of a collective whole.[13]

More than just a pragmatic approach to making sense of the variety of themes and content, BCMI was conceived as a series of environments with different atmospheres. Together they created an extraordinary and compelling image that was beyond a collection of everyday objects with normal significance and ordinary formal organization. The intention was a blend of the depictive, the abstract and the psychological that would evoke wonder and delight from the viewer – a visual language that connected emotionally with visitors and provided an antidote to the alienation that many people were feeling with the modern period and particularly just after the war. Historian Fiona MacCarthy puts it bluntly: 'exhibition designer James Gardner camouflaged austerity in BCMI. He made a post-war fantasy [and the public] was convinced.'[14]

Regardless of how literal or even artificial the displays of BCMI may appear, in retrospect the exhibition provided a much needed moment of escape from the context of the outside world. By borrowing from the visual techniques of surrealism, Gardner disguised and distorted conventional exhibition wisdom and provided a platform for visitors to expect a different interpretive experience of content and narration. The legacy of BCMI has been instrumental in challenging the established notions of what a museum experience can be, in providing an exhibition with a sense of place just as memorable or impactful as the content itself, and in defining exhibition design as a critical spatial practice.

NOTES

1 James Gardner, *James Gardner the ARTful Designer: Ideas off the Drawing Board* (London 1993), pp. 127–28.
2 J. Clark, A. de la Haye, with J. Husley, *Exhibiting Fashion: Before and After 1971* (New Haven and London, 2014), pp. 24–25.
3 Gardner (cited note 1), p. 129.
4 The display of design at this time owed much to the research of the Bauhaus and, in particular, teacher and artist Herbert Bayer, who wrote 'Fundamentals of Exhibition Design' in 1937. In this essay, Bayer argued that the great possibilities of exhibition design rested on its universal application, which included diagrams, lettering, words, photography, architecture, painting, sculpture, tone, light and film. For Bayer, new exhibitions should not be based only on style and the object. The exhibition space was available to the individual eye and should obtain its forms from the qualities of the eye itself: 'Fundamentals of Exhibition Design', *PM* 6, no. 2 (December 1939–January 1940).
5 Charles D'Orville Pilkington Jackson, 'Surrealism Comes Alive in the Art of Display', *The Studio* (August 1948), vol. 136, no. 665, p. 56.
6 http://www.tate.org.uk/learn/online-resources/glossary/s/surrealism, paragraph 2.
7 Thomas Mical (ed.), *Surrealism and Architecture* (New York and London 2005), p. 256.
8 Ibid., p. 20.
9 Dawn Ades, *Dada and Surrealism* (London 1974), p. 124.
10 J. Clark, A. de la Haye, with J. Horsley (cited note 2), p. 25.
11 Sylvia Pollak, 'Round the Exhibition', *Architectural Design & Construction* (October 1946), p. 270.
12 See *The Surreal House*, exh. cat., Barbican Art Gallery (London 2010).
13 See *Postmodernism: Style and Subversion 1970–1990*, exh. cat., V&A (London 2011).
14 Fiona MacCarthy , 'All Things Bright and Beautiful', *Design in Britain: 1830 to Today* (London 1972), p. 146.

SIMON GARFIELD

Mass Observation and Britain Can Make It

> We went on to the Britain Can Make It Exhibition and had no difficulty getting in. The exhibition begins like a peep show, or a fairy grotto in a fair, and the illusion is not dispelled for some time, in fact until one goes below into a miniature exhibition hall. There, set out in cases like shop windows, are tempting goods with no price labels but merely warning signs on most saying 'Available later' or 'Available soon'. Some indeed were marked 'Available now', but where was not shown ...
> —*Mass Observer from Sheffield, Thursday, 3 October 1946*

Everybody mentioned the queues: the queues were three deep; the queues were ridiculous; the queues were entertained by buskers and trinket sellers and food vendors [193]; the queues were not quite as bad as feared. The queues, of course, were a sign of popularity: here at last, the people said, was an exhibition for *us*.

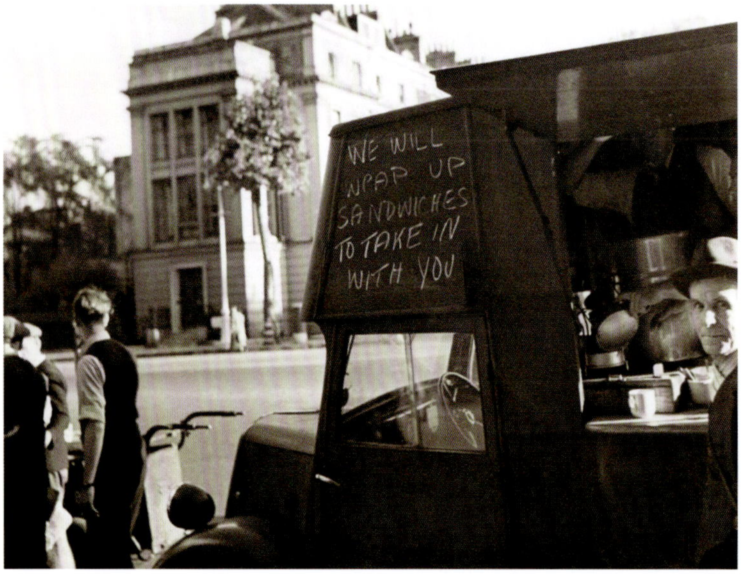

193. Street vendors outside BCMI offering sandwiches to take into the exhibition

But what would they say when they got home? Was the queuing worth it? Would they add to future queues by recommending the show to their friends? There were many ways to find these answers, but one organization did not even have to ask for them – they arrived automatically through the post. From its birth in 1937, Mass Observation (M-O) attempted to discover how ordinary people in Britain lived their lives. Initially relying on paid observers to snoop around pubs and other public gatherings and note what they heard, M-O soon supplemented its work with questionnaires and diaries sent in regularly by hundreds of eager volunteers. Many of the diarists visited the V&A at the end of 1946 to see the biggest show in town, and their musings – wholly subjective, rigorously unscientific, pointedly cynical – provide one of the most reliable snapshots of what visitors genuinely thought about their trip. This was not officialdom reporting back; this was sore-heeled literate punters, and their verdict was not wholly flattering.

'Britain can make it – one wonders,' wrote one observer, a 36-year-old single woman living near Slough (M-O knew the identity of their writers, but it guaranteed anonymity in any use of their material):

> The display is wonderful – wonderful, fascinating, terrible. Here is the Brave New World. The Shape of Things to Come, the sweeping lines, the pastel-coloured walls, the curves and candy pillars, fabric swathed ceilings, the cunningly concealed and directed lighting, the air conditioning and central heating, glitter of plastics and light metals – all that one expects and fears of the underground world of tomorrow. I was dazzled and dazed in this modern temple.

One London man, of retirement age, found the entrance queues and the crowds inside equally challenging: 'I could not see much of the Space Ship, as one had to look through a round opening in the wall to view a large section of the moon and a space ship apparently just going to land on its surface' [**185, 186**]. On his way to the Space Ship he found the waxwork female figures modelling new clothes 'hideously slim and lanky,' and he imagined that the modern child must have far stronger nerves than when he was young, as the Children's Toy Room contained bold 'gollywogs' and 'great glaring-eyed stuffed figures'. Apart from this, 'there were thousands of items which did not interest me at all.'

Other Observers were similarly unimpressed. 'Off to South Kensington for Britain Can't Get It,' wrote a housewife from Sheffield on 18 November 1946. 'Oh my my what a tedious business. Miles and miles, and we had already walked miles and miles that morning. Made one mad to see so much only to be looked forward to.' But the show was clearly influential. Three months later, the same woman noted, 'I have

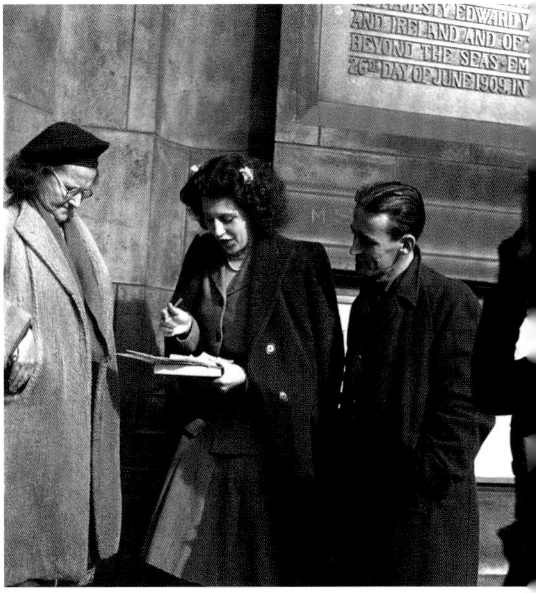

194. A Mass-Observation investigator questions visitors about their visit to BCMI

started to make a padded quilt like one I noticed at the "Britain Can't Get It" exhibition. It is already pram-size.'

On 16 December 1946, a 59-year-old female 'nurse companion' from Bristol noted the comments she had received from a friend in London. 'In my opinion it ought to be called "Britain Can Keep It". I never saw such a collection of tawdriness before. Everything is very "arty" .... There is nothing there that you cannot see in any good class store, and set out to much greater advantage.'

Not everyone was disillusioned with what they saw, but many had questions, some of them Zen-like in their profundity. 'Why so many eggcups?' asked a 46-year-old housewife from Yorkshire. 'Some of the things were interesting and I thought the layout very good. Didn't think much of the clothes .... The nicest things apart from the pottery and glass ... was a sailing boat in the Sports section, and a model of a catamaran in the last section. Also, the bicycle was interesting' [see **6, 184, 182**].

No one spent more time in the queue than Mass Observation itself [**194**]. Over a number of days, fifteen observers questioned visitors about why they had come, what they hoped to see, and what they thought of the show as they emerged. The organization then compiled its own report, complete with primitive info-graphics, suggesting that more women attended the exhibition than men, that three-fifths of visitors were under forty, and that half the visitors spent between two to three hours at the show. It was a fairly dry dossier, with visitors regarded as a curious tribe. 'Even on Sunday afternoon, when police had to control the crowd before the exhibition shut, there was joking between police and crowd throughout.' And when inside, 'Plastic curtains are touched by almost everyone passing them'. Public interest in each section of the exhibition was duly noted: 'Some attention to bags, none at all to umbrellas'. Without doubt, the biggest hit had been 'Shopwindow Street'. Visitors speculated to what extent the exhibits would fit into their own lives: they were impressed with the compactness of objects, and when it came to fabrics, the big question was 'Will it go in our room?' Concluding, M-O found that four out of five enjoyed their visit, 'but the overall impression is relief that it is all over'. And why then had they come at all? Most replied that they had 'an interest in future living', but there was also a strong sense that it was the 'done thing', and that not to come would be 'to lose face'. The ultimate word came from a 45-year-old 'milk roundsman'. He had come to Britain Can Make It 'just to see what Britain can make'. Clearly, the earnest social survey had met its match.

For further information on the Mass Observation Archive, please visit: www.massobs.org.uk

DIANE BILBEY

Reflections on Britain Can Make It

Finally, I would like to say 'Well done, British industry and British designers. You've made a great come-back; more strength to your elbow and good luck to you all'
—*Sir Stafford Cripps, President of the Board of Trade*[1]

Congratulations from leading industry figures flooded in to S.C. Leslie (Director of the CoID) following the close of *Britain Can Make It* (BCMI) on 31 December 1946.

What the experts said

The Chairman of Condé Nast and founder of British *Vogue*, Harry W. Yoxall, wrote, 'It is very well done indeed. True, the display is in most cases superior to the merchandise, but that is Industry's fault, not the Council's. You have dangled a juicy carrot in front of the manufacturer's nose, and I hope he will strain forward to absorb it.'[2] In his reply, Leslie politely suggested that perhaps more direct encouragement was needed, and asked that *Vogue* might like to help in this, and 'make some reasoned survey of the goods and apportion blame as well as praise. It will not hurt the export trade and it will help us get under the skin of those sections of industry which ought to be thinking again.'[3]

John Gloag, a prominent writer on design and architecture, wrote effusively to 'My dear Leslie', after his visit to BCMI on its opening day:

> that it is without exception the finest Exhibition that has ever been put on in this country or ever been put on by British interests in any foreign country. With this we finally get rid of the Victorian hang-over in design .... I think you have done a fine job and deserve the thanks of everybody who has the interests of good industrial design at heart, and as I have personally devoted the last 25 years to trying to improve this branch of human activity, I am personally most grateful to the man whose great gift for organization has achieved such an outstanding demonstration of the latent genius of Britain for design.[4]

The Managing Director of Josiah Wedgwood & Sons Ltd, The Hon. Josiah Wedgwood (great, great, great grandson of the original Josiah Wedgwood) wrote:

> I do congratulate you most sincerely on your achievement with [the] BCMI Exhibition – and perhaps you will feel this congratulation is the more sincere because it comes from a critical potter .... Your gate-money and the fainting queues of the colour-starved millions are more than proof of the success of your own organization, timing, & publicity and of your colleagues' excellent showmanship.[5]

At the close of the exhibition, *Punch* tried to analyse its unprecedented appeal, asking:

> How can we account, then, for the phenomenal success of the 'Britain Can Make It' Exhibition? .... I'm sure the principal reason is that our eyes have become so accustomed to drab colours and austere cuts that we just haven't been able to resist this first post-war opportunity to gaze into a brilliant shop window – as colourful as any of the old Christmastide bazaars.[6]

Misha Black – who as part of the Design Research Unit contributed to BCMI, including the 'What Industrial Design Means' display – also tried to explain why the British public was so attracted to the exhibition, concluding that 'Nothing seems able to quench people's eagerness for ceremony, festivity and exhibitions .... Nearly a million and a half people queued for hours on the hard Kensington pavements to inspect furniture and household goods which they were debarred from buying.'[7]

But it was not all round critical acclaim. A heated debate took place between Sir Thomas Barlow (Chairman of the CoID) and F.C. Hooper (Director of Business Training at the Ministry of Labour and National Service) when they met to discuss the important issue of 'Can Britain Make It Well Enough?' for a BBC radio programme broadcast on 24 January 1947, less than a month after the end of the exhibition. The discussion highlighted the failings of BCMI, and questioned whether it was the great success it was deemed to be. The adjudicator in the debate was Macdonald Hastings, a prominent journalist of the time, who succinctly remarked: 'I thought it was universally agreed that this was one of the most brilliantly designed exhibitions as an exhibition that's been put on and it was a credit to this country, but one of the criticisms was surely that the designs themselves although good were not new'. It is true – but understandable considering the lack of time – that pre-war designs were included, but these were confined to designs that were going back into production.

The discussion also covered the more fundamental issue as to whether or not BCMI had any meaningful impact on the public's awareness of design. On this point, Barlow was keen to point out how the Council had commissioned Mass Observation (M-O) to conduct investigations into what people thought about the exhibition, and had gained positive endorsement: 'Eighty per cent of the people who were asked said it was good, seventeen per cent expressed some disappointment, and two per cent said it was bad, but in addition, a number of people were interviewed in their own homes, and they all said that as a result of this exhibition, they had made certain alterations in their domestic internal arrangements, or had marked down certain commodities for purchase at the earliest possible opportunity.'[8]

In 1947, Noel Carrington – book designer, editor and publisher – wrote an article for *Graphis*, the influential Swiss graphic design journal, in which he pondered the post-war situation in Britain with regards to industrial design: 'How has it come about that Britain has suddenly emerged as the first country with a fully fledged State organization for the improvement of industrial design?'[9]

Continental investigations

Recent findings by Dr Anne Sudrow, a research fellow at the Center for Research on Contemporary History, Potsdam, perhaps provides us with some answers to this question. This transition was not as sudden or unplanned as Carrington surmised, it was much more calculated, and has direct links to some of the designers working on BCMI.[10]

Between July and December 1946, nine British design experts were dispatched to Germany on a Government mission organized by the British Intelligence Objectives Sub-Committee (BIOS). The idea had been put forward by Stafford Cripps, President of the Board of Trade, and by the CoID. The selected experts were divided into four teams. Each member was a prominent exponent in the field of industrial design, and their objectives were all-encompassing – to learn as much about the state of German industrial design as possible –and to 'investigate the place of the designer in German industry'. Nikolaus Pevsner, the renowned historian of art and architecture, chosen for his research and writings on British design history, headed one team.[11] Other teams included people who had direct connections to BCMI – the artist and designer Enid Marx, who contributed an essay on furnishing fabrics for *Design '46*, and who also designed some of the furnishing fabrics displayed; the textile designer Margaret Leischner, whose many fabric designs were included in the display of furnishing fabrics; John Beresford-Evans, who had worked on the glass display in the 'What the Goods are made of' and on the display of cameras, binoculars, thermometers and barometers at

BCMI [see **41**], and Allan Walton, a high ranking member of the Council of Industrial Design, who sat on the Design Committee and on the 1946 Exhibition Policy Committee. Other members of the teams included the esteemed historian the Hon. Margaret Lambert; Edward O'R. Dickey from the Ministry of Education and Fellow of the British Institute of Industrial Art; Hendrik A. Nieboer, Member of the Society of Industrial Artists; and Miss Lorna Hubbard of the CoID.[12]

Given unprecedented access, the teams travelled across Germany, visiting manufacturers and training institutions in the British, American and French Zones. Their remit was to investigate a wide range of products including furniture, plastics, textiles, pottery, glass, cutlery, leather goods, wallpaper, and kitchen and bathroom equipment. Their detailed report entitled *Design Investigation in Selected German Consumer Goods Industries* had not been revisited since its completion until re-discovered by Dr Sudrow in the archives of the Imperial War Museum Duxford. Its findings are illuminating, Dr Sudrow writes: 'it must be regarded as the single most comprehensive and significant source on the practice of industrial design in Germany around the middle of the twentieth century'. It also offers the opportunity to assess:

> Why, only one year after the end of the most destructive war humankind had until then experienced, was the British state interested in – of all things – the design of consumer goods in the country that had started this war? .... At home, the British government was also grappling with the economic consequences of war. In the face of the scarcity of basic goods and materials, consumer goods were still rationed until 1952 – much longer than in West Germany, where rationing ended in 1949. Why then were the British authorities so interested in the industrial design of German consumer goods after World War II?[13]

Perhaps in light of these findings, the BCMI exhibition must also now be viewed as part of a greater, more ambitious and far-reaching plan by the Board of Trade and CoID, and one that needs further examination.

Next steps?

The concept and spirit behind the BCMI exhibition did not end on its closing night, New Year's Eve 1946. Plans for a travelling version, albeit on a reduced scale, were seriously considered by the CoID, but rejected because of the anticipated costs and continued scarcity of raw materials. 1947 would also see the opening of the British Industries Fair in London and Birmingham, attracting the same manufacturers, buyers and public audience.

Scotland Made It Too

But what did result was another significant exhibition whose aim was to highlight the value of British design. Organized by the Scottish Committee of the CoID – and called *Enterprise Scotland* – this exhibition entirely devoted to showcasing 'post-war goods designed and made exclusively in Scotland' took place from 25 August to 30 September 1947 at the Royal Scottish Museum in Edinburgh [**195**]. Alongside the exhibition, a cultural element in the form of a Festival of Music and Drama featured renowned orchestras and theatrical companies, including the Vienna and Liverpool Philharmonic orchestras, Glyndebourne Opera, The Old Vic Company and Sadler's Wells ballet. This became the foundation of the Edinburgh Festival.

Although *Enterprise Scotland* was only a third of the size of its London counterpart, it included some significant specialist elements, including a large section on shipbuilding and tourism. Other themes were directly taken from BCMI, including displays on printing and packaging, toys, pottery and glass, jewellery, fabrics and toys, together with a suite of furnished rooms. Many of the same designers that worked on BCMI were involved, including John Hutton, Alec Heath, Bruce Angrave and Raoh Schorr. Basil Spence was again appointed in the role of Architect and Co-ordinator for Design and Lay-out, with James Gardner as Chief Designer for Displays. As well as a section on 'Scotland Today' at the very centre of the exhibition – displaying 'everything from Ships to Tartans' – there was a further display representing 'Scotland Yesterday', providing a 'symbolic treatment of the culture, crafts and traditions of the past'. 'Scotland Tomorrow' (designed by Edwin Calligan) was the final, motivational section of the exhibition, where 'Scotland's plans and ambitions for the future' were visualized in theatrical displays. Other sections focused on 'The Country' and the economic opportunities that tourism offered to Scotland.[14] Unsurprisingly, similarities between BCMI and *Enterprise Scotland* were many. The 'Hall of Pinnacles' or Commodity Hall in *Enterprise Scotland* was clearly inspired by Robert Goodden's sports display for the 'Weekend' section at BCMI [see also **6**].

As was the case for BCMI, the inclusion of art, in the form of murals and sculpture, was a prominent feature of *Enterprise Scotland*. The 'Hall of St Andrew' included a large engraved glass screen by John Hutton, illustrating five key figures in Scottish industrial history: James Watt, Henry Bell, Lord Kelvin, J.B. Neilson and Robert Foulis. A giant figure of St Andrew (made of aluminium sheet) by the Scottish artist Walter Pritchard, and 'suspended from the roof by piano wire', was described by some as 'forbidding, by others extraordinarily impressive'. The contemporary journal *Display* headed up its article enthusiastically, that 'Scotland Made It Too'.[15]

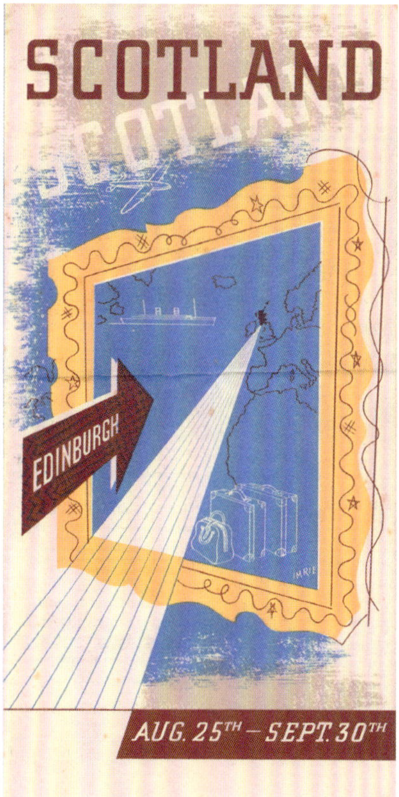

195. Promotional leaflet for the *Enterprise Scotland* exhibition that took place in Edinburgh in 1947. The cover design is by Archibald (Archie) Imrie, who was also the Special Designer for the 'Ships' section of the exhibition, and executed the ships mural.

Festival of Britain

One of the legacies of BCMI was that it provided a practical 'testing ground' for ideas that would shape the styling and design of the Festival of Britain five years later. When asked about the Festival of Britain and its debt to BCMI, James Gardner acknowledged that those involved in 1951 were basically the same team of designers as in 1946 – just more of them.[16] As well as Gardner and Spence, Misha Black (and his colleagues in the Design Research Unit), John Hutton, Robert Goodden, Wells Coates and Barbara Jones were amongst those who worked on both exhibitions. Ralph Tubbs, a pre-war assistant to the architect Ernö Goldfinger, was given the modest task of designing the Nursery School in the 'Things for Children' section at BCMI. For the Festival of Britain, however, he designed the iconic 'Dome of Discovery' for the South Bank Exhibition.

Design legacy

On the 40th anniversary of BCMI in 1986, an exhibition at the RCA and accompanying book reflected on the question, *Did Britain Make It?*.[17] Its timing allowed for direct parallels to be made with the austerity of 1946 and the straightened economic climate of 1986, which saw the highest rate of unemployment since the war. The political climate just over seventy years on from BCMI has remarkable similarities with 1946 and 1986. Austerity remains a toxic watchword in British politics and concerns over Britain's future, post-Brexit, are daily headline news.

As one of the main aims of BCMI was to educate the British public about design, James Gardner had a tough job on his hands. Forty years on, when Gardner was asked if he believed people attending the exhibition *thought* about design, he said he did not. On the question of had Britain made it,

> He [Gardner] was pretty sure that things would have been worse if there had not been a Britain Can Make It exhibition, but he didn't think that it had had the impact it should have done at the time. 'In Britain you can't just do it in one show – conservatism is too dug in'.[18]

Indeed, Bowden's revolutionary bicycle, one of the most talked about objects displayed at BCMI, did not make it into production in Britain, but was later manufactured in the USA. Other products, such as Ernest Race's BA 3 chair, were successful, despite being revolutionary in the use of materials and in manufacturing processes, and daring in design. And other seemingly insignificant items of domestic life displayed at BCMI made subtle changes to the future without even knowing it. Arnold Machin's *Taurus*, a large, striking pottery figure with zodiac decoration, was ahead of its time. Perhaps conservatism was not so dug in.

Wally Olins, the branding and corporate identity guru commented:

> Nobody in 1946 could conceivably have predicted the overwhelming success of the design business 40 years later. There was no precedent for it. And it didn't happen overnight. Although BCMI was a revelation to a visually starved public – it seemed to be about a Utopian world. I cannot think that it made any real lasting impact in a popular sense.[19]

But perhaps the impact on design was subtle. What BCMI accomplished was the raising of awareness of design and the elevation of the professional status of the designer. It was here, after all, that, whenever known, the name of the designer of each product was printed alongside that of the manufacturer in the accompanying catalogue. The aim of providing an explanation as to how things were designed was also attempted (if not altogether successfully) in the 'Birth of an Eggcup' display. The *Journal of the Royal Society of Arts* reported shortly after the close of BCMI that 'The effect of the Exhibition on the design policies of manufacturers, exhibitors and non-exhibitors alike, had been striking. From reports received it was evident that, so far as British manufactures and traders were concerned, Industrial Design is "on the map" to a greater extent than ever before'.[20] As a fourteen-year-old schoolboy, Terence Conran visited *Britain Can Make It* – a key event in his decision to become a designer. One personal story that demonstrates the tremendous impact of the exhibition on the landscape of design in Britain.

The commitment of those involved in BCMI in carrying through the objectives of the Board of Trade and CoID is best summed up by its designer, James Gardner, who wrote in his autobiography *The ARTful designer*, that, whilst working on the exhibition, 'The "easy riders" hadn't yet taken over, and we all worked as though we were still at war'. Wally (Walter) Cole worked closely alongside Gardner as his assistant. Gardner remembered, 'Wally and I kept up a routine of working sixteen hours a day until, near the end, the effort of climbing three shallow steps would set my heart racing'.[21]

Given that over 1.4 million people turned out to see *Britain Can Make It*, it is surprising that it is not a significant part of British cultural consciousness. Why has it been largely forgotten by the British public? Comparisons with the Festival of Britain are perhaps a little unfair. Unlike the Festival of Britain, *Britain Can Make It* was not a national event. It was focused on one location – London – and there is no physical, let alone architectural, legacy as was the case with the Festival of Britain and the Royal Festival Hall on the South Bank. Photography in the exhibition by the general public was not permitted, nor was sketching, to safeguard designers' and manufacturers' copyright. There is no colour footage to captivate a modern audience, and, other than CoID publications, there were no BCMI branded souvenirs for visitors to take home.

The monumental efforts required to put on such a show while large areas of London and cities across the country were still laid to waste by bombing, and material supplies were scarce, make staging the exhibition a truly remarkable feat. At the close of the exhibition, all objects were returned to manufacturers and displays torn down and discarded. Although Leigh Ashton, Director of the V&A, showed some interest in obtaining 'one or two features' that were displayed in the exhibition, it appears that the V&A did not – at the time – seek to acquire a single object from the exhibition for its permanent collections.[22] It was as if BCMI had been air-lifted into the V&A and just as quickly cleared out.

Perhaps the most tangible legacy was left to the British economy by the trade buyers from abroad, who placed orders to British industry estimated at between £25,000,000 to £50,000,000.[23]

NOTES

1 BBC Written Archives Centre, closing statement in his radio interview on BCMI by Sir Stafford Cripps, broadcast 27 September 1946, Home Service, Film T101.

2 Design Council Archive at the University of Brighton Design Archives (hereafter DCA): 910/ Britain Can Make It. Letters of appreciation. Letter from Harry Yoxall to S.C. Leslie, Brighton, dated 3 October 1946.

3 Ibid., Letter from S.C. Leslie to H.W. Yoxall dated 4 October 1946.

4 Ibid., Letter from John Gloag to S.C. Leslie, Brighton, dated 24 September 1946. Gloag followed this praise up with a letter to the editor of *The Times*: John Gloag, 'Britain Can Make It', *The Times* (28 September 1946), p. 5.

5 Ibid., Letter from Josiah Wedgwood to S.C. Leslie, October 1946.

6 'The Great Exhibition', *Punch* (16 October 1946), p. 305.

7 Misha Black, 'Exhibition Design', in Misha Black (ed.), *Exhibition Design* (London 1951), p. 18.

8 'Questions in the Air. Can Britain Make It Well Enough?', BBC Written Archives Centre (24 January 1947), Film 201.

9 Noel Carrington, 'State and Industrial Design in Great Britain', *Graphis* (1947), vol. 3, no. 18, pp. 108–09.

10 Anne Sudrow, *Nikolaus Pevsner u.a. Geheimreport Deutsches Design. Deutsche Konsumgüter im Visier des britischen Council of Industrial Design (1946),* Deutsches Museum Abhandlungen und Berichte, Neue Folge, 28 (Göttingen 2012). Dr Sudrow's book includes an English translation and reproduces a copy of the original report. I am grateful to Stephen Walton, Imperial War Museums, for his help with locating a copy of the report. The full title is B.I.O.S. Final Report No. 952, item no. 22. 'Design Investigation in Selected German Consumer Goods Industries' (1947). The existence of the report was not secret as it is cited in the *Council of Industrial Design Annual Report 1946–1947* (London 1947), pp. 16–17.

11 Pevsner had published his critical *An Enquiry into Industrial Art in England* in 1937.

12 Allan Walton was also a member of the Scottish Committee of the CoID. Other members on the BIOS mission included Lorna Hubbard, member of the CoID; Hendrik Albert Nieboer, a designer; E.M. O'Rorke Dickey, Ministry of Education; and Margaret Lambert, an historian of German history and politics.

13 Anne Sudrow (cited note 10), pp. 106–07.

14 Promotional leaflet for *Enterprise Scotland*. For further information on Enterprise Scotland, see 'Scotland Made It Too!', *Display* (1947), pp. 8–19; W.O. Hutchinson, 'Can Scotland Make It Too?', *Art & Industry* (December 1947), pp. 183–91; *A Pictorial Review of Scottish Industry as displayed in the Exhibition Enterprise Scotland* (Edinburgh 1947); *Enterprise Scotland 1947* catalogue (Edinburgh 1947); Henrietta Goodden, *The Lion and the Unicorn. Symbolic Architecture for the Festival of Britain 1951* (London 2011), pp. 31–37.

15 'Scotland Made It Too!', *Display, Design & Presentation* (1947), pp. 8–19.

16 'Giles Verlarde interviews James Gardner', in Penny Sparke (ed.), *Did Britain Make It? British Design in Context 1946–86* (London 1986), p. 13.

17 The exhibition was called 'Make or Break. Design and Industry in 1946', held at the Royal College of Art 9 September–24 October 1986 and Manchester City Art Gallery 8 November–7 December 1986). Penny Sparke (cited note 16).

18 Giles Verlarde (cited note 16), pp. 12–13.

19 Penny Sparke (cited note 16), p. 59. Although not directly relevant to BCMI, in 1983 the first of a series of exhibitions highlighting British goods took place. Initially held at the Kensington Centre, like BCMI these *Better Made in Britain* exhibitions (BMIB) focused on promoting British manufacturing.

20 'General Notes', *Journal of the Royal Society of Arts* (17 January 1947), p. 149.

21 James Gardner, *The ARTful designer. Ideas off the drawing board* (London 1993), p. 131. Walter Cole worked for the CoID in 1946–47 and went on to buy and reopen the Rye Pottery in Sussex with his brother Jack in 1947.

22 Letter from Leigh Ashton to S.C. Leslie dated 1 August 1946. Leslie wrote: 'The exhibition is proceeding really well and does I think, the greatest credit to the architect and decorator. I find it very well done and am equally pleased that as it is to be torn down again most of it is rather unsuitable for permanent installation. However, there are one or two features which I shall seriously have to consider trying to obtain and no doubt we can come to some arrangement about this if it goes to Glasgow it will need considerable adaptation.' V&A Archive, MA/28/75/1. Any objects relating to BCMI that are now in the V&A's collections have all been acquired retrospectively. For information on the uneasy relationship between the CoID and V&A, particularly with regards to Ashton's lack of interest in collecting examples of modern design for the Museum, see Anthony Burton, *Vision and Accident. The Story of the Victoria and Albert Museum* (London 1999), p. 204.

23 Royal Society of Arts (cited note 20).

Index

Pages with illustrations in bold

A.E. Gray & Co. Ltd **98**, 99, **155**
A.H. Cornwall & Sons Ltd 223
Abbatt, Marjorie 179, 180
Abbatt, Paul 179, 180
Abbatt Toys 179, 181
Ackery, W.M. 227
Acme 57
Adie Brothers 86
Advertising Creative Circle 214
Adys Road Secondary School 183
Age of Austerity 1945-1951 (Sissons and French) 19, 27
air-conditioned bed 117, 221, 227, **222**, 234
Air Ministry 28, 36, 63, 68
Aldridge, John 174, 177
Alexandra, Princess 58, 59
Alexandra Palace, London 226
Allen Bowden Ltd 228
Allen Solly & Co. Ltd 188
Alphabet and Image 120, 121
aluminium 26, 42, 61, 65, 66, 69, 72, 73, 108, 112, 120, 127, 135, 136, 209, 221, 226, **228**, 245
Aluminium Development Association 120
The Ambassador 123, 125, 129
Angrave, Bruce 203, 245
Anny Lewinter Ltd 126
Appleby, H. 90
Appleton, Leslie 229
Architect and Building News 19, 27, 38, 54, 59, 81, 141, 143, 144, 161, 163, 177, 221, 225, 231
Architects' Journal 25, 28, 77, 81, 170
Architectural Design & Construction 81, 107, 115, 171, 230, 237
Architectural Review 176, 177
Ardern, Miss M.L. **55**
Armstrong & Co. Ltd 223
Art & Industry 41, 95, 103, 137, 144, 203, 249
Art Deco 1910-1939 exhibition (London 2003) 58, 59
Art in Industry exhibition (London 1935) 19, 27
Art Workers Guild 194, 197
The ARTful Designer (Gardner) 25, 28, 62, 68, 81, 137, 167, 171, 237, 247, 249
Artists' International Association 65
Arthur Sanderson & Sons 173, 176, 177
Arts Council 32, 65, 163, 167
Ascher Ltd 137, 156, 163
Ashford, F.C. 117, 222
Ashton, Leigh 24, 28, **36**, 37, 38, 131, 137, 155, 196, 248, 249
Asprey 90
Attlee, Clement 32
Ayrton, Michael 176, 177

B. Cohen & Sons Ltd 153
BA 3 chair 135, **136**, 137, 209, 246
BBC (British Broadcasting Corporation) 20, 21, 27, 28, 32, 54, 55, 57, 58, 59, 62, 68, 80, 81, 108, 114, 115, 135, 169, 171, 219, 224, 226, 231, 242, 249
Bainbridge, John 167, 170
Baker, Ethelwyn 126
Ballard, J.G. 29, 32
Barbican Art Gallery 235, 236, 237
Bardfield wallpapers 174, 177
Barker, John 198, 199, 203
Barlow, Sir Thomas 194, 231, 242, 243
Barr, Edward 107, 115, 212, 218
barrage balloon **33**, **62**, 63
Barronia Metal Works 181
Bawden, Edward 119, 153, 157, 174, 175, 177, 213
Bayer, Herbert 234, 237
Bayes, Kenneth 118
The Baynard Press 158
Beamish, Major Tufton 44
Beaton, Cecil 173
Beck & Pollitzer 13, 115, 203
Belk, Walter 85, 87
Bell, Vanessa 78, 165, 169
Belle, James Cleveland 127, 128, 129
Belton, T.A.L. 135, 136
Bennison, Geoffrey 123, 126
Bentley, Nicolas 141, 142, 143, 147, 148, 162
Beresford-Evans, John **69**, 71, 243
Bernays, Walter 43, 45
Bethnal Green Museum (now V&A Museum of Childhood) 146
Betjeman, John 140, 141, 143, 144, 218
Bevan, Aneurin 32
Bew, John 98
Bicknell, Peter 135
Biggs, Madeleine 182
Bimini, **91**, 92, 93
Birkinshaw, Franklin 29
Birmingham Cooperative Society 123
Birmingham Jewellers and Silversmiths Association 85, 88
Black, Misha 27, 28, 49, 118, 171, 200, 202, 203, 221, 242, 246
Blackfriars Road, London **31**
Blackshaw, Mrs E.M. 126, 129
The Blitz 33, 34, 35, 39, 40, 63, 65, 76, 205
Bloxidge Bros. of Birmingham 92
Board of Trade 17, 23, 27, 39, 45, 65, 68, 80, 81, 82, 86, 89, 97, 124, 125, 142, 145, 151, 157, 158, 194, 197, 202, 203, 205, 206, 207, 208, 218, 241, 243, 244, 247
Bodleian Library, Oxford 36
Bolsoms 204
Bombee 205
Bonella, Raph 12–13, 22, 28
Bon Marché 125, 127
Booth, David 99, 100
Boucheron 89, 90, 91
Bourne, Stafford 161, 163
Bourne & Hollingsworth 93, 124
Bowden, Benjamin (Ben) 227, 228, 246
Braddell, Dorothy (also known as Darcy) 23, 28, 99, 105, 142, 149
Bratt Colbran 155
Bray, Phyllis 133, 134, 135, 137
Brearley, Harry 86
Brenner Sports Ltd 124, 129
Britain Can Make It
1. 'From War to Peace' 24, 57, 60, **61**, **62**, **63**, **64**, **65**, **66**, 67, 68, 70, 75, 76, 163, 168
2. 'What the Goods are made' of 24, 66, 68, **69**, **70**, **71**, **72**, **73**, **74**, **75**, 76, 117, 243
3. 'Shopwindow Street' 24, 73, 75, **76**, **77**, **78**, **79**, 80, 81, **83**, 90, 94, **95**, 101, 107, **164**, 165, 169, **171**, 174, **178**, **218**, 240
4. 'Dressing the Goods' 24, **116**, **117**, 118–21
5. 'Heat, Light, Power' 24
6. 'Furniture and Furnishing Fabrics' 24, 204, 205, 206, **207**, **208**, 209
7. 'Furnished Rooms' (Things in their Home Setting) 19, 23, 24, 25, 28, 56, 98, 99, **100**, 103, 106, 135–46, **147**, **148**, **149**, **150**, 151, **152**, **153**, **154**, 155, 156, **157**, 158, 159, **160**, **167**, 170, 174, 204, 205, 208, 224, 225, 245
8. 'Radio etc.' 24, **109**
9. 'Dress Fabrics I' 24, 122–29, 204,
10. 'Tea Lounge' (Rest Lounge) 24, 26, 130, **131**, 132, 133, 134, **135**, 136, 137, 165, 204, 209, 210, 234
11. 'Women's Dress' 24, **122**, **123**, **124**, **125**, 126–29, 202, 234
12. 'Things for Children' 24, 57, 122, 127, 132, 169, **178**, 179–84, 186, 239, 246
13. 'Men's Clothes' 24, 185–86, **187**, **188**, **189**, 190–91
14. 'Dress Fabrics II' 24, 122–29, 204
15. 'Great British Designers' 24, 192, **193**, 194–97, 215, 218
16. 'What Industrial Design Means' 24, **198**, **199**, **201**, **202**, **203**, 242
17. 'The Council of Industrial Design' 24, 25
18. 'Travel Goods' 24, 59, 166, **168**
19. 'Utility and Other Furniture' 204–05, **206**, **207**, **208**, 209
20. 'Weekend' (Garden & Hand Tools, Sports) **22**, 24, **170**, 245,
21. 'Books and Printing' 24, 25, 49, 115, 210, **211**, **212**, **213**, **214**, 215–18
22. 'Designers Look Ahead' (Hall of the Future) 22, 24, 28, 49, 75, 117, 219, **220**, **221**, 222–31, **234**
Britain Can Take It film 39, 45
British Hospitals Association 184
British Industries Fair exhibitions (London and Birmingham 1947) 244
British Institute of Industrial Art 244
British Intelligence Objectives Sub-Committee (BIOS) 243, 249
British Film Institute (BFI) 27, 42, 75
British Landscape in oils exhibition (London 1942) 37
British Movietone News 18, 27
British Overseas Airways Corporation (BOAC) 62
British Pathé 27, 54, 56, 59, 68, 81, 182, 184, 223
British Plastics 27, 44, 45, 59, 75, 137
British Pottery Manufacturers' Federation 96, 103
Bromley School of Craft and Industrial Design 127
Brown, Mr 166
Brumwell, Marcus 118, 121
Bryanston Biscuits 118
Building Centre, London 197, 205, 207
Bullers Ltd 98
Bunyard, Claud 88, 93
Burton's 190

CC41 mark 205
C. & J. Clark 182
Calligan, Edwin 245
Cambridge University Press 214, 218
Camouflage 22, 28, 49, 74
Caravan-trailer 221, **231**
Carrington, Noel 27, 42, 45, 121, 162, 165, 171, 199, 215, 218, 243, 249
Carter Tiles 150, 154, 162
Casson, Sir Hugh 21, 27, 28, 109, 143, 162
Cast Courts (V&A) 34
catamaran 221, **229**, 231, 240
Cecil Drayson 89, 92

Central Institute of Art and Design 173, 194, 197
Central Office of Information 17, 21, 27, 42
Central School of Arts and Crafts 93, 127
Chad Valley Company 180
Charles C. Whitlock 187
Charles Kuperstein Ltd 126, 129
Chermayeff, Serge 145, 146
Cheverton, Charles 36
Children's Outfitter Magazine 182
Chilprufe 182
Chippendale, Thomas 192
Christie-Tyler Ltd 224
Chrysler Motors 48
Church & Co. 188
Churchill, Winston 17, 82
Clark, Sir Kenneth 23, 28, 38, 155, 156, 163, 195, 222, 231
Clark, Lady 173
Clark, May 91
Clifford, Alexander 81, 166
Clinch, Herbert 205
Coates, Wells 27, 28, 229, 246
Cockerell, Catherine 91
Cole & Son 155, 159, 173, 174, 175, 176, 177
Collins, Jesse 121
Condé Nast 241
Conran, Sir Terence 9, 10, 247
Contemporary Lithographs Ltd 159, 160, 163
Cooke, S.D. 66, 68, 70, 73, 75
Cooksey, Sergeant-Major R.A. 56
Cooper, Susie 96, 97
Co-operative Wholesale Society 209
Coram, Robert (known as Maroc) 26
Cornishware 98, 99
Cossor 55
Cossor Model 464 radio 55
Cotton Board 127
Council of Art and Industry (CAI) 127
Coxon, Dr Wilfred F. 80, 81
Crane, Walter 172
Crawford, W.S. 46, 48, 49, 50, 185, 186, 187
Creke, J.W. 86
Cresta Silks Ltd 158
Cripps, Sir Stafford 17, 18, 20, 22, 27, 28, 39, 41, 43, 44, 45, 82, 97, 103, 144, 145, 194, 197, 232, 241, 243, 249
Crosbie, William 78, 165, 166, 169, 171
Crown Films Ltd 42
Curry, Leonora 128, 129
Curwen & Sons 188

de Normann, Eric 28, 37, 38
Daily Herald Modern Homes exhibition (London 1946) 105, 106, 115
Daily Herald Post-War Homes exhibition (London 1945) 104, 105, 115
The Daily Mail 28, 40, 45, 55, 57, 58, 59, 61, 68, 81, 191, 222
The Daily Mail Ideal Home exhibitions 43, 146, 151
The Daily Telegraph 59, 65, 68
Daks Simpson 49, 185, **186**, 188, 191
Dalì, Salvador 234
Dalton, Dr Hugh 111
Darwin, Robin 84
Day, Lewis F. 172
Deans 181
Debenham & Co. 122, 129
Decca Record Co. Ltd 79, 109
Delanghe, Angele 127
Denby, Elizabeth 28, 99, 150, 154, 162, 205
Design '46 (CoID) 20, 21, 51, 59, 81, 87, 92, 93, 96, 103, 115, 117, 118, 120, 121, 124, 129, 137, 171, 177, 183, 184, 187, 191, 192, 197, 215, 216, 218, 243
Design and Industries Association 142, 146, 180, 181, 197
Design and the Designer in the Dress Trade (report) 127
Design for Better Living film 19
Design Investigation in Selected German Consumer Goods (H.M. Stationery Office) 244, 249
Design Quiz (CoID) 6, 28, 104, 105, 106, 107, 108, 109, **110**, 111, 112, 114, 115, 177
Design Research Unit (DRU) 69, 109, 110, 114, 115, 118, 142, 171, 194, 200, 203, 221, 227, 228, 242, 246
Designers in Britain book, 1947 51, 118, 121, 171
Dickens, Ronald 83, 87, 93
Dickey, Edward O'R. 244, 249
Disobedient Objects exhibition (London 2014–2015) 58
Display, Design & Presentation 27, 28, 43, 45, 63, 68, 72, 75, 80, 81, 105, 114, 115, 137, 161, 163, 168, 171, 191, 198, 203, 219, 223, 231, 245, 249
Display Presentation (Pick) 68, 168, 171
Douglas, Thomas 26, 130
Dowler, Miss C.M. 81, 111, 112, 184
Drayson, Cecil 89, 90, 92
Drew, Jane 98, 99, 149
Driver, Mr 13
Duchamp, Marcel 234, 235
Dufaylite 230
Duncan, Mary 158
Duncan Miller Gallery, London 49
Dunlop Rubber Co. Ltd 63, 74
Dutton, Norbert 103, 201, 203
Dyson's 29, 30

East London Group 134, 137
Eckersley, Tom 49, 213
Edinburgh Festival 245
Edinburgh Weavers (Morton Sundour) 155, 159, 209
Edward Johnston exhibition (London 1945) 37
eggcup 49, 142, 198, 199, 201, 202, 203, 240, 247
Eisler, Georg 154, 162, 163
electric cooker 221, **227**
Elephants in the Attic (Gardner) 22, 28
Elizabeth, Queen 26, 42, 45, 53, 55, 139, 143, 229, 230
Elkingtons 86
Ellis, Clifford *166*
Ellis, Rosemary *166*
Emberton, Joseph 187, 191
Eno's **48**, 49
Enterprise Scotland exhibition (Edinburgh 1947) 245, 249
Entwistle, C. 143
Ercol (Furniture Industries Ltd) 208, 209
Ercolani, Lucian 208, 209
Eric Gill sculpture exhibition (London 1941) 37, 38
Ernest Brown & Philips Ltd, Leicester Galleries 153, 162
Ernest Race Furniture Ltd 135, 136, 137, 209, 246
Evans, Joan 69, 71, 87, 243, 250
Evening Standard 42, 45, 227, 231
Everett's Advertising 46
An Exhibition of Historical & British Wallpapers (London 1945) 173, 177
Exposition Internationale du Surréalisme (Paris 1938) 234, **235**
Exposition surréaliste d'objects exhibition (Paris 1936) 234

Faculty of Royal Designers for Industry 194
Farnham Castle 74
Fashion and Fabrics Overseas 125, 126, 129
Feibusch, Hans 132, 133, **134**, 137, 165, 166
Fellowes, Mrs Reginald 123
Fennemore, Thomas Acland 173, 174, 177, 196, 197
Ferramola, Floriano 33, 34
Ferranti Ltd 224, 225
Festival of Britain exhibition (London 1951) 13, 15, 21, 22, 58, 110, 165, 176, 209, 212, 246, 247, 249
Films of Fact (Paul Roth Productions) 27, 42
Firth Brown 86
Fitton, James 213
Flat-screen television projector 221, **224**, 231
Fleetwood-Hesketh, Peter 139, 140, 141, 142, 144
Fortnum & Mason 182
Fougasse (Cyril Kenneth Bird) 213
Fox's of Bexley 90, 91
Frankfurt kitchen 149
Franklin, Miss 178
Fraser, Eric 119, 194
Frazer, Austin 200
Frederick Starke Ltd 124, 129
Freedman, Barnett 22, 109, 118, 119, 180
Freitag 92
French, Philip 19, 27
Fry, Maxwell 28, 98, 149, 150
Furnishing to fit the family (CoID) 50, **143**, 154, 162
Furniture Industries Ltd 208

G.F. Glenny 187
G. Music & Sons of Hatton Garden 92
Galerie Charles Ratton 234
Galerie Beaux-Art, Paris 235
Gallagos, Isabel 173
Gamage, Leslie 65
Games, Abram 49, 166, 213
Games and Toys 183, 184
Gardiner, Miss 125
Gardner, (Leslie) James 9, 12, 14, 15, 21, 22, 25, 26, 27, 28, 49, 52, 62, 63, 66, 68, 69, 70, 71, 72, 73, **76**, 77, 78, 79, 80, 81, 83, 87, 95, 106, 107, 108, 115, 126, 129, 132, 133, 134, 137, 140, 142, 144, 165, 166, 167, 171, 178, 179, 180, 181, 182, 183, 201, 203, 212, 218, 227, 232, 235, 237, 245, 246, 247, 249
Garland, Madge 93, 124, 125
Gaunt, William 227, 231
Geffrye Museum of the Home, London 146, 151, 204, 205
General Election 1945 18, 27
General Electric Co. Ltd 65, 68, 130
Gense of Eskilstuna 86
Geological Museum, London 34
Georg Jensen 86
George VI, King 20, 26, 27, 53, **105**, 171, 204
George Webb & Sons Ltd 188
Gibberd, Frederick 149
Gilbert, Geoffrey 190, 191
Gill, Eric 37, 38, 215
Gladwin Ltd 86
Glenny, G.F. 187
Gloag, John 40, 45, 115, 117, 241, 249
Goldfinger, Ernö 181, 246
Goldsmiths' Company 83, 87, 88, 92, 93
Goodden, Robert 22, 27, 28, 73, 137, 173, 177, 245, 246, 249
Goodhart-Rendel, H.S. 172, 177
Goodwin, Miss 42
Gordon Russell Ltd 9, 24, 41, 45, 115, 137, 144, 151, 155, 194, 195, 205, 206, 209, 222, 224
Grant, Duncan 78, 165, 169
Graphis 27, 118, 121, 137, 243, 249
Gray, Milner 27, 28, 118, 121, 163, 194, 227, 228, 250
Great Exhibition (London 1851) 18, 23, 28, 58, 194
Greenly, Henry 181
Greenwood, Arthur 130, 131
Greenwood, Joan 56
Groag, Jacqueline 137, 209
Groag, Jacques 207

H.A. Byworth & Co. 90
H.G. Stone (Chiltern Toys) 182
H.K. Furniture Ltd 135
H.M. Stationery Office 39, 47, 218
Haines, Judy 29
Haley, Mrs 57
Hall, D. Clarke 160
Handley, Tommy 31
Hardie, Martin 36
Harling, Francis 91, 92
Harper's Bazaar 93, 124, 126, 128, 129
Harris Lebus 204, 205, 209
Harrison, Mary 110, 115
Harrison, Molly 146

Harrods 180
Hart, Nat 175, 176
Hastings, (Douglas Edward) Macdonald 242
Havinden, Ashley 6, 7, 9, 17, 20, 22, 26, 27, 42, 46, 47, 48, 49, 50, 51, **185**, 186, 187, 188, 189, 190, 191, 213
Hawes, Miss 128, 129
Haycock Press 47
Heal, Ambrose 146
Heal, Christopher 207, 208
Heal's Ltd 95, 98, 103, 112, 130, 131, 207, 208, 209
Heap, Anthony 190
Heath, Alec 245
Helios Ltd 158
Henrion, F.H.K. 28, 49, 213, 226
Hepplewhite, George 192
Hepworth, Barbara 48
Heriot-Watt College, Edinburgh 214
Heron, Patrick 158, 163
Heston Aircraft Co. Ltd 65
Hill, John 154, 155, 156, 175
Him, George 27, 125, 166, 168, 171, 213
Himsworth, Joyce 86
Hoff, Max (alias I.T. Pallas) 186, 187, 191
Hogan, James 102
Holbein exhibition (London 1943) 37
Holford, Marjorie 143
Holland, James 165, 171
Holliday & Browne Ltd 188
Holoplast 209
Home Service 20, 28, 55, 57, 59, 62, 68, 169, 171, 219, 249
Honey, W.B. 95, 96, 97, 103
Hooper, F.C. 242
Horder, Mary 182
Hornchurch Chronicle Childrens' Corner 178
How They Fly exhibition (London 1939) 22, 28, 180
How to understand modern painting exhibition (London 1945) 37
Howitt's of Hatton Garden 90, 91
Hoy, Agnete 98
Hubbard, Lorna 244, 249
Hughes, George Ravensworth 83, 85, 87, 93
Hunter & Hyland 108, 115
Hussey, Walter 157, 163
Hutchinson, Stanley 36
Hutton, Clarke 215
Hutton, John 22, 68, 75, 165, 166, 221, 229, 230, 245, 246

IKEA 151
IWM Duxford 244
Illingworth, Leslie 45, 142, 143, 144
Imperial Chemical Industries (ICI) 120, 127
Imrie, Archibald (Archie) 245
Incorporated Society of London Fashion Designers (INCSOC) 123, 127
International Model Aircraft Ltd 180
International Surrealist Exhibition (London 1936) 77, 81
It's That Man Again (ITMA) 31

J. Lock & Co. Ltd 188
J. & P. Coates 118
Jaeger & Co. Ltd 182, 188
James Jobling & Co. 101
James Powell & Sons 102
Jarrett and Rainsford 85
Jarvis, Alan 28, 45, 51, 115, 214
Jeffcott, Ian 43
Jeffrey & Co. 172
Jewellery and Silverware Working Party 82, 83, 84, 86, 87
John Lewis Partnership 92, 95, 98
John Line & Sons 175, 176, 177
Johnson-Marshall, Stirrat 74
Jones, Barbara 165, 166, 167, 169, 171, 179, 246
Joseph Bourne & Son 99
Josiah Wedgwood & Sons 96, 97, 99, 100, 101, 192, 222, 242, 249
Juda, Hans 123, 125

Katz, Bronek 200
Kearley, Mr N.E. 55, 81
Keith, Howard 135, 137
Kelly, Irene **56**, 59
Kennedy, Warnett 221, 223, 230
Kestelman, Morris 63, 65, 163, 165
Keyes, Phyllis 97
Kiddicraft 181
KLM 49

L. & R. Morley 187
L.S. Mayer 92
La Dell, Edwin 158
Lachasse 126
Lambert, Margaret 244, 249
Lampl, Fritz 92
Last, Nella 224
Latimer, Clive 112, 115, 209
Lavers, R.S. 193, 196, 197
Le Witt, Jan 166, 168, 171, 213
Leach, Bernard 97, 98, 103
Lebus, Herman 204, 205, 209
Leicester Galleries, London 49, 153, 162
Leischner, Margaret 243
Leith, Patricia (Pat) 135, 137, 178
Leslie, Samuel Clement (Clem) 18, 23, 27, 28, 32, 41, 42, 45, 51, **56**, 57, 65, 68, 97, 103, 131, 137, 140, 141, 142, 144, 161, 163, 171, 192, 193, 194, 195, 197, 203
Lewinter, Charlotte 126
Lewis, Walter 214, 218
Liberty **49**
Lillywhites 188
Line Drawing for Reproduction (Havinden) 48, 51
Lines, Walter 179
Lines Brothers 65, 66, 179
Lipski, Tadeusz 180
Litfaß, Ernst 170
Livery Companies 193
London Can Take It film 39
London City Guilds 193
London County Council (LCC) 146
London Model House Group (LMHG) 124, 128

London Press Exchange 46
Lott's Bricks 181, 182
Lowndes, George 146
Lund Humphries 118
Lund Humphries Gallery, London 49

MARS Group 25, 28
M.P. Greengross 89
MacCarthy, Fiona 237
Macdermott, Elvie 94, 97, 103
Mackinnon, Duncan 158
Mackinnon, Lana 158, 163
Machin, Arnold 100, 101, 246
Maclagan, Eric 34, 35, 36, 38
MacLeod, Mr 169
MacManus, Frederick 99, 155, 156, 158
Magna, Reginald 229
Mappin & Webb 85, 86
Marx, Enid 158, 243
Mass Observation (M-O) 7, 27, 54, 59, 80, 81, 98, 102, 103, 110, 115, 141, 144, 151, 153, 161, 162, 163, 176, 177, 182, 184, 188, 191, 194, 197, 199, 201, 202, 203, 207, 209, 224, 226, 228, 238, 239, 240, 243
McKnight Kauffer, Edward 48
Mecca Cafés 131
Medd, David L. 149, 159
Melless, Lance Sergeant John 36
Mercer, F.A. 41
Mesling, C.A. 41, 45, 115, 231
Metal Box Ltd 120
Meynell, Sir Francis 155, 195, 214, 215, 218, 222
Milk Marketing Board 48
Milne, John A. 19
Ministry of Information 21, 28, 36, 39, 60, 163
Minns, Eden 224
Mitchell, Arnold 182
Mitchell, Private 54
Moholy-Nagy, László 48, 186, 191
Mondrian, Piet 48
Monotype Corporation 46, 211, 214, 215, 217, 218
Montacute House, Somerset 33, 35, 36
Moorcroft 98, 131, 137
Moore, Henry 43, 48, 137, 154, 156, 158, 163
Mooring, Able Seaman Herbert 36
Morgenstern, Milan 181
Morison, Stanley 48, 217, 218
Morris, Gabriel 170
Morris, Geoffrey 186
Morris, William Morris 192, 195, 216
Morris column 170
Morrison, Herbert 40
Mortimer, Raymond 87, 174, 177
Morton, Alastair 159, 209
Morton, Digby 125
Morton, George 36
Mosely, Edna 98, 143, 148
Mostyn, Rita 55, 56, 59
Moviegram Ltd 22, 28
Mr Designer **15**, **25**, **26**, 28, 49, 69, 70, 75, 113, 118, **119**, 219
Mr Therm 15

Müller, Fritz 97, 103
Munro & Co. Ltd 188
murals 7, 63, **64**, 65, 68, 75, 78, **123**, **125**, 126, 132, 133, 134, 135, 137, 142, 153, 160, 163, **164**, 165, **166**, **167**, **168**, **169**, **170**, **171**, 178, 179, 182, 187, 191, 195, 197, 198, **199**, 210, **211**, **221**, **229**, 230, 245
Murphy Radio Ltd 224
Murray, Keith 97, 102

NHS Bill 32
National Art Library 34, 36, 37, 38
National Insurance Act 32
National Savings 18, 19, 27, 30
Natural History Museum, London 34
Neale, Mary 182, 184
Neurath, Otto 171
Newman, Greta 126
Newman, W.H. 51, 93
Newnes Practical Mechanics 67, 230, 227, 231
New Statesman and Nation 87, 174, 177
New Towns Act 32
Nicholson, Ben 48
Nicholson, Christopher 224, 225
Nicoltoys 181
Nieboer, Hendrik A. 244, 249
Nonesuch Press 214, 218
Norsk Fokemuseum (Norway) 151
Norvic 182
Nossiter, Dorrie 91
Nunn, Kenneth **55**, 56
nylon 75, 127, 128, 219, 221

O'Neill, Olive 126
Odeon and Gaumont-British cinemas 47
Odney Pottery 98, 103
Old Master drawings from the Witt Collection exhibition (London 1943) 37
Olins, Wally 247
Oman, Charles 83
Orwell, George 32
Oved, Sah 91

Page, Hilary 181
Panting, Phyllis (known as Ann Seymour) 124, 125
Paris International Exhibition (Paris 1937) 186
Pasmore, Victor 154, 155, 160, 162, 163
Paul Rotha (Films of Fact) 42
Pedigree 181, 182
Penrose, Roland 166
The Penrose Annual 51, 118, 191
Percival, Michael 227
Perspex 26, 49, 61, 66, 67, 75, 77, 92
Peter Robinson Ltd 182
Pevsner, Nikolaus 101, 103, 195, 196, 243, 249
Phillips, Frank 219
Pick, Beverley 61, 62, 63, 68, 76, 168, 171
Picture Post 18, 27, 124, 191, 227
Pilkington Jackson, C.D.O. 117, 233, 237
Piper, John 159, 160, 163
Pirelli Glass Ltd 102

Plimpton Engineering 181
Pollak, Sylvia 81, 107, 115, 169, 171, 230, 235, 237
Pompa, Philip 175, 176
Porter, H.B. 126
Post War Export Trade Committee 17
poster 9, **16**, 19, 26, 42, **46**, **47**, 46–51
Poster Services Ltd 47
Postmodernism: Style and Subversion, 1979–1990 exhibition (London 2011) **236**, 237
Pountney & Co. 99
prefabs 30, 42
Price, Dennis 56
Printing and Allied Trades Association 117, 119
Pritchard, Walter 245
Product Design Unit 73
Punch **26**, 28, 43, 45, 80, 81, **130**, 137, 142, **143**, 144, 159, 163, 186, 190, 191, 203, 242, 249
Purchase Tax 67, 84, 86, 89
Pyrex 101
Pytrams 126, 127

Quennell, Marjorie 146
Quiz coins 78, 105, 107, 108, 115

RFD Co. Ltd 62
R. Pringle & Sons Ltd 126
Race, Ernest 135, 136, 137, 209, 246
Raphael Cartoons 35, 38
Rationing 17, 18, 27, 29, 30, 65, 132, 189, 190, 204, 205, 213, 217, 218, 244
Ravilious, Eric 77, 97, 99, 100, 177
Read, Herbert 25, 28, 96, 103, 114, 115, 118, 121
Recording Britain exhibition (London 1944) 37
Redux 112, 209
Regent Street Christmas lights 62
Rembrandt Hotel, London 13, 58
Richards, J.M. 77
Richard Shops 49
Richardson and Bailey 126
Riddell, Dr G.L. 117, 119, 121
Rima (Wholesale) Ltd 124, 126, 129
Roberts, John 36
Roberts and Belk 85
Robertson, Howard 95
Robinson, Basil (Robbie) 36
Romac Industries Ltd 57
Romac portable 106 radio **56**, 57, 63, **67**
Rose, Yootha 182
Rose & Blairman Ltd 126
Rowntree, Kenneth 158, 160
Royal Academy, London 19, 27
Royal Air Force (RAF) 12, 35, 38
Royal College of Art, London 34, 84, 93, 127, 249
Royal Designers for Industry (RDI) 27, 194
Royal Mint 108, 115
Royal Scottish Museum, Edinburgh 245
Royal Society of Arts, London 19, 24, 27, 28, 45, 129, 171, 194, 247, 249
rubber chair **62**, **63**, 65
Russell, Audrey 57, 59, 169, 171
Russell, Gordon 9, 24, 41, 45, 115, 137, 144, 151, 155, 194, 195, 197, 205, 206, 209, 222, 224
Russell, R.D. 150, 151, 153
Russell, W.H. 155
Ryan, John M. 117, 121
Ryder, R. Dudley 18, 27, 66, 80, 81, 115, 137, 161, 163, 165, 166, 171, 192, 193, 197, 203

S.H. Benson 46
Sadler, Arthur 189, 191
Sandbrook Witting & Co. Ltd 188
Sangster, Margaret 48
Saxon Mills, G.H. ('Bingy') 48
Scanlon, Robert 168, 170
Scarfe, Laurence 165, 210, 211
Schleger, Hans (known as Zero) 49, 120
Schlesinger, S.B. 89, 92
School Prints Ltd 158, 163
Schorr, Raoh 78, 127, 129, 245
Schütte-Lihotzky, Margarete 149
Science Museum, London 35, 197
Scott-James, Anne 93, 124, 128
Scottish Mac Fisheries Group 120
Scottish National Gallery of Modern Art, Edinburgh 49, 50, 51
Selfridge's 180
Selwyn, Arnold 92, 93
Semphill, Lady 180, 183, 184
Settle, Alison 128, 129, 195, 197
sewing machine 221, **226**, 231, 234
Seymour, Ann (Phyllis Panting) 124, 125
Seymour Products Ltd 223
Shaw, Norman 172
Shelf-Appeal 117
Shell-Mex House 22, 28, 180
Shell Motor Oil **120**
Shinwell, Emanuel (Manny) 43, 143
Sign and Display Trades Union 166, 171
Simpson, Alec 185
Simpson, Simeon 185, 186
Simpsons of Piccadilly 48, 185, 188
Sissons, Michael 19, 27
Sitwell, Sacheverell 173
Skeen, Henry 175, 176, 177
Skillen, Hugh 182
Skinner, Freda 181
Smail, J. Cameron 214
Smith, Albert 74, 117, 121
Smith, Sir Ben 30
Society of Industrial Artists (SIA) 51, 118, 121, 171, 187, 203, 244
Society of Mural Painters 65, 166
Society of Woodworkers 166, 171
Southam 85
space-ship 221, 227, **230**, 231, 234, 239
Spear, Ruskin 39, 40, 45
Spence, Basil 12, 21, 22, 26, 27, 49, 52, 69, 179, 219, 220, 226, 227, 245, 246
Spurr, Elizabeth 156, 157, 163, 175
Standard Emergency Furniture 205
Standard Ware 97, 103
Stanhope, Eileen and Eric **56**, 57, 59
Stead, Albert 123
Steer, Philip Wilson 36, 38
Steiner of Grosvenor Street 126, 127
Stevens and Williams 102
Stone, Reynolds 119
Stone, Robert E. 85, 86
Stuart Bell, D. 178
The Studio 137, 233, 237
Sudrow, Dr Anne 243, 244, 249
Sunday Pictorial 190, 191
The Sunday Times 93, 126, 129, 163
Sunderland Daily Echo 111, 115
Surreal House exhibition (London 2010) 235, **236**, 237
Susie Cooper Pottery 96, 97
Sutherland, Graham 65, 155, 156, 157, 158, 159, 160, 162, 163, 173, 176, 177
Swift, W.A. 187

T.G. Green & Co. 98, 99
T.H. & J. Daniels Ltd 198
Tailor and Cutter & Women's Wear 128, 129
Tanner, Betty 177
Tarr, John 214, 215, 218
Taurus the bull **5**, **100**, **101**, 246
taxi 221, 227, 228, **229**, 231
Tennyson, Sir Charles 28, 41, 45, 155, 177, 185, 191, 201, 203
Thoday, Leonard 130, 131
Thomson, Mrs 132
The Times 28, 36, 38, 45, 46, 54, 59, 63, 65, 67, 68, 103, 140, 144, 153, 162, 169, 171, 190, 191, 216, 218, 249
The Times Furnishing Company 145
Tin Research Institute 73
Tomalin, Miles 139, 141, 144
Tomrley, Mrs C.G. 115, 231
Tomtopia 31
Tootal 188
Torode, Mr 166, 171
Trethowan, Harry 95, 103
Trevelyan, Julian 155, 156, 163
Tri-ang 180
Tubbs, Ralph 28, 246
Tunnard, John 152, 153, 162

Union des Artistes Modernes, Paris, 1933 49
United Glass Bottle Manufacturers Ltd 101, 102
Utility Furniture Advisory Committee 205
utility products 7, 123, 124, 128, 129, 131, 137, 182, 189, 190, 204, 205, 206, 207, 208, 209

Van Dyck exhibition (London 1941) 37
Van Gogh, Vincent 154, 163
van Vlymen, D. 28, 45, 46
Vaughan, Reginald (William) 200, 221, 227, 228
Vaughan-Thomas, Wynford 62, 68
Victoria and Albert Museum (V&A) 2, 6, 9, **11**, 13, 15, 17, 21, 23, 24, 25, 26, 28, **33**, **34**, **35**, **36**, 37, 38, 46, **47**, **53**, 58, 59, 77, 78, 83, 95, 126, 129, 132, 134, 146, 166, 167, 173, 186, 196, 232, **233**, **236**, 239, 248
Victory over Japan exhibition (London 1945) 60, 62, 68
Vogue 67, 91, 93, 124, **126**, 129, 173, 180, 241
Voysey, C.F.A. 172, 192

W. Anderson & Sons Ltd 188
W.J. Bassett Lowke 181
W.S. Crawford 46, 48, 49, 50, 185, 186, 187
Wadsworth, Edward 120, 121
Wakeley and Wheeler 85
Wales, Prince of 58, 59
Wall Paper Manufacturers Limited (WPM) 173, 176, 177
wall radio 221, 223, **230**, 231, **234**
wallpaper selection committee **174**
Walton, Allan 231, 244
War Artists' Advisory Committee 39, 157, 163
War Economy Standard 217
Warde, Beatrice 215, 216, 217, 218
Warerite Ltd 134, 135, 137
Waring and Gillow 145, 146, 151
Warner, Sylvia Townsend 30, 32
Warner Brothers 39
Watercolour drawings and sculpture exhibition (London 1942) 37, 38
Wakouwa 180
Webster, Miss R. 108
Wedgwood, Hon. Josiah 222, 242, 249
Weiss, Otto 126
Welch Margetson & Co. Ltd 188
Weldon, Fay 29, 32
Westwood, Wiltshire 35
Whalley, Philip 195
Whitaker, Mrs 13
Whitefriars 102, 103
William Britain 181
William Mallinson & Sons Ltd 73
Williams, George E. 228
Williams-Ellis, Clough 133
Williams-Ellis, Susan, 154, 162
Williamson, A.H. 102, 177
Windsor chair **208**, 209
Winsor & Newton 121
Winterthur Museum, Garden and Library (USA) 151
Withers, Audrey 93, 124, 125, 129
Wolfe & Co. 90
Women and Beauty 124, 126, 127, 129
Wood, Christopher 154, 159, 163
Woods, W. Joseph 226
Woolton, Lord 111
Woolworth's 92
Worcester Royal Porcelain Co. 99
Worth 126
Wray, Miss 125
Wright, Hilton 143, 162

Yeats, J.B. 154, 155
Yerbury, F.R. 193, 196, 197
Yorke, F.R.S. 13, 109, 115, 130, 156, 157, 175, 177
Yoxall, Harry W. 241, 249

Zero (Hans Schleger) 49, 120
Zoo-Zag puzzle **179**
Zwemmers 160, 163

Contributor biographies

Martin Andrews is a printing historian, exhibition designer and lecturer in the Department of Typography & Graphic Communication, University of Reading. He has lectured widely in this country and abroad and published a number of books including a biography of Robert Gibbings and most recently a book about the artist and lithographer, Rena Gardiner.

Diane Bilbey is a freelance researcher and writer on art and 20th-century design. She has worked at the V&A in the Sculpture Department and published on its collections of British sculpture and plaster casts. Most recently she has worked on the Art UK Sculpture Project.

Edwina Ehrman is a Senior Curator at the Victoria and Albert Museum and a specialist in 19th century fashion. Since joining the museum in 2007 she has curated three exhibitions, *Fashioned from Nature* (2018), *Undressed: A Brief History of Underwear* (2016) and *Wedding Dresses 1775*–2014 (2014) and from 2009–2013 she was the lead curator for The Clothworkers' Centre for the Study and Conservation of Textiles and Fashion, the V&A's fashion archive at Blythe House in West Kensington. Her previous publications include *Fashioned from Nature* (2017), *London Couture 1923–1975: British Luxury* (2015) of which she is co-editor and a contributor, and *Wedding Dress: 300 Years of Bridal Fashions* (2011).

Simon Garfield is the author of 19 acclaimed books of non-fiction including *Mauve, Our Hidden Lives* and *In Miniature*. His study of Aids in Britain, *The End of Innocence,* won the Somerset Maugham prize, while *Just My Type* and *On The Map* were New York Times bestsellers. He was a staff feature writer on the *Independent* and *Observer*, and was formerly editor of *Time Out London*. He lives in London and St Ives, Cornwall. www.simongarfield.com

Alexandra Goddard was Exhibitions Manager at the British Library. After receiving a Masters degree in Early Sienese Painting from the Courtauld Institute she worked at the V&A as a curator in the Sculpture Department. Following that she was Assistant Keeper – Interpretation and Exhibitions at The Geffrye, Museum of the Home where she developed a strong interest in 20th-century and contemporary interiors and social history with particular expertise in the Utility Furniture scheme and the post-war period.

Alun Graves is a Senior Curator in the Sculpture, Metalwork, Ceramics and Glass Department of the Victoria and Albert Museum, London. He has responsibility for the collections of 20th- and 21st-century ceramics, and has published frequently in this field. He was a key member of the project team in the redevelopment of the V&A's Ceramics Galleries, opened in 2009/10, and has curated many V&A displays and events.

Andy Groarke is co-director of London-based architectural practice, Carmody Groarke, which he established with Kevin Carmody in 2006. The practice has developed a reputation for working internationally on a wide range of arts, cultural, heritage and residential projects. They have won several architectural competitions and awards for completed projects, as well as the Building Design Architect of the Year Award in 2018. With over 20 years' experience as an architect, Andy has lead the development of a number of award-winning projects, including the 7 July Memorial, Maggie's Merseyside at Clatterbridge and Studio East Dining. Andy has lectured and taught extensively in the UK and overseas at institutions including the Royal College of Art and the Bartlett School of Architecture (University College London). He is currently Visiting Professor at the University of Sheffield and was Visiting Professor at the University of Stuttgart, 2015–2016. He is currently leading delivery of the Windermere Jetty Museum in the Lake District and the Special Exhibition Gallery at the Museum of Science and Industry in Manchester.

Catherine Howell was formerly Collections Manager at the V&A Museum of Childhood where she worked from 1991 until 2017. She was the lead curator of the toys and games collections and curated several of the Museum's major exhibitions, including ones on teddy bears, magic and board games. She has written on various aspects of childhood, presented papers at conferences and seminars both nationally and internationally, and has contributed to television and radio programmes on the history of toys.

Paul Jobling has published widely on the intertexuality of word and image in the fashion media. His books include *Advertising Menswear: masculinity and fashion in the British mass media since 1945* (2014), awarded Outstanding Academic Title of 2014 by the American Libraries Association (*Choice*, January 2015); *Man Appeal: advertising, modernism and masculinity* (2005); and *Fashion Spreads – word and image in fashion photography since 1980* (1999). He has taught at several institutions in the UK and Europe, and is currently professor in fashion studies at Parsons New School, Paris.

David Kynaston has been a professional historian since 1973 and is the author of some 20 books. His corporate histories include studies of the Bank of England, Financial Times and of HSBC; his four-volume history of the City of London during the 19th and 20th centuries was published between 1994 and 2001; and he is currently engaged on a multi-volume history of post-war Britain, of which *Austerity Britain, Family Britain* and *Modernity Britain* have so far appeared, taking the story to 1962. He is an honorary professor at Kingston University.

Clare Phillips is a Curator in the Department of Sculpture, Metalwork, Ceramics and Glass at the Victoria and Albert Museum, with particular interest in jewellery from the late 19th century to the present. Her publications include *Jewels and Jewellery* (2000, revised 2008 and 2019) and *Bejewelled by Tiffany* (ed., 2006). She has contributed to a wide range of V&A exhibitions and books, and curated the Tiffany exhibition at the Gilbert Collection, Somerset House in 2006.

Paul Rennie is a design historian and specialist in the history of graphic communication in Britain. He is subject leader in Contextual Studies in Visual Communication at Central Saint Martins, London.

Katrina Royall researched poster propaganda of the Second World War for her MA degree and has maintained an interest in the subject for many years. Having worked at the V&A in various departments she developed a strong interest in the V&A during the war, realising that little had been written about this subject. She has undertaken research for an architectural company specializing in historic buildings, and is currently Treasurer of the Cornish Buildings Group.

Gill Saunders, Senior Curator of Prints at the Victoria & Albert Museum, has written extensively on 20th-century and contemporary art and design. Her publications include *Apocalyptic Wallpaper* (1997), *Wallpaper in Interior Decoration* (2002), *Prints Now: Directions and Definitions* (2006) and *Bawden, Ravilious and the Artists of Great Bardfield* (2015; with Malcolm Yorke). She was co-curator of *Walls Are Talking* at the Whitworth Art Gallery, Manchester in 2010, and editor of the accompanying book. She is currently researching the V&A's poster collection for a forthcoming publication. She publishes, lectures, and broadcasts on a wide range of print-related topics.

Margaret Timmers was formerly Senior Curator of Prints at the V&A. She devised the V&A's exhibitions and publications *The Power of the Poster* (1998), and *A Century of Olympic Posters* (2008), and also edited *Impressions of the Twentieth Century: Fine Art Prints from the V&A Collection* (2001). With support from Arts Council England she has recently worked as a Consultant Curator at the V&A, helping to develop the UK Subject Specialist Network for Posters. She is currently contributing to the publications *The Poster: A Visual History* (2020) and to *Shell Art and Advertising* (2020).

Eric Turner has been a curator in the Metalwork Section of the Victoria & Albert Museum, London, since 1976. His particular areas of expertise are associated with the collections of post-1880 metalwork and Sheffield plate. He has contributed to many publications, several conferences and exhibitions on these subjects both in Britain and abroad. He has been a Freeman of the Goldsmiths' Company since 2002 and was made an Honorary Member of Contemporary British Silversmiths in 2017.

Lesley Whitworth has been involved with the University of Brighton Design Archives since its inception. She is Deputy Curator and a Senior Research Fellow whose published work embraces many aspects of the early history of the Council of Industrial Design (afterwards Design Council), including its advocacy on behalf of the emergent profession of designer, and on behalf of the consumers of designed goods too. Exhibition design and the role of the curator are emergent areas of her teaching.

Jonathan M. Woodham is a Professor Emeritus of Design History having been teaching, researching and publishing in the history of design for more than forty years. He has presented keynote addresses in twenty-eight countries across five continents, been a specialist peer reviewer for national and international research councils, and has published well over a hundred peer reviewed books and articles. The latter include *Twentieth Century Design* (1997) that has sold nearly 60,000 copies worldwide and has been translated into Korean and Chinese and the *Dictionary of Modern Design* (2005) now part of Oxford Reference online (significantly updated and expanded in 2016); all are published by Oxford University Press. Woodham is now completing a substantial monograph on *British Design 1915–2015: Empire, Welfare State and Enterprise*.

Select bibliography

Design '46. Survey of British Industrial Design as displayed at the "Britain Can Make It" exhibition (London: Council of Industrial Design, 1946)

Britain Can Make It Exhibition Catalogue (London: Council of Industrial Design, 1946)

Penny Sparke (ed.), *Did Britain Make It? British Design in Context 1946–86* (London: The Design Council, 1986)

Patrick J. Maguire and Jonathan M. Woodham (eds.), *Design and Cultural Politics in Postwar Britain: The Britain Can Make It Exhibition of 1946* (London and Washington: Leicester University Press, 1997)

PRIMARY ARCHIVAL SOURCES

Design Council Archive, University of Brighton Design Archives

Britain Can Make It Exhibition records 1945–1948, V&A Archive of Art & Design (AAD)

BBC Written Archives Centre

The Mass Observation Archive, University of Sussex: massobs.org.uk

Photographic credits

Every effort has been made to trace and credit all known copyright and reproduction right holders; the publishers apologise for any errors or omissions and welcome these being brought to their attention.

Frontispiece, pages 11, 15 Design Council Archive, University of Brighton Design Archives; **BRITAIN CAN MAKE IT** 1 © Victoria and Albert Museum, London. Reproduced by kind permission the Estate of Ashley Havinden; 2 © Victoria and Albert Museum, London; 3 Reproduced by kind permission the Estate of Ashley Havinden; 5, 6, 8–9 Design Council Archive, University of Brighton Design Archives; 10–11 Punch Cartoon Library / TopFoto; THE **V&A 1939–46** 13–16 © Victoria and Albert Museum, London; 17 Design Council Archive, University of Brighton Design Archives; **NAMING THE EXHIBITION** 18 Design Council Archive, University of Brighton Design Archives; **ASHLEY HAVINDEN** 19, 21 Design Council Archive, University of Brighton Design Archives; 20 © Victoria and Albert Museum, London. Reproduced by kind permission the Estate of Ashley Havinden; 22 © Victoria and Albert Museum, London. Reproduced by kind permission the Estate of Ashley Havinden, and GSK; 23 © Victoria and Albert Museum, London. Reproduced by kind permission the Estate of Ashley Havinden, and Liberty Fabrics; 24–27 © National Galleries of Scotland. Reproduced by kind permission the Estate of Ashley Havinden; **STATISTICS, STATISTICS** 28 © Victoria and Albert Museum, London; 29–32 Design Council Archive, University of Brighton Design Archives; **FROM WAR TO PEACE** 34–40 Design Council Archive, University of Brighton Design Archives; **WHAT THE GOODS ARE MADE OF** 41–50 Design Council Archive, University of Brighton Design Archives; **SHOPWINDOW STREET** 51–56 Design Council Archive, University of Brighton Design Archives; **SILVERWARE** 57–60 Design Council Archive, University of Brighton Design Archives; **JEWELLERY** 61, 63 © Bedford, Lemere & Co, courtesy of The Goldsmiths' Company; 62 Design Council Archive, University of Brighton Design Archives; **POTTERY AND GLASS** 64, 68 Design Council Archive, University of Brighton Design Archives; 65 © Victoria and Albert Museum, London. Reproduced by kind permission the Estate of Susie Cooper; 66 © Victoria and Albert Museum, London. Reproduced by kind permission Portmeirion Group; 67, 70 © Victoria and Albert Museum, London; 69 © WWRD. Photograph © Victoria and Albert Museum, London; 71 © Victoria and Albert Museum, London. Reproduced by kind permission O-I Glass Limited; **DESIGN QUIZ** 72, 76 © Victoria and Albert Museum, London; 73–75, 77, 79 Design Council Archive, University of Brighton Design Archives; 78 Reproduced by kind permission Scott Brownrigg; **DRESSING THE GOODS** 81, 82, 85 Design Council Archive, University of Brighton Design Archives; 84 Reproduced by kind permission the Estate of Barnett Freedman; 86 Reproduced by kind permission Shell Brands International AG; 87 Reproduced by kind permission Winsor & Newton; **WOMEN'S FASHIONS** 88–90, 92 Design Council Archive, University of Brighton Design Archives; 91 Vogue © The Condé Nast Publications Ltd.; **TIME FOR TEA** 93 Punch Cartoon Library / TopFoto; 94–98, 100–01 Design Council Archive, University of Brighton Design Archives; 99 © Victoria and Albert Museum, London. Reproduced by kind permission the Estate of Hans Feibusch; 102 © Victoria and Albert Museum, London. Reproduced by kind permission Ernest Race Furniture; **INTRODUCTORY STORY** 103 © RIBA Collections; 104 Design Council Archive, University of Brighton Design Archives; 105 Punch Cartoon Library / TopFoto; 106 Reproduced by kind permission the Estate of Nicolas Bentley; **THE FURNISHED ROOMS** 107–09 Photograph © Geffrye Museum of the Home; 110–12, 114, 115 Design Council Archive, University of Brighton Design Archives; 113 © Victoria and Albert Museum, London. Reproduced by kind permission the Estate of Margarete Schütte-Lihotzky; **DISPLAYING ART** 116 –19, 122 Design Council Archive, University of Brighton Design Archives; 120 © RIBA Collections; 121 © Victoria and Albert Museum, London. Reproduced by kind permission the Estate of Kenneth Rowntree; **THE MURALS** 123–26, 128 –30 Design Council Archive, University of Brighton Design Archives; 127 Photograph courtesy of Brian Webb. Reproduced by kind permission the Estate of Barbara Jones, and Nicholas Anderson; **WALLPAPER** 131, 134–36 © Victoria and Albert Museum, London; 132–33 Design Council Archive, University of Brighton Design Archives; **THINGS FOR CHILDREN** 137–38 Design Council Archive, University of Brighton Design Archives; 139–44 © Victoria and Albert Museum, London; **RULED BY FASHION** 146, 148–50 Design Council Archive, University of Brighton Design Archives; 147 Reproduced by kind permission Daks Simpson and the Estate of Ashley Havinden; **GREAT BRITISH DESIGNERS** 151–52 Design Council Archive, University of Brighton Design Archives; **BIRTH OF AN EGGCUP** 153–57 Design Council Archive, University of Brighton Design Archives; **UTILITY AND OTHER FURNITURE** 158–59 Photograph © Geffrye Museum of the Home; 160–63 Design Council Archive, University of Brighton Design Archives; **BOOKS AND PRINTING** 164 © Victoria and Albert Museum, London. Reproduced by kind permission the Estate of Laurence Scarfe; 165–67, 169 Design Council Archive, University of Brighton Design Archives; 168 © Victoria and Albert Museum, London. Reproduced by kind permission Estates of Jan Lewitt and George Him; 170 © Victoria and Albert Museum, London; **DESIGNERS LOOK AHEAD** 171 © Historic Environment Scotland: Sir Basil Spence Archive; 172–77 Design Council Archive, University of Brighton Design Archives; 178 © Victoria & Albert Museum, London; 179 © RIBA Collections; 180–86 Design Council Archive, University of Brighton Design Archives; **SURREALIST STRATEGIES** 187–88 Design Council Archive, University of Brighton Design Archives; 191 © Luke Hayes. Barbican Art Gallery, London; 192 © Luke Hayes. Victoria and Albert Museum, London; **MASS OBSERVATION** 193–94 Design Council Archive, University of Brighton Design Archives